CHEMISTRY AS A GAME OF
MOLECULAR CONSTRUCTION

CHEMISTRY AS A GAME OF MOLECULAR CONSTRUCTION

The Bond-Click Way

SASON SHAIK
The Hebrew University, Jerusalem, Israel

Contributors: Dr. Racheli Ben-Knaz Wakshlak,
Dr. Usharani Dandamudi and Ms. Dina A. Sharon

Published by John Wiley & Sons, Inc., Hoboken, New Jersey
Published simultaneously in Canada

For general information on our other products and services or for technical support, please contact our Customer Care Department within the United States at (800) 762-2974, outside the United States at (317) 572-3993 or fax (317) 572-4002.

Wiley also publishes its books in a variety of electronic formats. Some content that appears in print may not be available in electronic formats. For more information about Wiley products, visit our web site at www.wiley.com.

Library of Congress Cataloging-in-Publication Data:

Shaik, Sason, 1947–
    Chemistry as a game of molecular construction : the bond-click way / Sason Shaik.
        pages cm
    Includes bibliographical references and index.
    ISBN 978-1-119-00140-9 (pbk.)
 1. Chemistry–Textbooks.   I. Title.
    QD31.3.S47 2016
    540–dc23

                                            2015024292

Printed in the United States of America

10 9 8 7 6 5 4 3 2 1

Dedicated to my loved ones, Sara and Yifat-Sela Shaik

CONTENTS

FOREWORD

BUILDING

Two ways merge into the sensation of beauty: one is an appeal to the senses, an immediate visceral appreciation that this is right, and the second is a coaxing into play of our intellectual faculties as they ponder what it is that makes a work of nature or artifice different, yet similar to all that is in our experience.[1]

So it is with chemistry. There are two ways in to the beauty (and danger) of chemistry—I could call the first the "smells, stinks, and colors" way, and the second the "molecular architecture" way. I could also call one Oliver's way, referring to the way Oliver Sacks was attracted to the physicality of macroscopic matter transforming. Which he recounts masterfully in his *Uncle Tungsten*, a contender (with Primo Levi's *The Periodic Table*) for the best introduction to chemistry ever written. And there is Sason's way, the delight in the macroscopic molecule, assembled and taken apart, that pervades every page of this book. I think one needs both, especially in our time.

Chemists are a productive bunch. To the zillions of molecules of nature (your DNA is a molecule; it's ever so slightly different, that is, it is a different molecule—thank God—from my DNA) chemists have added in ~200 years of frenetic activity around 70 million new compounds, which heal, occasionally harm, and mostly do the things we want. That is why someone made them in the first place—dye that cotton denim, and if tomorrow you change your mind and prefer it turquoise, make it greener. Clean up that water, shrink a tumor, make things stick together when you want them together, and loosen that bond tomorrow.

---

[1] Immanuel Kant, a great philosopher, wrote this (I paraphrase to save you from the original flowing German).

We do build, and as I said, we are prodigal and talented at doing so. We build the world of complex function—an antitumor agent such as taxol, a new biodegradable polymer—by assembling simpler pieces. Not that nature necessarily has those simple pieces lying around; we often must get the building blocks for our tinker-toy extravaganzas from commensurately big natural molecules in the petroleum and coal resources of the world. The disassembly of these large molecules also often provides the energy we need to make chemical reactions go. It may also happen (and increasingly will do so) that the natural building blocks are small—the methane of natural gas, the CO_2 recycled from combustion. Then our task is different; instead of disassembly, we must learn to activate those small molecules, propel them on their chemical way.

In this book, you will learn to love another kind of construction, the click and clack, so natural, of a bond or bonds forming and breaking. The tearing apart and building up of molecules you will see is a symbolic one. That's usually not the way nature (by dint of the complexifying mechanisms of natural evolution) or chemists (working from cultural evolution, shaped by history) make the molecules that color, house, help plants grow, heal, and hurt. You will learn this conceptual way, beautifully crafted to help us fathom how the complexity of effect and function of the natural and synthetic worlds, a complexity absolutely necessary to allow a child to dance, can be created. This too is a way to build—and what is built is understanding.

ROALD HOFFMANN

PREFACE

This book presents a new way of teaching chemistry. It was written after more than 20 years of teaching introductory chemistry to chemistry majors, and about 9 years of teaching a course called "Chemistry as a Game of LEGO." This book was based on material taken from the introductory course I lectured to chemistry majors and then to social scientists and students of humanities. This experience strengthened my feeling that one could teach chemistry to anyone at most ages and project the immense beauty of this science. *This is what this book is all about.*

During these years, I was extremely fortunate to have teaching assistants and demonstrators who shared my enthusiasm. One of them is Racheli Ben-Knaz (at present, Ben-Knaz Wakshlak, in short, Racheli). Others were the demonstrators Calvin DeLano and Yohann Aouat, who were magicians with a flair for entertaining and staging the "magic of chemistry."

With Racheli, who contributed the demonstrations included in this book, I had a special rapport. She was a student in all the courses that I have given at the Hebrew University since 2003, and she always excelled. But more importantly, her hunger and enthusiasm for knowledge and wisdom made her an ideal student and then a wonderful teaching assistant. During the 3 years we taught the course "Chemistry as a Game of LEGO," we used to meet frequently to converse about the course material. Our dialogue has become a tool for grooming the ideas and communicating also to students. It also gave us the feeling that we are not only preparing lectures but also staging them so we could impress upon the audience the beauty of our science.

Since I keep a diary, I have documented key points from these conversations. After some experiences in writing trialogues with Roald Hoffmann, I realized that the discussions with Racheli could be interesting for the teachers and students of this textbook. Based on the points I had, I reconstructed, with due freedom, the discussions as conversations on contents of each lecture.

Another person who will appear in these conversations is Usharani Dandamudi (Usha for short). Usha was an excellent postdoctoral fellow in my group for over 4 years (2009–2014). She too has innate hunger and enthusiasm for knowledge and wisdom. Her subtle understanding of chemistry, dedication to research, and curiosity gradually attracted her to the book that I have been writing starting June 2013. She finally joined us and took care of all the artwork, in addition to reading the entire book and providing insightful feedback. You will see Usha's name in conversations on the contents of some lectures.

A third person that became instrumental in the completion of this book is Dina Sharon, a Princeton graduate, who visited my group for 9 months as a Fulbright Post-Graduate Fellow, after receiving her bachelor's degree. Dina then extended her time in lab for several months after her Fulbright term ended. In her hands, almost every detail in the art items and their captions was thoroughly investigated. She found errors that both Usha and I overlooked, and if the art items are now flawless (I hope so), it is due to Dina's incredible editing talents.

The conversations that precede the lectures not only tell about the contents of the lecture, but are also intended occasionally for advising the teacher (e.g., if a certain lecture requires a slide presentation, demonstration tutorial, etc.), and at times providing also background material. These "corridors" into the lectures are generally lighthearted and you may find at least some of them fun to read. An example follows right away, in the dialogue Racheli and I conducted after having started to get some referee reports on the book from the various publishers.

I owe Racheli, Usha, and Dina a debt of gratitude for their unwavering enthusiasm and assistance during the project. Dina and Usha were hugely important in shaping the book during the final stages of its production. My research associate David Danovich has been helpful in proofreading the references. I owe a debt of gratitude to Assaf Friedler, Santiago Alvarez, Philip Ball, Zvi Rappoport, Michael Michman, David Avnir, Carol Parish, Helmut Schwarz, Mansoor Niaz, and Roald Hoffmann, who have read some or all of the lectures. Roald participated in the conversation on the contents of Lecture 7, and he was very encouraging throughout the project. His reading of entire book and comments were extremely useful. His foreword made me blush … Michael read the book and his message in Hebrew, "the book is a bomb," sounded divine to my ears. He also designed some of the art in the book. Carol's and Helmut's comments were equally uplifting. Discussions with Avinoam Ben-Shaul on solvation were enlightening. Zehava Cohen drew some of the beautiful artwork in the book. W. B. Jensen has kindly made available his book of cartoons and permitted us to use it at will.

Thanks are owed also to my colleagues Assaf Friedler and David Avnir at the Hebrew University, who urged me ceaselessly to teach the LEGO course. Ms. Ester Ben-Shoan, the department's administrator, has been a source of major support for the course. She also attended it successfully in 2013. Calvin and Yohann made enchanting contributions to the lectures, and their demonstrations enlivened the lectures.

PICTURE P.1 The author and the three contributors to the textbook (Racheli in the top left photo, Usha in the top right photo, and Dina in the bottom photo).

I owe special thanks to my Editor, Anita Lekhwani from Wiley Hoboken, who has been a positive force in my decision to start and finish the project. Her assistant Cecilia Tsai was very helpful in getting permissions, which is the single most frustrating part of putting a book together. Thanks go also to the rest of the Wiley team, Paula, Katrina and Shalini who jointly managed the production.

Every writer of a book has ups and downs. Let me confess mine. In many moments, throughout the time between June 2013 and April 2014, I had the feeling that the completion of the book would require a miracle. This feeling was intensified when I left Israel for two sabbatical months as an Alexander von Humboldt Fellow, and I started arranging many seminars and trips. As the schedule during and after my sabbatical looked rather hectic, it was clear to me that the book would be delayed indefinitely. But miracles do happen! Away from my usual place, a great peace of mind engulfed me, and despite the relatively little time I had to devote to the book, I still managed to complete the four final chapters with ease. Who said that changes make coping difficult? I owe a debt of gratitude to my Berliner hosts Martin Kaupp and Helmut Schwarz.

Finally, the textbook contains problem sets (and answers in the ending chapter of the textbook) for every lecture. These problems may be helpful for the teacher to prepare the tutoring sessions and "home" assignments. Any teachers or students who wish to communicate with the author on aspects of the book are welcome to do so by e-mail (sason@yfaat.ch.huji.ac.il and sason.shaik@gmail.com).

SASON SHAIK
The Hebrew University
Jerusalem, Israel

A CONVERSATION ON THE TEXTBOOK AND ITS INTENDED READERS

Sason: I am ready for your interview, Racheli. Let's see what you have in mind …

Racheli: Let me provoke you gently, Sason. Why should someone start using this book? After all, time is precious, and chemistry is not overly cool.

Sason: Elementary, my dear Racheli. Let me remind you, if you already forgot what a colleague wrote on the book proposal: "I am convinced that any book, which achieves teaching the most important concepts of general chemistry, in a simple and understandable manner, is highly needed."

This is what the book is all about. It takes the reader into a fascinating journey that shows the universal aspect of chemistry: it argues quite playfully, I think, that we are made of chemical matter, the same matter as any other chemical matter, and hence the book makes a point as to why knowledge of chemistry is so essential to Mankind.

There is a unity of matter from H_2 to the DNA molecule, and chemists have achieved this amazing feat in the 300–350 years of chemistry. In this context, chemistry is a central pillar of human culture. Why not learn about this culture?

I am really provoked now Racheli. I have to introduce the textbook to the prospective readers, and tell them maybe what is so special about it.

Racheli: Oh Sason, no need to preach here. Why not let the book speak for itself?

Sason: Now, now Racheli, you are flipping from being provocative to optimistic … This is the sign of youth … but I am old-fashioned and I believe in telling "my captive audience" what's in store for them …

Racheli: OK Sason, since you insist …

Sason: When I started my academic position as a Lecturer in Ben-Gurion University, in 1980, my first teaching assignments were courses in "general chemistry." Initially, I used a few textbooks for my teaching. Some of these started with the quantum model of the atom and the quantum revolution that ushered it … Others started from quantitative topics like units (energy, force, distance …), moles, pKa, pH, strong and weak electrolytes, the kinetic theory of gases, etc.

The molecule was of course there all along, but how to construct it from atoms, how to understand the three-dimensional structure of molecules, and what is beautiful and enchanting about their structure? All this was reached quite late; in some cases you had to go through 300–400 pages of the book to find this topic. *I felt that the beauty of chemistry ought to be revealed right at the onset of the course* to attract the students and motivate them to be engaged in the rest of the course.

Racheli: So, are you criticizing the traditional teaching in chemistry?

Sason: Yes and no. All the traditional texts I saw are very well-written, but they are quite advanced and they teach chemistry to chemists, which is sort of preaching to the believer. Considering that in the general public the term Chemistry evokes a buzz of negative words like pollution, disasters, and whatnot … I think **it is important to start by highlighting how beautiful chemistry is**.

Racheli: Are you teaching Chemistry or Poetry, Sason?

Sason: So young and so sarcastic … Beauty serves a good purpose, Racheli.

By the way, after teaching Ist year chemists at Ben-Gurion University, I already noticed that many of the first year students lacked a background in chemistry and did not know how to construct simple molecules. Given that, it made little sense to me to start with advanced fundamental and quantitative aspects of chemistry, even for chemistry students.

Racheli: I think you are right. I find the same lack of knowledge even among advanced students in chemistry.

Sason: Indeed, this was when I realized that I had somehow to cut the viscous wad that held the student audience from the molecule and its beauty.

I decided to start the general chemistry course by teaching how to construct molecules. I invented a playful approach which I called "a LEGO principle" based on simple pairing rules (the octet/duet rules), and I slowly proceeded to show them how to predict the three-dimensional (3D) structures of these molecules, and the architectural developments which occur as 3D objects form the macroscopic matter we see and feel. I reserved the more quantitative and physical aspects of chemistry for the end of the course. In hindsight, the seeds of these ideas were planted during my postdoc at Cornell where I heard my teacher, Roald Hoffmann, explaining to us his "isolobal analogy." I will talk about it in Lecture 9.

Racheli: This must have been a strange approach to the students of chemistry …

Sason: On the contrary! I felt that the communication with the majority of the students improved, as between the years 1982 and 1992 I appeared a few times on the list of the best teachers. One of those years I was on the same list with the writer Amos Oz, which is not too bad for a chemistry teacher …

Racheli: Did you talk to other teachers? Chemistry teachers tend to be conservative.

Sason: You got me there, Racheli. I never gave teachers an in-depth exposure to the method. One day I should …

Racheli: Yes, you should!

But what on earth made you start teaching the course to humanities and social science students here at the Hebrew University?

Sason: Racheli, you are jumping too far ahead. I am old-fashioned …

Racheli: So be it: tell us your entire story …

Sason: In 1992, when you barely finished elementary school, I moved to the Hebrew University (HU) in Jerusalem. And, lo and behold, I was asked again to teach general chemistry.

Racheli: You must have been bored, teaching the same course for so long.

Sason: No, Racheli. I have this self-inflicted habit of rewriting my courses every year, and this is what I did when I started my general chemistry teaching at the HU.

Racheli: Yes, but it is still the same course!

Sason: Not really. Rewriting the course made its first part into a grand scheme of constructing the material world. I taught how to construct and comprehend molecules, starting from H_2 all the way to DNA. I even taught energy and structure as a game of construction from building blocks, like LEGO or TINKERTOY. I wanted to show the unity of matter from the chemical perspective.

Racheli: I know this because I took your course in 2003. But you still did not answer my question.

Sason: Be patient Racheli, we are almost there …

Well, after one round of teaching, I decided that I should start the course with a motivating lecture. I called this lecture. "Chemistry, Love and the Price of Diamonds." You may recall it.

Racheli: Yes. This was a fun lecture. You also argued that **knowing chemistry is also a journey of Mankind into his material essence, and a process whereby we learn to appreciate the limitations of matter and become its benevolent master**, in the spirit of Primo Levi, the chemist and writer.

Sason: Yes, Racheli. This lecture has become subsequently a basis for the public lecture I have given plenty of times to all kinds of audiences, including teachers, high school students, scientists, and lay audiences. It has served also a basis for an essay I wrote in 2003.[1]

Racheli: What about my question???

Sason: I got carried away, Racheli …

Racheli: As usual … So?

Sason: Well this is how it happened. My colleague, David Avnir, who was also your PhD adviser, had been in charge of the curriculum at the HU in 1993. One day, he invited me to talk about teaching, and he suggested that I teach in the Mount Scopus Campus a course to social science and humanities students. I agreed to do this.

Racheli: Weren't you concerned to teach chemistry to such an audience?

Sason: I was … Even though I let myself be tempted into giving such a course, still I was very concerned and had many sleepless nights.

Racheli: Sason, sleepless nights aren't a solution. How did you resolve your worries?

Sason: I constructed initially a course, which was basically storytelling about chemists and alchemists, such as the story of the pathological science of "polywater."[2] I barely touched any chemistry …

Racheli: So what made you change it completely to the format that is now the basis of this textbook?

Sason: I learned something from a clever student …

At the end of that course one of the brightest students (from an elite program in the Institute of Psychology), approached me and said: "It is clear you like teaching, but it is also clear you were afraid of teaching us any chemistry."

Her statement was arresting! Immediately I realized what a serious mistake I had made. I did not dare to teach chemistry!

Racheli: Well this is not so good if you realize that you chickened out …

Sason: In the following year, I decided to use all the material of the first half of the general chemistry course and teach it to humanities and social science students on the Mount Scopus Campus. In the next year, I had 200 or more students. The course became a hit …

Racheli: Yes. I saw somewhere that you were voted as one of the best teachers in the HU in 1995. But, assuming we do not want to get indulged in patting ourselves on the back, tell us why did the course become a hit?

Sason: Well, one of the fun features of the course was the demonstration sessions, which revived "the magic of chemistry"—one material disappears, and another one appears …

At the HU, we had a laboratory technician, named Calvin DeLano. One day, it was intimated to me that Calvin is a wonderful demonstrator, and I immediately decided to enlist him for the course on Mount Scopus.

Racheli: This is the same Calvin who joined us in 2010 when you decided to re-teach the course. Right? By then he was nearing his 80s. It is like the return of the Jedi …

Sason: Yes, it is the same Calvin. Calvin was a "magician," and his demonstrations were a lot of fun. He also loves teaching.

Racheli: Sason, how can you love a course so much and then quit it as you did, in the 1990s? This does not make much sense. Does it?

Sason: Well, first I left for a sabbatical in Rochester, New York. And when I came back, I recalled the Herculean effort of giving this course. Each week, Calvin and I would schlep all the chemicals in his car, and drive to Mount Scopus, and then carry all the chemicals to the lecture room …

So after teaching a few more times, I decided to stop schlepping chemicals, and teach other courses.

Racheli: I am still waiting to understand what is it that made you go back to teaching this Herculean course?

Sason: I love chemistry and teaching, and I guess I am easily tempted. We get like that as we grow older …

In 2009, I was approached by the Vice-Rector, Oded Navon, who asked me to teach this course to social science and humanities students in the new program called "Cornerstones." He promised me I could have budgets for a teaching assistant and a demonstrator. I promised him an answer within a week.

Guess what I did all that week?

Racheli: You rewrote the course …

Sason: No, no. Be serious … I spent time convincing the Jedi Calvin to join me.

Calvin was apprehensive, because he was living now in the north of the country (Netanya). The prospects of having to travel to Jerusalem and then schlep chemicals were not too attractive.

Racheli: So how did you convince him to join?

Sason: First, I had a great idea to look for you and offer you to join as a teaching assistant. You too hesitated initially and then decided to join me, so I must have been quite convincing or lucky. I just hope I did not tell you it was going to be easy …

Racheli: I do not recall how you actually convinced me. But what attracted me for sure were the prospects of a unique experience to teach such a course to humanities and social science students. As you know, I play the piano and sing, and I still study chemistry. So it looked natural to me that anyone, even staunch artists, should learn chemistry.

Sason: When you agreed, I rushed to Calvin and told him that you will be joining us old Jedi. Then the three of us met, and the rest is history. You and Calvin made a wonderful team. You wanted to learn his tricks, and he was eager to demonstrate.

Racheli: Yes, sometimes we must have looked like the Marx brothers … It was fun!

Sason: More like the Three Stooges (from the American vaudeville slapstick comedy acts).

It was fortunate that you joined us, because when Calvin had to leave us, you managed to recruit another magician, Yohann Aouat, who was doing his PhD with Avnir and Gad Marom. We had two absolutely exciting years teaching this course, which grew from 80 to 200 participants. The memory of this wonderful time was behind my decision that it is time to write the material as a textbook.

Racheli: Sason, I hate to interrupt your enthusiasm for reminiscing, but isn't it time you told the readers something about the structure of the book?

Sason: Thank you Racheli for interrupting my endless digressions. This is a **textbook** for teaching chemistry in a likeable and fun manner. It is integrative, and the topics are meshed together by association rather than by systematics.

It opens the course with a wild introductory lecture, which starts with a presentation about love, lust, hate, etc., based on my essay,[1] and argues that all is chemistry and we are chemistry too … This is followed by fun demonstrations of the magic of chemistry.

Racheli: But this appetizing lecture is not chemistry teaching. What do you actually teach in the textbook?

Sason: I teach constructing molecules. This is done here in a playful way using atom connectivity and combining atoms into bonds using the "click word." For example, consider two H atoms, having each a single connectivity, and "click"—there is a bond …

Racheli: Mind you, atoms have many electrons. Which ones do you click?

Sason: Only the valence electrons, and not all of them. I teach how to "read" their numbers from the periodic table. Then applying the upper bounds octet/duet rules results in a <u>table of connectivity</u> that enables the student to construct simple molecules, which obey the "nirvana rule."

Racheli: Yes, Nirvana. How on earth did you come up with this term as a chemical rule?

Sason: Being playful did not hurt any teacher. I was seeking one term that can replace the duet and octet rules, and I came up with the "nirvana rule"; the "profound peace of mind" of the atom. The nirvana rule enables me to construct in a single stretch H_2, H_2O, H_2O_2, NH_3, CH_4, N_2, P_4, CO_2, etc.

Racheli: How many more molecules can you make without getting bored?

Sason: To avoid boredom, I usually tell the students/readers about the significance of the molecules we make, something about history, and sometimes funny stories (e.g., the discovery of phosphorus by Hennig Brand by extraction from urine).

This is also an opportunity to present concepts such as a single bond, a double bond, a triple bond, a lone pair, and the large molecular variety that comes with a connectivity of two or more.

Constructing SiO_2 is an example of a large molecule, an infinitely large one, made from an infinite array of $SiO_{4/2}$ units. This is a corridor for presenting the terms "polymer" and "stoichiometry."

Racheli: Sason, do not forget that after you demonstrated that $SiO_{4/2}$ is an "infinite" molecule, you told the students about glass coloring and you read a recipe from the time of the Assyrian King Ashurbanipal (seventh century BC), explaining how to color glass. Then came the demonstrations by Calvin/Yohann. And I followed on this with a demonstration of making glass from sol-gel ("water glass") by adding gradually some acid, and coloring it by adding a colorant. I recall that you hardly let us finish the experiment before you told the class about the great alchemist van Helmont, who made "water glass" from sand and then reconverted it to sand by usage of acid. For van Helmont, this was a proof that the entire universe was made of the "element water."[3]

Sason: Racheli, you know I love stories! I am very glad you remember them so well.

Anyway, back to the main story, the construction of SiO_2 shows that building molecules from atoms can be tedious when the molecules get large. The game of construction has to be scaled up. This is done by using as a basis molecules like H_2O, NH_3, and CH_4, which we already built by the "click" method, and then by plucking off bonds ("unclicking") we form new molecular fragments with specified connectivities such as CH_3 with a connectivity of one, CH_2 with a connectivity of two, etc. Using these large modular fragments, we then proceed to make molecules using the advanced kit. In this process, we start with H_2, and we end with polymers, RNA, DNA, and transition metal complexes. It's all there in the table of contents and in the lectures that follow.

Racheli: Don't forget; by tutoring I cover for your dizzying pace of lecturing …

Sason: Absolutely! Chemistry without tutoring does not sink in.

Racheli: And … What about 3D?

Sason: Goodness. You are a slave driver Racheli!

Once we complete the grand construction of the molecular worlds, I will move on to discuss the 3D structure of molecules, using the famous valence shell electron pair repulsion (VSEPR) rules.

Now we can teach also isomerism such as geometric and positional isomers of C_2H_4, benzene, and *chiral* isomers, how Nature makes use of *cis-trans* geometries (retinal, Tamoxifen), and chirality. Mind you, the 3D rules can be applied also to longer chains of hydrocarbons. In brief, we show that molecular architecture can be reconstructed based on a few simple structural elements.

Racheli: Molecular models help to visualize these 3D objects. Internet sites (e.g., http://www.3dchem.com/) can also be useful here. There is even a magnetic molecular kit, called Snatoms, in which the atoms/fragments actually "click" during the bond making (https://www/youtube.com/watch?v=He30D8M5fNc). Are you telling this to your readers at some point?

Sason: Absolutely. You tempt me to keep giving advice to the teachers …

Racheli: You make me feel guilty. Do you want me to stop reminding you?

Sason: Heavens, no! I am used to this affliction by now.

Racheli: So here is another reminder: will your book cover electronegativity, ionic bonds, or transition metal complexes?

Sason: Sure. Electronegativity and polarity of bonds provide me with an opportunity to talk about the hydrogen bond and its role in the architecture of the chemical matter: water, ice, proteins (silk, hair, nails, muscles, tendons), the "double helix," etc. This is the point where I turn back to the genetic code and show that it is based on hydrogen bonding.

Racheli: Have you forgotten ionic bonds?

Sason: How could I?

Racheli: Please explain briefly if possible, how you intend to teach these bonds?

Sason: I invented the "click-clack" method to deal with ionic bonds. We first "click" the atoms according to their connectivity to form covalent bonds such as "Na-Cl," and then we "clack" by shifting the bond pair exclusively to Cl, to form Na^+Cl^-. This allows the students to make ionic molecules with the correct stoichiometry and charges.

Subsequently, we can easily explain the formation of ionic solids by close packing of charged spheres. Then we speak about solubility of these solids and show that some of them like Na^+Cl^- are soluble, while others like $Ca^{2+}CO_3^{2-}$ are not. This gives me the opportunity to tell about the role of ions and ionic material in living systems (firing during neurotransmission and skeletal construction).

Racheli: This was not too brief … Anyway, what about transition metals? I know it is tough …

Sason: I am not chickening out. I have a special lecture (Lecture 9) on bonding in transition metals using the "click" method. I teach about hemoglobin, cytochrome P450, Ziegler–Natta catalysts, and whatnot …

The colors of transition metal complexes will give me a chance to connect matter and color or light. I really want to stress how and why we see the world through a narrow window that is like a tiny slit in a huge cathedral.

Racheli: You forgot … the usage of "light" to characterize all the bonds existing in a molecule and even gauge the bond lengths and bond angles (spectroscopy and X-ray diffraction).

Sason: Nothing is forgotten, Racheli! As I plan to show, molecular dimensions and energies can also be described using the LEGO principle, as combinations of typical building blocks.

Racheli: I always felt that addressing the negative aspects of chemistry is an important lesson for students. Did you include this address in your textbook?

Sason: This will be done in Lecture 10, where I remind the students about the two faces of chemistry—the beneficial versus the destructive, like the Roman god Janus.

The course ends with a popular lecture (Lecture 11) entitled "Chemistry is Everything … ," which is followed by demonstrations that emphasize the magic and fun of chemistry.

Racheli: You have not said even one single word about quantitative aspects of chemistry. As you know, Chemistry is not only poetic, but it is also a quantitative science. Aren't you concerned by criticism that you are not being systematic, and that chemistry without a mole is not chemistry?

Sason: I am prepared for some criticism such as "be more systematic," "introduce terms like conformations," "teach the mole," "teach pK_a," etc. But I am not concerned because I think any teacher who will be convinced to use this book will see that this criticism is not particularly valid. The foremost aim is *to project the beauty of chemistry* and to make the students familiar with molecules, which make up this beauty.

This book will not "harm" the students but will make them better prepared for the future standard chemistry course, by having gone through the fundamental microscopic construction process, and having some appreciation of the immense beauty of chemistry.

Racheli: Let me insist. Why aren't you being systematic?

Sason: Elementary! Through the "click" construction,[4] the book enables the students to sample various advanced topics in chemistry, which are traditionally taught only later during the chemistry major curriculum. As such, I prefer to teach in an associative manner and not in a systematic one. It is much more interesting for the students. They will eagerly learn quantitative aspects in advanced courses after they learn to love chemistry.

Racheli: Here is a tougher question, what about the mole?

Sason: Not a problem! I am going to teach balancing equations using microscopic constitution: *The principle of conservation of the number of atoms on the two sides of the chemical equation.*

At some point, I shall explain the mole in historical and practical contexts—for example, by calculating how much CO_2 is produced by a car per year. You won't believe how much!

Racheli: What do you plan to do for advanced topics, like quantum theory?

Sason: I try to tell the readers about more advanced issues at the end of the chapters, in special sections, which I call "retouches."

Racheli: What, what … ?

Sason: In my childhood, when I was an amateur photographer, we used to "develop" photos on a paper that was covered by a silver salt. The developed picture would contain black areas of metallic silver and white areas where all the silver was washed away during the process. This was never perfect, and the silvery black part would contain small white spots, stains, scratches, etc. It would need to be refined to make it perfect.

This act of refinement is called "retouching." So the retouching sections in our book will attempt to complete the white stains left by me in the text and refine the knowledge for the advanced students.

Racheli: That was a long digression, Sason. What I still do not understand though is if your retouching will be on the level of advanced courses?

Sason: No!

The retouches are meant to offer "gentle touches" of topics that belong to more advanced courses and thereby make the students aware of these topics.

Racheli: I am afraid to ask more questions, but I really have an urgent one: Who after all is your audience in this textbook?

Sason: Remember Racheli, the material in this book constitutes the first part of the general chemistry course I taught in the University, and then it was adapted for the course I taught for humanities and social science students. As such, the book is written in a conceptually modular form. It will fit very well either *as a preliminary part of a general chemistry course*, wherein the teacher augments the material with more physical and quantitative topics including the structure of the atom and periodicity, or as *a preparatory course for chemistry majors and scientists* (biologists, earth scientists, etc.) who a lack background in chemistry. As is, or with some deletions at the discretion of the teacher, it is suitable as a textbook for chemistry for humanities and social science students, chemistry for agriculture students, economics students, for hotel-school students. And, last but not least, it is very suitable for teaching chemistry in high schools, for those who wish to take credit points of chemistry and "not jump into the quantitative pool."

Racheli: I recall that one of the referees of the book proposal wrote: "Maybe the chemical industry would use this as a tool to improve the image of chemistry."

Sason: To that I say, why not? Industry will have much to gain if such a book becomes popular!

Racheli: Yes. Let's dream it will come to pass.

Sason: Racheli, I must say I enjoyed our interview. I think I will adapt it to usher all the lectures, instead of writing tables of contents. Do you mind?

Racheli: This is a heavy responsibility. Will I be the only interviewer?

Sason: I think Usharani Dandamudi, my postdoc, will take over when you tire. And I have some surprises for you and the readers. There will be some more interviewees …

REFERENCES AND NOTES

[1] S. Shaik, *Angew. Chem. Int. Ed.* **2003**, *42*, 3208.

[2] F. Francks, *Polywater*, MIT Press, 1981.

[3] P. Ball, *Elegant Solutions: Ten Beautiful Experiments in Chemistry*, The Royal Society of Chemistry, Cambridge, UK, 2005, pp. 11–21.

[4] Our "Click" word for making a bond should not be confused with the now popular "Click Chemistry," which is reviewed in: M. V. Gill, M. J. Arévalo, O. Lòpez, *Synthesis*, **2007**, 1589.

LECTURE 1

MOLECULAR BLUES

1.1 CONVERSATION ON CONTENTS OF LECTURE 1

Racheli: Knowing you, Sason, in your first lecture you are going to preach to the students about the motivation to learn chemistry, using material from the essay you published in 2003.[1] You are probably going to use a presentation and bring molecular models to wow your students by the beauty of these molecules and by the dizzying pace of your lecture …

Sason: Absolutely so, Racheli, your memory is good. I intend to show the beauty of chemistry as much as I can … And yes, the presentation is available upon request. Regarding the "dizzying pace"—no worries. This is an appetizer lecture that will not be included in the final exam.

Racheli: Do not forget the role of the fun demonstrations (*ca.* 30 min.), which show the magic of chemistry, "one material disappears and a completely new one appears!"

Sason: How can I? Chemistry without demonstrations? No way!

Racheli: When do you plan to tell about the development of the atomic hypothesis?

Sason: Not during the lecture! I will have however some stories about the fathers of the atomic hypothesis in the Retouches section at the end of the lecture.

Racheli: It is annoying to have only fathers … I wish we would have known **of mothers too.**

Chemistry as a Game of Molecular Construction: The Bond-Click Way, First Edition. Sason Shaik.
© 2016 John Wiley & Sons, Inc. Published 2016 by John Wiley & Sons, Inc.

Sason: We are not really orphans ... There is of course Marie Curie and others. But this aside, I am eager to start lecturing about the huge importance and beauty of chemistry.

1.2 THE UNIVERSAL ASPECT OF CHEMISTRY

The term "chemistry" evokes immediately a buzz of words, like air and water pollution, ozone hole, chemical disasters, terrible stuff ... In other words, the public image of chemistry is poor! In fact, little seems to have changed in this respect since Gabriel-François Venel depicted in 1753, in Diderot's *Encyclopedia*, the "miserable state of the chemical community" as being "isolated in the midst of the greater people hardly curious of its business."[2] All chemists are painfully aware of this public image of chemistry,[3] and most of them feel that this is a gross mistake that overlooks the essence of chemistry. Unfortunately, even though it is absolutely necessary to show what a wonderful science chemistry is, and what great achievements it has made in deciphering the constitution of this universe, the popularization of chemistry[4] is still scant. Thus, instead of explaining or apologizing for the negative issues of chemistry, I wish to make a 180° turn and take you on a fascinating journey that shows the universal aspect of chemistry.

> ☎ Chemistry is the window given to Mankind to glimpse into its material essence.
> ☎ Chemistry is the subconscious journey of the human race into the mysteries of his material self ... And is hence a central pillar of human culture.

Let us talk then about the connection of chemistry to love, sadness, addiction, the price of diamonds, and about Viagra—a molecule more precious than diamonds. It is through its relation to fascinating human attributes as love, addiction, sex, dreams, pleasure, etc., that chemistry can be more easily appreciated in its universal sense. It is this universality that makes chemical knowledge, on the one hand, an enchanting story, and on the other, an educational dictum for us as inhabitants of this universe.

1.3 LOVE, ADDICTION, PSYCHOLOGICAL BALANCE, ETC.

Let us start with falling in love or romance. It has been known for some time that the "love life" of insects is regulated by small molecules called pheromones.[1] This is true also for mammals.[1] But, what about love in humans? Well, humans may not be an exception to this *chemical control* of love. Thus, even though the chemical details of the emotional map are not yet worked out precisely, it is becoming gradually evident that emotional life is written in our genes and is coded by specific molecules made in the brain.[5] Indeed, our brains are chemically wired.[1,6] Some of these neurochemicals are neurotransmitters, others are neuromodulators that regulate the

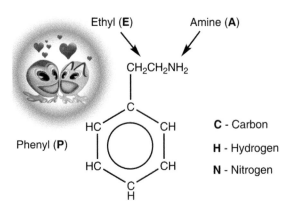

SCHEME 1.1   The molecule PEA and the meaning of the acronym. The constituent atoms, C, H, and N, and their names are shown alongside the molecule. Phenyl (**P**) is the hexagon with C and CH groups.

action of neurotransmitters, and others are hormones, etc. These neuromolecules act sometimes on their own, other times with neuromodulators and other neurotransmitters, and still other times by creating secondary messenger molecules that trigger the emotions.

Psychiatric research suggests that the brain chemical PEA,[1] in Scheme 1.1, may well be the neurotransmitter and neuromodulator of libido and interpersonal energy; its administration increases emotional warmth, affection, sexuality, and feeling of physical energy. It was even implicated in the "runner's high," which is the feeling of wellness that accompanies physical exertion.[1] PEA is a molecule, and its name is simply the acronym of the chemical groups in the molecule, as indicated in Scheme 1.1. The hexagon, made of five CH groups and one C atom, is called by chemists *Phenyl,* the little chain, CH_2CH_2, is called *Ethyl*, and the NH_2 tail is called *Amine* (put together, phenylethylamine).

We shall learn in the next lecture to construct molecules from atoms. At this point you should not be concerned about the excessive jargon I am using at times; this is a scientific language, and we shall learn it as we go along. Right now, let us return to "love."

While PEA may certainly function together with other neurotransmitters (e.g., dopamine, noradrenaline, and serotonin[7] in Scheme 1.2), its action in the emotional domain seems to be unique, and it has indeed a specific site (so-called receptor) located in the *amygdala*, the brain's emotional command center.[1] Unique also are the facts that PEA has a very short lifetime (minutes), and is degraded by a specific form of the enzyme MAO (monoamine oxygenase), the MAO_B form. The short lifetime suggests that PEA has a special biodynamic role, possibly associated with excitory effects triggered in a short time slot. By contrast, other neuroamines (serotonin, noradrenaline, and dopamine) have long lifetimes (hours) and are degraded by the other form of the enzyme, MAO_A.

The effects of PEA on human behavior have led to the hypothesis (henceforth the "psychochemical hypothesis") that PEA is associated with the event of "falling in

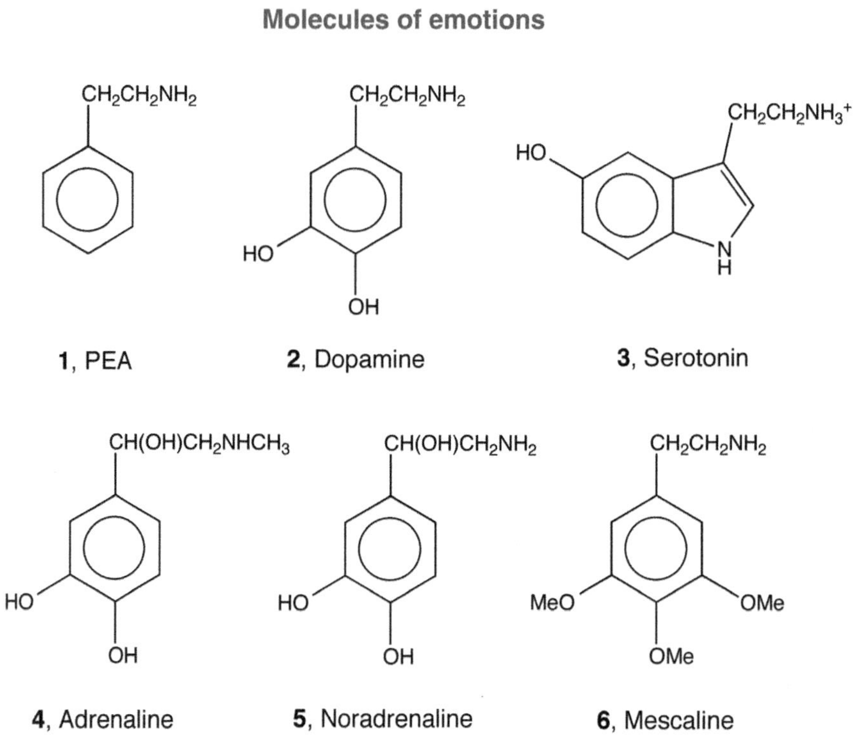

Molecules of emotions

$CH_2CH_2NH_2$

1, PEA

$CH_2CH_2NH_2$

HO

OH

2, Dopamine

$CH_2CH_2NH_3^+$

HO

N
H

3, Serotonin

$CH(OH)CH_2NHCH_3$

HO

OH

4, Adrenaline

$CH(OH)CH_2NH_2$

HO

OH

5, Noradrenaline

$CH_2CH_2NH_2$

MeO

OMe

OMe

6, Mescaline

SCHEME 1.2   Molecules of emotions.

love."[8–10] Even though this idea is very simplistic and may still be in the realm of speculation, nevertheless, *the molecular regulation of emotions is well established.* As such, it is interesting to go over the "psychochemical hypothesis," since it might contain some seminal truth considering the role of PEA in regulating emotional impact.

What is the role of PEA? Walsh[8a] gives a vivid description of the action of PEA: When we meet someone who attracts us "the whistle blows at the PEA factory." Namely the brain starts to generate PEA like crazy! Note that unlike other mammals, where "falling in love" is triggered by scent, I am confident that none of you has ever seen a guy who rushed to smell the gal he was suddenly attracted to, or vice versa. In humans, the attraction is triggered by sight and mediated by PEA. A mere glimpse is required to activate the sensation of romantic love.

The most famed instance of *love at first sight* is the biblical story of David and Bathsheba. The Bible tells it in a succinct manner:

"From the roof he saw a woman bathing. The woman was very beautiful …"

That's it! There is nothing more to explain because we understand from our own experience what King David must have gone through. He was flooded with PEA at the sight of Bathsheba. The synthesis of PEA in the brain and its relay into the entire nervous system are implicated in the generation of excitement at the sight of the subject of love, and the great longing when the lover is not around.[8a–d,11] Thus, PEA is a messenger of the brain that activates the entire body (assisted by quite a few other companion neurotransmitters), causing faster heartbeat, dry mouth, sweaty palms, and a general feeling of intoxication, alertness, and high. PEA is the arrow of Cupid. It goes to the heart and alters reality.

By the way, PEA is also present in chocolate, Nutrasweet (in which the main ingredient is aspartame), and diet soft drinks, and its concentration rises with smoking marijuana,[8d] etc. However, all these sources of PEA do not produce the kick of the brain-PEA. They fail to do so, in part because of their fast degradation by the enzyme MAO_B. Thus, even though PEA can cross the blood–brain barrier (which protects the brain from outside-incoming chemicals), most of the external PEA is degraded.[1] Externally administered love potion may exist in Shakespeare's *A Midsummer Night's Dream*, while in reality, our chemical system safeguards zealously the exclusivity of the emotional response.

The "psychochemical hypothesis"[8–11] assigns roles to other neurotransmitters, which figure in our psychobiological energy. One such family of neurotransmitters is the endorphins produced in the *hypothalamus*, the brain stem, and the *pituitary gland*.[12] Endorphins are neuropeptides (β-endorphin, the most potent of the known endorphins, contains 31 fragments, called amino acid residues) that serve as the natural painkillers of the body. They act on specific receptors on the membranes of our cells, so-called *opiate receptors* (the same receptor as for marijuana).[8e] In so doing, endorphins produce analgesia (pain relief) and create a sense of wellness.

However, according to the "psychochemical hypothesis,"[8d] endorphins also play a major role in maintaining durable relationships. Thus, over the course of time after falling in love, the body requires larger doses of PEA to maintain the same initial kick, and we gradually "fall out of love."[8a–c]. Had this been the entire story, love would have probably consisted merely of romance or lust. Fortunately, the brain, being a versatile chemist, makes also endorphins, which fix us with durable partnerships. Thus, whenever we see our partner, the brain squirts a dose of endorphin that permeates the entire body,[8d] which in turn gets imbued by a sense of security, calmness, and wellness. The sharp, almost physical pain that accompanies the loss of a loved one is thought to be caused by the sudden arrest of the endorphin effect. *We are addicted to our own brain molecules.*

There is in fact a whole science, called "brain chemistry," which has been an active part of chemistry for a long time. Figure 1.1a shows a caricature of Einstein's head with some of the neurotransmitters that are produced in the brain by instant chemical reactions. Most of these reactions are rather simple, and are taking place in the nerve cells in our body. Figure 1.1b, taken from the weekly journal *Chemical & Engineering News*, shows brain organs wherein molecules of emotions are made and then dispatched to the entire body. One of these is the almond-like organ called *amygdala*, and was mentioned above as an emotional center. When we say "made," we mean that there is *a chemical reaction that converts a precursor molecule to the neurotransmitter*. We shall talk about chemical reactions, since they form the basis of the magic of chemistry mentioned in the outset of this lecture.

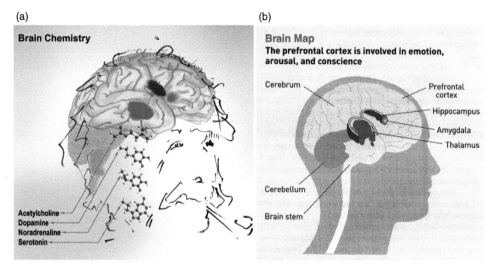

FIGURE 1.1 (a) Brain chemistry in the human head. Although I am using Einstein's head, these molecules exist in all brains (designed by M. Michman and executed by Ms Z. Cohen; copyrighted by the author). (b) A map of emotional organs in the brain. Adapted with permission from *Chemical & Engineering News*, June 2, 2003, p. 36.

Some of the additional molecules of emotions, which are even more widely important than PEA (**1**), are depicted in Scheme 1.2 alongside PEA to highlight the molecular similarity. The molecule dopamine, **2**, is manufactured in the brain in an area called the *ventral tegmental*, and is released in the frontal cortex (see Figure 1.1b). This molecule is involved in reward-seeking behavior, learning, drug addiction, and development of dependencies, including trained conditioning and willed control.[1] According to the "psychochemical hypothesis,"[8] dopamine regulates the desire to pursue pleasure, to seek danger, and it is the motivator to achieve, to win, and to desire. It also regulates the repetition of those actions that bring about the sensation of pleasure/satisfaction. With little dopamine, there is little joy of life, little sense of adventure and experience,[8] while too large quantities or enhanced dopamine absorption by nerve cells leads to addictive behavior. It appears now that addictive behavior is associated with long-term chemical changes in the communication between the nerve cells (the communication transpires via spaces called synapses).[1]

Serotonin, **3**, is formed by a simple chemical reaction deep in the brain stem (in an organ called the *raphe nuclei*).[1] It is the "mood molecule,"[8d] responsible for the psychological balance, and is the innate antidepressant "drug." It is also the "brake" against impulsiveness, causing us to "think first and only then act." Problems in the serotonin cycle are associated with chronic ailments, such as pain, depression, Alzheimer's disease, strokes, Parkinson's disease, as well as in character traits, which reflect the lack of the biological brake and restraint,[9] such as inclinations leading to impulsive murdering tendencies and premeditated ones.

Figure 1.2 shows positron emission tomography (PET) scans of a "normal brain" (left-hand side) versus a "murderer's brain" (right-hand side).[9] Though not

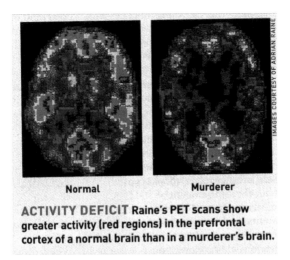

Normal   **Murderer**

ACTIVITY DEFICIT Raine's PET scans show greater activity (red regions) in the prefrontal cortex of a normal brain than in a murderer's brain.

FIGURE 1.2   Positron Emission Tomography scans show greater brain activity in the prefrontal cortex (red color) of a normal brain on the left-hand side versus a murderer's brain on the right-hand side. Adapted with permission of Adrian Raine, the University of Pennsylvania.

measuring serotonin activity explicitly, PET scans nevertheless shed insight into the remarkable impact of serotonin. When serotonin activity is abundant or highly reduced in the brain, this phenomenon is indicated by glucose consumption, or lack thereof, in the prefrontal cortex in PET imaging.[10] (Yes, the brain "eats" a lot of glucose, a sugar, to stay active.) The red areas in Figure 1.2 are the active ones, while the dark areas are inactive/less active. It follows therefore that the "normal brain" in Figure 1.2 is flooded with serotonin activity compared with a "murderer's brain" (on the right), where this activity is hugely depleted, marking the acute serotonin deficiency (which has been reported in the literature[9]). Indeed, acute serotonin deficiency is a biological cause of aggression and violence.

Less acute deficiencies are responsible for depression, schizophrenia, craving for carbohydrates and sweets, etc. Normal levels cause a balanced personality, while excess is associated with anxiety. The permanent state of anxiety and fearfulness was recently proven to be associated with a hyperactive *amygdala* (the emotional command organ of the brain)[5] produced by inefficient back-transport of serotonin to its mother cell.

Serotonin (or its precursor tryptophan, or serotonin-promoting molecules) can be found in eggs, meats, grains, bananas, tomatoes, and pasta. Even though you might be wondering now if this is what makes your Italian friends so jovial, actually eating any one of these foods does not necessarily increases serotonin levels in the brain, because of serotonin's fast degradation by the enzyme MAO_A.

Near serotonin, Scheme 1.2 shows adrenaline, **4**, and noradrenaline, **5**, also known as epinephrine and norepinephrine, respectively. These are the stress hormones, which prepare the body for strenuous activity (*the fight or flight instinct*). Adrenaline, released from the *adrenal medulla* (adrenal gland), acts to raise the level of glucose in the blood and thereby provides the surge of energy needed for performing demanding tasks or facing dangers. Noradrenaline is a neurotransmitter, which is also released in stressful situations along with adrenaline. However, it is implicated (in conjunction with PEA and dopamine) also in maintaining the sensation of vitality, high degree of attention, and focus.

In **6**, we show mescaline, a hallucinogenic drug. The structural similarity of mescaline to the innate drugs is apparent (in the Retouches, you can find related drugs ephedrine, pseudoephedrine, and methamphetamine). It raises thoughts about the "reality" of emotions, independently of their unique molecular effectors.

1.4   THE CHEMICAL MECHANISM OF NEUROTRANSMISSION

The discovery of chemical neurotransmission in the late 1950s has led to a paradigm shift from an "electrical brain" to a "chemical brain."[1,6] It is instructive to say a few words on the mode of action of neurotransmitters. Thus, *the neurotransmitters are effectors of information flow between nerve cells*, and they serve as the means by which the brain commands and receives information from the body, and reacts to external stimuli. At any given time, 100 billion nerve cells may be engaged in this "conversation," which is conducted by two prototypical mechanisms, fast ion "firing" and slow biochemical cascade, and combinations thereof.[1] Cell communication is an exceedingly rich and a fast-developing topic that cannot be elaborated in this lecture.[6,11] Perhaps the easiest mechanism to illustrate, albeit very briefly, is the mode of cell communication that is triggered by the neurotransmitter molecule serotonin and mediated by an electrical impulse ("firing"), as exemplified in Figure 1.3.

As shown in Figure 1.3, the nerve cells are segregated by tiny gaps, known as synapses.[1,6,11] One can see that the presynaptic cell ending contains pockets which store serotonin molecules (symbolized as red spheres). When an electrical signal (a flux of positively charged potassium, K^+, ions in the serotonin case) reaches the cell ending, it swells and its pockets release the neurotransmitter molecules. These in turn cross the gap and attach themselves to receptors in the membrane of the next cell, the postsynaptic cell. The serotonin receptor is an ion channel made from a bundle of a few proteins; for example, the key receptor of serotonin contains five protein bundles.[12,13] The attached neurotransmitter gates the channel in the postsynaptic cell by causing a structural change that opens the pore of the receptor.[12a,b] This opened channel permits now the ions to flow through it and this causes another neurotransmitter to cross the synaptic gap, leading another cell to fire ions. In a fraction of a second, billions of neurons exchange serotonin and fire ions, and this brings about a psychological balance. We still do not know how to describe the expression of this serotonergic networking as a "psychological balance." But we know all too well this integrated experience to be able to feel it or diagnose its absence. Chemically, what we know for sure is that *the driving force of this wondrous pacifying effect is the serotonin-gated ion firing*.

As shown in the top part of (A) in Figure 1.3, the neurotransmitter molecules, which completed their job, must then return to their mother cells (the presynaptic cell) by recrossing the synaptic gap and attaching to the mother cell. To avoid cell

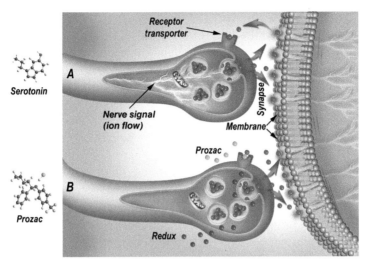

FIGURE 1.3 Drawings of the serotonin and Prozac molecules (left side), and a schematic representation of the events in the serotonin cycle using drawings A and B: In (A), a neuron ending of a presynaptic cell (on the left) and its pockets full of serotonin (the red spheres) shown near the membrane of the postsynaptic cell (on the right). An ion current (the golden-colored lightning) passes through the presynaptic cell. The pockets open wide, and the serotonin molecules are released into the synaptic gap. These serotonin molecules cross the synaptic gap and attach to receptors in the postsynaptic cell, causing ion firing, which releases serotonin to another cell, etc. The top part of (A) shows a serotonin molecule that completed its action going back and being taken up by the mother cell. Part (B) shows the action of Prozac (symbolized by the green spheres). It blocks the receptor in the mother cell and does not permit the serotonin to be taken up again. Those "rejected" serotonin molecules may bind again to the receptor in the postsynaptic cell. Thus, a small amount of serotonin can achieve enhanced ion firing. Some medications, like Redux (blue spheres), home on the presynaptic cell and cause ion firing that stimulates serotonin release. The figure was created by the artist Z. Cohen and is copyrighted by the author.

damage,[14] the enzyme MAO_A makes sure to destroy any stray serotonin, not on a receptor or inside a pocket in a transporter cell (neurotransmitters are replenished in the brain by specific chemical reactions).

A recent exciting discovery is associated with the mechanisms of return of serotonin molecules to their presynaptic cells.[5] Serotonin is assisted to recross the synaptic gap by molecules called transporter proteins, which are in turn encoded by a gene that has two forms (alleles). One of these alleles produces only about half of the transporter protein as the other allele, thus resulting in a more frequent serotonin neurotransmission.[5] This higher frequency is expressed in the brain as a hyperactive *amygdala* (the emotional command of the brain), and is associated with the psychological state of permanent anxiety and fearfulness. *This is a striking connection between different forms of a single gene, transportation efficiency of a single brain chemical, and different patterns of human response to emotion-laden stimuli.*

We already mentioned that serotonin deficiency is a major cause of depression, which is the most widespread disease in the western world. Once I heard from a renowned neuroscientist[15] that the Phoenicians (an ancient people who used to live along the shores of Lebanon and who founded the superpower Carthage that was later destroyed by the Romans) had a magic solution for the treatment of depression. They used to collect all the depressed members of the community, put them in wrecked ships, and send them to an unknown destination. When these members did not return, the Phoenicians assumed that they were healthy and content wherever they settled.

In modern times, the solutions are a bit more elaborate. The understanding of the chemical essence of neurotransmission enabled chemists to synthesize drugs that function by homing on the elements of the neurotransmission mechanism. For example, as shown in (B) in Figure 1.3, Prozac, an antidepressant drug (symbolized by green spheres), homes on serotonin by blocking the receptor in the presynaptic cell and causing the small amount of serotonin to go back to the postsynaptic cell and induce refiring, etc. Note from Figure 1.3 that Prozac shares some chemical similarity with serotonin. Redux, an antiobesity drug (symbolized by blue spheres), homes, on the other hand, on the pockets of the transporter cell and causes enhanced transportation of the serotonin (Redux has been taken off the market due to undesirable side effects). Many are the options of tweaking the cycle of a neurotransmitter.

Similar mechanisms are found for other neurotransmitters. Figure 1.4 shows the dopamine cycle with cell-to-cell communication via synaptic gaps and ion firing.[8, 11] On the right-hand side, we can see that the mechanism of addiction to cigarette smoking combines two chemical events: (i) nicotine enhances dopamine release, and (ii) another molecule in the cigarette smoke, presumably ammonia (NH_3; see blue spheres in Figure 1.4) intensifies the release of dopamine, while other additives block the enzyme MAO_B. These events enhance the action of dopamine, which leads to the sensation of pleasure, alertness, and satisfaction, and eventually to addiction (recall the movie *The Insider* with Russell Crowe). Indeed, in the basis of human behavior, there lies an enticing simplicity of chemical origins.

1.5 MOLECULES OF PLEASURE, WELLNESS, AND PAIR BONDING

Let me mention now the neurotransmitter oxytocin that is formed in the *pituitary gland* and plays a major role in the "psychochemical hypothesis."[8, 16] Oxytocin is shown in **7** in Scheme 1.3. This molecule is made from many atoms, and for this reason, it is represented using three-letter codes for its amino acid building blocks (e.g. Tyr, Cys, etc.) rather than showing all the atoms in these amino acids. These building blocks make both our proteins, which are huge molecules, and smaller molecules called peptides, which is what oxytocin is. Later in the textbook, once we construct the amino acid building blocks, we shall learn how to piece together them as proteins and peptide molecules.

Oxytocin is one of the first peptides which was synthesized in the laboratory and proved thereby the chemical origins of psychobiological "energy flow." In laboratory

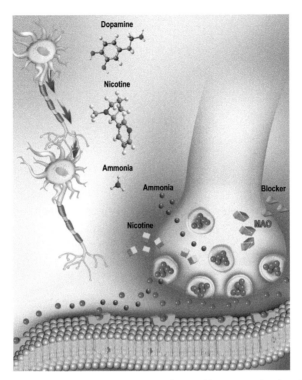

FIGURE 1.4   The dopamine cycle occurs by cell-to-cell communication (left-hand side) as dopamine crosses the synaptic gap between the cells. A schematic description of addiction to smoking is shown on the right-hand side, in which the region of the synaptic gap between neuron cells is enlarged. The nicotine (greenish cubes) and ammonia (NH_3; blue spheres) stimulate dopamine (red spheres) release, and another molecule (symbolized by pink pyramids) blocks the enzyme MAO (MAO_B). The molecules dopamine, nicotine, and ammonia are shown in the middle of the figure. The figure was created by the artist Z. Cohen and is copyrighted to the author.

animals, oxytocin was established as a promoter of sexual behavior and pair bonding. By extrapolation, it has been hypothesized[8, 16] that oxytocin is one of the molecules that are responsible for human "pair bonding"[16] and is behind the human urge to touch, caress, and reassure. According to this hypothesis, humans with a healthy production of oxytocin tend to be also cuddly.

One can simply rub the back of one's neck to feel the pleasing sensation caused by oxytocin release. The sensation gets much stronger when your partner does it for you, presumably because oxytocin secretion is an evolutionary chemical mechanism of seeking and getting addicted to "togetherness." In women, oxytocin is thought to play additional important roles: During nursing, oxytocin stimulates the flow of milk during suckling, and the surge of oxytocin causes pleasure and relaxation and thereby reinforces the mother–child bond. During lovemaking, the surge of oxytocin causes orgasm in females. Oxytocin thus regulates our "chemical commitment" and its consequential wellness.

Molecules of pairing, pleasure and wellness

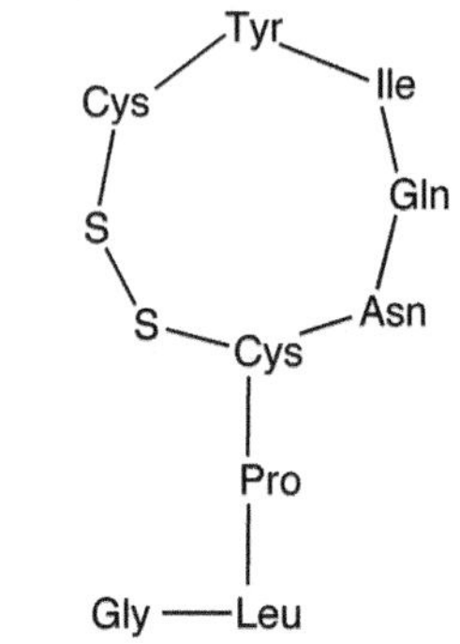

7, Oxytocin   **8**, Nitric oxide

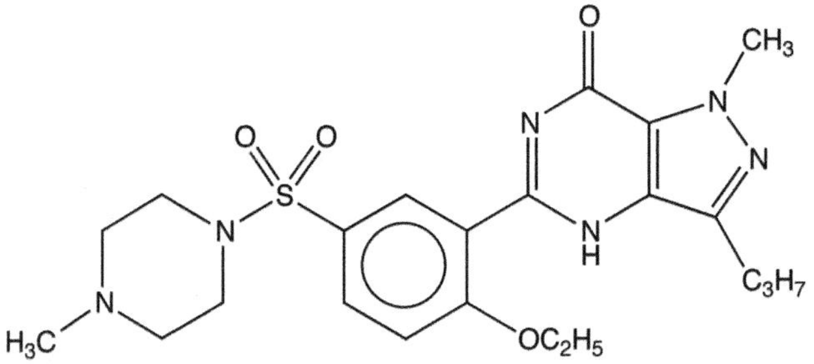

9, Viagra

SCHEME 1.3   Molecules of pairing, pleasure, and wellness: Oxytocin **7** and nitric oxide **8**. Shown also is Viagra **9**.

Males' orgasm[17] requires an initial erection, which is chemically triggered by the "natural Viagra," shown in **8** in Scheme 1.3—a little molecule made just of two atoms, N and O, so-called nitric oxide (NO).

NO is an important neurotransmitter responsible for the regulation of blood flow, breathing, long-term memory, etc.[18a] So essential is it that a special enzyme, called nitric oxide synthase (NOS), exists for the sole purpose of degrading a molecule, called L-Arginine (a natural amino acid found in granola, cottage cheese, ham, etc.), and generating from it NO in the *endothelial* cells (the *endothelium* is the thin layer of cells that lines the interior surface of blood vessels and lymphatic vessels).[18b] The diffusion of NO to the smooth muscle cells, in the walls of the penile arteries of the erectable tissue, causes a cascade of chemical events,[18b] by the end of which the muscle cells are depleted of their positively charged calcium Ca^{2+} ions (Ca^{2+} ions stiffen the protein molecules which make up the muscles), and the muscles then relax and enable thereby the flow of blood into the penis and its eventual erection. NO is involved also in the mechanism of long-term memory (so-called also long-term potentiation, LTP) in the *hippocampus*.[12b–d] In memory, NO acts as a neuromodulator that helps the cell receptor[19] to increase the concentration of calcium ions within the cells, thereby causing repetitive firing patterns of the cell, which is the essence of LTP. One cannot avoid smiling at the recognition that erection and memory require opposite directions of Ca^{2+} flow. These two actions must not go too well together ...

By the way, NO is one of the causes of the "ozone hole," and one cannot escape the recognition that chemical matter is like the Roman god Janus with the two faces: It is beneficial where it is needed (e.g., in LTP), and it is harmful in higher amounts and where it is not needed (in our atmosphere). This is a general truism about chemical matter. We shall meet the "Janus effect" many times throughout the book.

The artificial compound called Viagra (also called Sildenafil) is the molecule shown in **9**, in Scheme 1.3. Its synthesis is sufficiently simple that a second-year undergraduate student who has completed a laboratory course of chemical synthesis might be able to produce good, homemade quantities of this molecule (I am joking. Do not try!). The molecule interferes with the mechanism of the NO-induced Ca^{2+} depletion mechanism described above. The excitement caused by Viagra has barely subsided, and chemists are already utilizing their understanding of the mechanism of the erection that is triggered by NO stimulation, and are fervently working to synthesize drugs that are "more than Viagra."[20] The elixir (the potion of immortality, one of the principal goals of alchemy) may well be within the reach of chemistry.

1.6 MORE CHEMICAL CONTROL

This amazing chemical simplicity behind the human attributes is further emphasized by the story in Figure 1.5.[21] It seems that the transport of the sperm to reach the egg is chemical and is controlled by molecules (presumably secreted by the egg) to which the sperm is attracted and is navigated by in its fateful journey. One molecule that was tested is called Bourgeonal (see Figure 1.5). When Bourgeonal gets attached to the receptors on the sperm, the sperm navigates quickly and directly to the egg. By

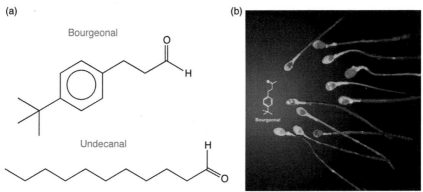

FIGURE 1.5 Bourgeonal and undecanal play roles in the selection of the future generation.[21] Downloaded from http://www.pm.ruhr-uni-bochum.de/pm2003/msg00090 .htm with permission of Prof. H. Hatt.[21a]

contrast, if the sperm attaches to the second molecule (see undecanal in Figure 1.5) that was tested, its navigation became inefficient. Since navigation toward the egg leads to its eventual fertilization, this experiment suggests that the ability of the sperm cells to get attached to very specific molecules like bourgeonal determines whether or not fertilization will occur. This may also be the manner whereby dysfunctional sperms are labeled such that the "healthy" ones are selected to fertilize the egg. It follows that *the future generation is chemically selected ...* As we shall study during this course, molecules are attracted to one another by very weak forces that can be extremely sensitive to the three-dimensional (3D) structure of the molecules and other properties (e.g., charges).

The above examples indicate that despite the biocomplexity, in the basis of the most wondrous human attributes, there lies incredible molecular simplicity. Molecular triggers, through weak molecular interactions and small structural changes, regulate information flow between the unconscious (the body) and the conscious (the brain) domains of the biosystem. Even learning and behavioral changes in one's own lifetime have chemical pathways.[1] In short, the entire emotional–sensory–intellectual system,[11, 22] which we term "human," has a molecular basis. This highlights the following central position of chemistry in human existence and culture:

☎ Mankind is made from chemical matter, the same matter that constitutes all chemical matter. Mankind is indeed the "salt of the earth."

☎ Mankind has been learning the secrets of manipulating and shaping chemical matter. This knowledge is chemistry.

☎ Chemistry is hence the window that Mankind uses knowingly and unknowingly to probe his own material being and come to terms with its limitations.

☞ It is the ultimate fate and quest of Mankind to widen the window ...

Philosophy tends to characterize self-cognition and free choice as human characteristics that form the great divide from the rest of the animal world. Some time ago, I read that rats dream about themselves and their whereabouts during the day while running in their mazes. While we cannot claim to know the imagery that might appear in the brain of sleeping rats, the signals recorded during the brain imaging are virtually the same as those while running in their mazes.[23] In this sense, a rat certainly has a form of self-cognition and awareness. In fact, the chemistry of emotions, memory, learning, etc., is universal, much as is the genetic code. Thus, the great divide between humans and the animal world is not any form of self-cognition and free choice. It is rather the self-cognition gained as Mankind journeys through chemical matter in the unconscious process of self-introspection, we call chemistry. This enables the human race to create a new evolutionary process driven by intelligence and knowledge rather than by "Natural Selection." Thus, whereas the details of the above story may very well be simplistic and new ones will replace quite a few of them, still the general lesson is with us to stay.

1.7 THE CHEMICAL MATTER

This is a good point to say a few words about the chemical matter. The central paradigm of chemistry is that:

> ☎ There exist atoms, which can form chemical bonds and thereby molecules.
> ☎ These bonds are directed in space and determine the molecular architecture, as well as the tendency of the molecule to interact with others, and to change by chemical reactions.

Let us talk about atoms very briefly. Chemists have discovered over the years materials, which they called "elements," meaning that they represent fundamental forms of matter and cannot be decomposed by chemical means to simpler entities. At the start of the nineteenth century, the English chemist Dalton hypothesized that matter was made of small entities called "atoms," which meant that the "elements" were made of atoms which were different from one element to the other. When physicists, like J.J. Thomson, Millikan, and later Rutherford, got into the picture, they probed the inner structure of the atom, and this atomic picture continues to evolve as science progresses. We confine ourselves here to the simplest possible picture (consult, however, the Retouches at the end of the chapter).

The atom is a neutral entity composed of electrons and protons, which are charged particles (the protons are positive, the electrons are negative) and neutrons, which are neutral particles. The electron's mass is 1/2,000 of those of the proton and neutron. Scheme 1.4 shows a simplistic cartoon of an atom as a sphere. The center is occupied by a nucleus (N) that packs together all the massive particles. The radius (R) of the atom is 1 Å (0.00000001 cm; see Retouches at the end of the chapter), which makes it a very small entity. However, the radius of the nucleus is 100,000 times smaller. If we imagine the atom as large as a football field, the nucleus will not even be the football, but rather a speck of dust on a shoe of the player standing in the center of the field. *All that "matters" resides in a null-space, while all the space is void, and the electrons "hover" over the surface of the deep. The chemical matter is empty and we, made from it, are sculpted voids.*

Why does not the void collapse unto itself? The architecture of the void is shaped by the behavior of this tiny entity called electron. In general, all chemical matter and its various interaction modes abide by the rules of the electronic behavior. In this sense, chemical matter is unique because the electron is an elementary particle and hence chemical patterns cannot be further reduced or reconstructed from bottom up (i.e. from more elementary constituents).

As an elementary particle, a few "laws" govern the electronic behavior. The "uncertainty principle" forbids the electron to be definitely located (on the nucleus) and have at the same time a definite energy level (or velocity). Therefore, the electron is delocalized in space and "hovers" over the nucleus. Thus, the uncertainty principle

The electron hovers over the surface of the deep

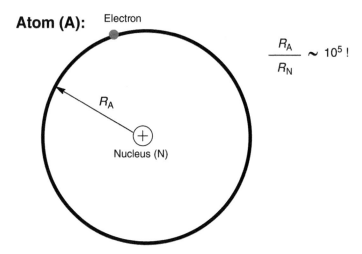

SCHEME 1.4   Relative dimensions of the atom (A) and the nucleus (N), given by the ratio of the corresponding radii, R_A/R_N. All the heavy particles (protons and neutrons) reside in the nucleus. As can be seen from the R_A/R_N quantity, the electron (shown as a red circle) hovers in a large distance around the nucleus.

gauges the size of the void in the atomic entity (you may want to read the Retouches at the end of the lecture).

Accepting that atoms do not collapse, why then do they not all combine into a single giant molecule, making all of us pieces of the same material continuum, a piece of a single universal entity? Here comes to play the "Exclusion Rule," which when applied to electrons imposes upon all atoms *a specific connectivity* that chemists discovered already in the nineteenth century and called it *valency*, and which in our course is simply the *atomic connectivity* in the bond construction process. It is the Exclusion Rule that underlies the division of matter into discrete molecules, and this is also the creator of the molecular diversity of the chemical matter (for further details, see the Retouches at the end of the lecture).

The outcome of these two laws is that chemistry is "a game of LEGO" with atoms and fragments, with specified connectivities. Take, for example, the molecule which is made only from hydrogen (H) atoms. As we shall learn in the next lecture, H is the simplest atom, it has only one electron, and according to the Exclusion Rule, it can accommodate only a pair of electrons and no more. As such, when two H atoms approach one another, then like in LEGO and similar games, the electrons "click" and pair up to create a bond, to form the H_2 molecule, shown in Scheme 1.5.

Other atoms or fragments have higher connectivities, and hence they form molecules with more atoms combined together by these electron-pairs. With more than three atoms bonded in a molecule, we start to have molecules with 3D shapes. When a molecule has many, many atoms, the structural elements become more complex and we may speak about an architecture of a molecule. Figure 1.6 provides a few examples just to illustrate the molecular shapes, which we shall learn during

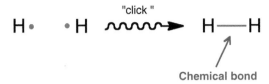

SCHEME 1.5 Two H atoms approach one another and their electrons (drawn as red circles) "click" to pair up and form a bond. The red line connecting the atoms represents the chemical bond.

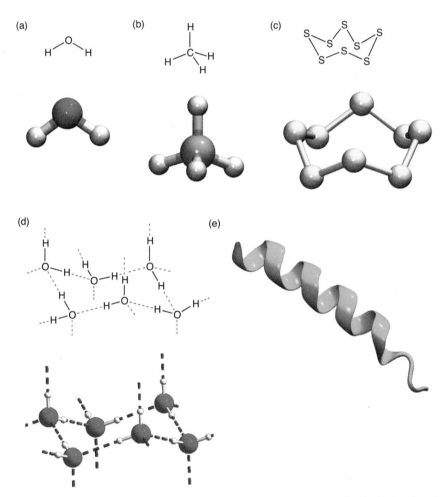

FIGURE 1.6 Molecules and their 3D shapes: (a) H_2O (water). (b) CH_4 (methane). (c) S_8 (elementary sulfur). (d) An assembly of six water molecules. (e) A small antimicrobial peptide derived from *Helicobacter pylori* (1OT0.pdb), drawn without atomic details. The term "pdb" means that the drawing was generated from data deposited in the "Protein Data Bank"; hence, pdb.

this course how to predict and understand. Each molecule is presented using a line drawing (a line represents a chemical bond) and a ball-and-stick drawing, wherein the atoms are represented as spheres and the bonds as sticks.

Thus, in Figure 1.6a, we show a molecule of water, made from an oxygen atom (O) and two H atoms—it is planar (lying in a single plane). The second molecule (Fig. 1.6b) is methane (used as fuel), involving one carbon (C) atom and four H atoms. This molecule is already 3D. In Figure 1.6c, we show S_8 made from eight sulfur (S) atoms and looking in 3D like a crown; this is the shape of the sulfur element as found in nature (yellow powder or crystal). The next example, in Figure 1.6d, shows six water molecules as they appear when water freezes to ice. This cluster looks like a chair, and the individual water molecules hold on to one another by a weak interaction called *hydrogen bonding*, which we shall discuss later during the course. Finally, Figure 1.6e shows a short protein molecule which coils like a helix, again because of internal *hydrogen bonding* that causes the molecule to coil. It is clear that the more complex 3D structures start to exhibit "architecture."

1.8   MOLECULAR ARCHITECTURE AND ITS EMERGENT PROPERTIES

1.8.1   Diamond, Graphite, and More

The molecular architecture creates spatial information, which is sensed by other molecules as well as by the surrounding light. Sometimes, the molecular object and its interaction scenario create a "combined property," which is not simply the sum of the individual properties of the atoms but is something "more" than that. This kind of property is called an "emergent" property. Let us look now at some examples where the molecular architecture creates new information that leads to "emergent" properties.

Figure 1.7 can serve as a demonstration of the emergent function of architecture. The figure shows three different molecules all made from carbon (C) atoms only. A mere inspection reveals that the molecules are different in their architectures as a result of different utilization of the fourfold connectivity of carbon ("valence"). C_{60} on the left, which looks like a soccer ball (football in Europe), is called buckyball or fullerene after the name of the famous architect Buckminster Fuller, who built geodesic domes made from hexagons and pentagons. Graphite in the middle is like a honeycomb made from layers of carbon hexagons. The distance between the layers is indicated as 3.40 Å (1 Å = 0.00000001 cm) and is longer than the bonds between the carbon atoms in the layer. Diamond on the right-hand side is also made from layers of hexagons, but these are not planar and are packed closer to one another compared with the graphite layers. In fact, the distance between the layers (1.54 Å) is the same as the length of the C–C bonds within the layers. Diamond is a mesh of C–C chemical bonds arranged in space.

"Real" graphite and diamond are shown at the bottom of the figure. Diamond is hard and translucent, while graphite is opaque and brittle. *The different architectures at the atomic scale determine the outcome of the interaction of the two compounds*

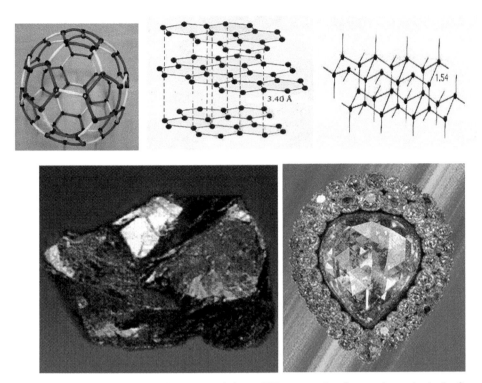

FIGURE 1.7 Molecular architecture of three different molecules made exclusively from carbon atoms. On top: structures of buckyball C_{60} (left), graphite (middle), and diamond (right). On the bottom: the "real materials," a piece of graphite (left) and the famous Topkapi diamond (right). The architecture of diamond transcends to beauty. Adapted with permission from Figure 1 of Shaik[1].

with light. This in turn creates an emergent property. Thus, graphite absorbs all the visible light and hence it is dark and opaque, while diamond reflects light and hence it looks translucent, winking, and scintillating. But much more than that: Diamond is *beautiful* and expensive, while graphite is *simple looking* (not to say *ugly*) and cheap. Some people will murder for diamonds and plan daring robberies, others will simply eulogize its beauty. No one will devote a single such thought to graphite.

The architectural differences of diamond and graphite transcend via interaction with light and with our biological light sensors into "beauty" or lack of it. This is a demonstration of how information stored in the architecture of matter is mediated by light through molecular processes of the biosystem to evoke an emotional–mental response. This is an emergent property of the architecture of matter through the information created by its interaction with light, and through the specific evocation of beauty mediated by the chemical mechanisms of neurotransmitters.

1.8.2 And There Was Light …

Our inner visual world is shaped by the ability of one molecule to exist in two distinct structures, and by the response of this molecule to light. The visual pigment

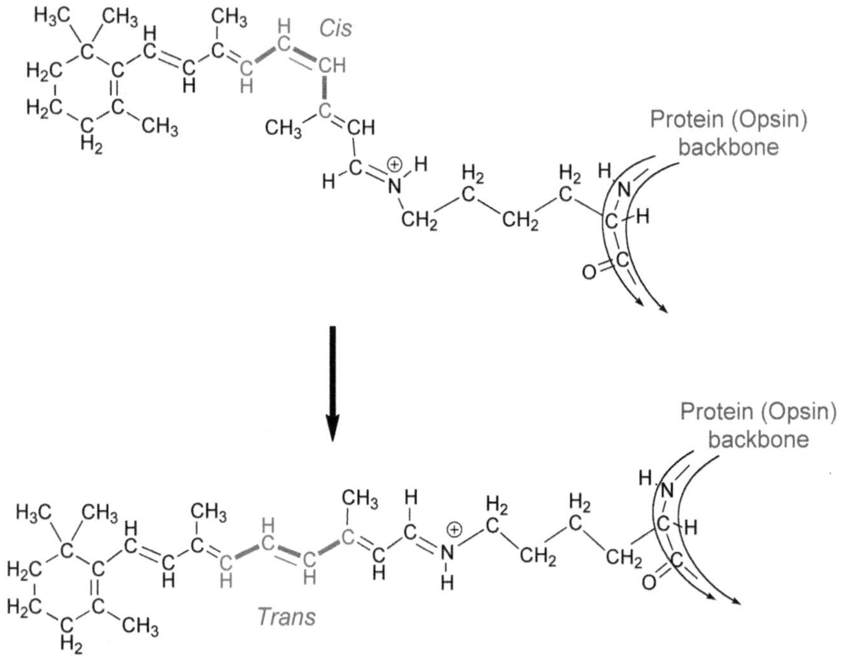

SCHEME 1.6   The light-sensitive retinal molecule (connected to a component of the opsin protein), shown in its resting state (in the upper component of the scheme) and in its state after absorbing visible light. Note that the structures differ in the geometry of the portion colored in red. In the resting state, this geometry is called by chemists *cis*, and, after light absorption, the geometry is called *trans*.

is the rhodopsin receptor, made from proteins, and is a member of the family of receptors we have in our cells (called GPCR: G-protein-coupled receptors).[13] The receptor contains a molecule called retinal, which is sensitive to "visible light," and is connected to the protein (called opsin), as shown in Scheme 1.6.

The retinal in Scheme 1.6 is made from a chain of carbon (C) atoms, which are bonded also to hydrogen (H) atoms. In the resting state, the chain has a structural motif which chemists call *cis* (scientists like to use Greek and Latin) that endows the molecule with a kinked shape. Upon absorption of visible light, the molecule gains sufficient energy to twist the *cis* region and give rise to the structure labeled as *trans*, which has a longer chain and detaches from the protein. This architectural change activates the protein, and the event triggers an amplification of the initial signal, which is transmitted as a neural signal and leads onward to the creation of an image.[24] Subsequently, another event of neurotransmission, yet to be elucidated, takes over and evokes an emotional response (oh, diamonds are beautiful!). Our cognition of shape and the innate sense of beauty are outcomes of the interaction of matter with light and of the motions of the atoms of matter. This is so much reminiscent of the chemical evocation of the emotional–sensory system that is activated by chemical mechanisms of the neurotransmitters and neuromodulators.

1.9 CHIRALITY, AND THE MAGIC BY WHICH MOLECULES RECOGNIZE OTHERS IN NATURE

Let me start with a few words about a very special architectural element in chemistry, called *chirality*. When we look at the mirror, do we really see ourselves? The answer to the question is no! To understand the reason, look at the cute snail and her mirror image in Figure 1.8a. They are clearly not identical. For example, the spiral on the snail's shell is counterclockwise, while on the shell of her mirror image the spiral is clockwise. You cannot possibly match a clockwise spiral with a counterclockwise

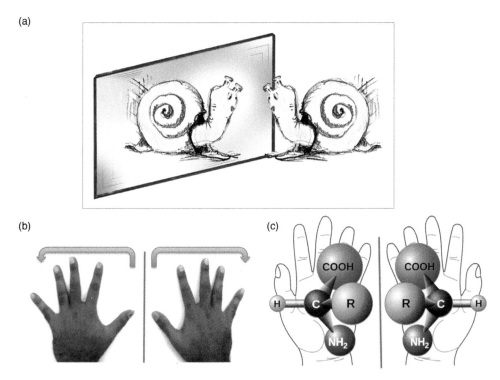

(a)

(b)

(c)

FIGURE 1.8 Chirality/Handedness: (a) A lady snail and her mirror image are not identical (note the direction of the spirals on their shells) or as chemists say, "non-superimposable." (b) The right hand "sees" in the mirror (the red line) the left hand and vice versa. Note that the directionality from the thumb to the little finger is clockwise for the right hand and counterclockwise for the left hand. The two hands are non-superimposable. (c) A molecule (an amino acid) with four different groups linked to a carbon atom resting on the left hand and "seeing" through the mirror the right hand, which is the mirror image of the left molecule. Like the two hands, these two molecules are non-superimposable. The molecules are "chiral," that is, having "handedness," and are therefore different. (The lady snail and her mirror image were designed by M. Michman and drawn by the artist Z. Cohen. Copyrighted by the author. The image containing the amino acids, which was created by NASA, was reused with modification. This image, which is in the public domain, was obtained from http://en.wikipedia.org/wiki/File:Chirality_with_hands.svg.)

one. Chemists call this mismatch of an object and its mirror image as being "non-superimposable."

As a whole, we are not superimposable on our image in the mirror, because of asymmetry in the human body, which makes any man/woman and their mirror images non-superimposable. Consider, for example, our hands. Figure 1.8b shows that a right hand is reflected through a mirror to a left hand. The two hands have an inherent difference, an internal direction (from thumb to the little finger) either clockwise or counterclockwise, and therefore the right and left hands are non-superimposable. If you try to rotate say the left hand in any fashion imaginable so it becomes "a right hand," you will find that this is impossible. The only way to superimpose them is "to cross the mirror, to the world on the other side." ... This kind of asymmetry is called "handedness" or *chirality*, which is the Greek word for handedness (as I said, chemists like Greek terms).

Molecules exhibit precisely the same phenomenon of chirality or handedness. Figure 1.8c shows a molecule made from a carbon atom (C) that is linked to four different chemical groups, which are also color coded. This molecule is held by the left hand with the gray group (H) pointing toward the thumb. The red line is a mirror plane. It is seen that the "mirror image" of the molecule that is nestled in the left hand is the molecule nestled in the right hand. The two molecules are "handed" or simply chiral. Using models of the two molecules will reveal that no matter how we rotate one of these molecules, it can never be superimposed over the other. With an experienced eye, one can see this non-superimposability by inspection of the two molecules in Figure 1.8c. These are two different molecules, even though the four groups attached to carbon are identical in the two molecules. What differs is the directionality going from the orange group to the lilac via the blue one. This direction is counterclockwise in the molecule resting in the left hand and clockwise in the one resting in the right hand. This is *molecular chirality*.

Nature makes routine use of molecular chirality. We spoke above about proteins, which are made from units called amino acids. In fact, the two molecules resting in the two hands in Figure 1.8c are amino acids. So, amino acids are chiral. Interestingly, all the proteins in living creatures are made exclusively from left-handed amino acids, and when the creature dies and its protein disintegrates, one finds both left- and right-handed amino acids. How does the living matter "know" to select only left-handed amino acids?

There is a simple demonstration, which I learned from my teacher Roald Hoffmann.[25] Approach one of your students and offer your left hand for a handshake. The student will stretch his/her left hand and the two hands will snuggle together in a perfectly fitted handshake. Now ask your student to shake your left hand with his/her right hand, and now the handshake will be cumbersome; the right and left hands cannot snuggle nicely in a handshake (Scheme 1.7).

Nature uses an analogous mechanism during the making (synthesis) of proteins. Thus, in rough terms, molecules that have a preference for left-handed amino acids transport the amino acids (handshake ...). These amino acid molecules are then arranged on another long molecule that serves as a template (the DNA/RNA molecules which will be seen later). When a left-handed amino acid is brought in, it snuggles

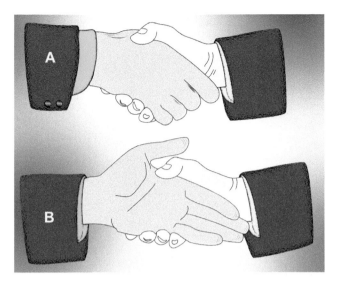

SCHEME 1.7   Two handshakes A and B—Right knows right from left. (Designed by M. Michman, drawn by the artist Z. Cohen, and copyrighted to the author)

nicely near the first left-handed molecule, while if a right-handed molecule is brought in, it fits less well (like the mismatched handshake shown in Scheme 1.7) and it can easily be rejected and replaced by a left-handed molecule. In this simple manner, nature "knows left from right" and generates protein molecules with components that have the same handedness/chirality. This is referred to as "homochirality" (uniform chirality). How did homochirality start in our universe? This is a hot and still-debated topic in science.

1.10   OUR GENETIC CODE IS CHEMICAL

Nature uses relatively weak interactions between molecules to store information and transfer it along to make specific molecules like on a book-printing template. This is the genetic code stored in the DNA molecule on the left-hand side of Figure 1.9, and it is based on the architecture of weak interactions that are called *hydrogen bonds* (these are not the "click-bonds" as were demonstrated in Scheme 1.5 by the "clicking" of the electron pair, and which will be taught in the next lecture). The hydrogen bonds in DNA are very specific and they couple the four bases of the DNA into two *faithful pairs*, which are shown on the right-hand side of Figure 1.9. These faithful couples, which are color coded in the left-hand drawing in the figure, serve to hold together two coils of DNA strands into "a double helix."[26] Since the bases holding the strands come only in faithful pairs, the two strands are necessarily complementary and they have perfect "knowledge" of one another. Therefore, each strand is a template for making the other. This happens during the division of the cell.

The DNA serves also as a template for the synthesis of proteins. There are 20 naturally occurring amino acids from which all proteins of living systems and plants

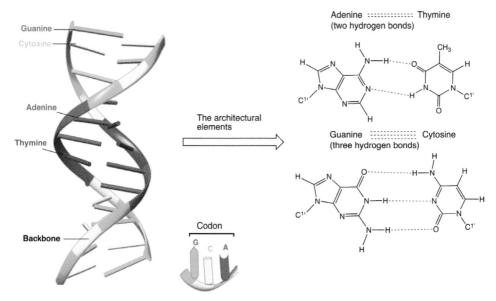

FIGURE 1.9   Left-hand side: The "double helix" of DNA is made of two strands of DNA molecules twisted around each other and connected by means of hydrogen bonds between the base fragments (color coded) that hang on the strands. Right-hand side: The hydrogen bonds (specified in dashed lines) occur between particular and consistent pairs of bases. At the bottom middle: three successive bases form a codon that recognizes a specific amino acid (in this codon, G is guanine, C is cytosine, and A is adenine).

are made by the code embedded in the DNA strand. It turns out that three successive bases recognize one of the amino acids. Such a codon is schematically shown in Figure 1.9 (bottom-middle of the figure). In this manner, the amino acids get organized following directions from a copy of a DNA strand called RNA (or a piece of it) in a special order that is determined by the sequence of codons, and are subsequently stitched into a protein molecule. The DNA code is "personal" and hence the resulting protein is also "personal." For example, hair is protein, and some of us have blonde hair, others dark, still others have curly hair, etc. In fact, all human characteristics, including character traits (e.g., a nervousness[5]) are "written" in the DNA, which is our "software" (some would say "hardware"). Thus, in a nutshell, genetics is a chemical code of *hydrogen bonding* templates, which are weak and transferrable interactions, which give rise to cell division and to the properties of the next generation.

1.11   CHEMISTRY AND ITS EMERGENT EXPRESSIONS

Again and again, we could see that in the basis of the human attributes, there exists chemical information of molecular architecture and molecular motions. This algorithm is everywhere, from our falling in love and being psychologically balanced … , all the way to the genetic code. Chemistry provides infinite ways by

1	1 H																	2 He
2	3 Li	4 Be											5 B	6 C	7 N	8 O	9 F	10 Ne
3	11 Na	12 Mg											13 Al	14 Si	15 P	16 S	17 Cl	18 Ar
4	19 K	20 Ca	21 Sc	22 Ti	23 V	24 Cr	25 Mn	26 Fe	27 Co	28 Ni	29 Cu	30 Zn	31 Ga	32 Ge	33 As	34 Se	35 Br	36 Kr
5	37 Rb	38 Sr	39 Y	40 Zr	41 Nb	42 Mo	43 Tc	44 Ru	45 Rh	46 Pd	47 Ag	48 Cd	49 In	50 Sn	51 Sb	52 Te	53 I	54 Xe
6	55 Cs	56 Ba	57-71	72 Hf	73 Ta	74 W	75 Re	76 Os	77 Ir	78 Pt	79 Au	80 Hg	81 Tl	82 Pb	83 Bi	84 Po	85 At	86 Rn
7	87 Fr	88 Ra	89-103	104 Rf	105 Db	106 Sg	107 Bh	108 Hs	109 Mt	110 Ds	111 Rg	112 Cn	113 Uut	114 Fl	115 Uup	116 Lv	117 Uus	118 Uuo

6	57 La	58 Ce	59 Pr	60 Nd	61 Pm	62 Sm	63 Eu	64 Gd	65 Tb	66 Dy	67 Ho	68 Er	69 Tm	70 Yb	71 Lu
7	89 Ac	90 Th	91 Pa	92 U	93 Np	94 Pu	95 Am	96 Cm	97 Bk	98 Cf	99 Es	100 Fm	101 Md	102 No	103 Lr

FIGURE 1.10 The periodic table: the atom is conserved in all chemical changes.

which a few fundamental mechanisms and the power of combinatorics can lead to a great variety of emergent qualities. This chemical machinery, its genetic origins, and its emergent emotional–sensory expressions[5] form the psychophysical connection (body = chemistry; spirit = thinking, emotions, motivation)[1,6c,11,27–29] that is gradually unfolding. We may be standing on a threshold of a new dawn, in which the gap between "matter" and "spirit" will be bridged. *Through chemistry we have created the alphabet; we now have to create the words, sentences, the poetry of science ...* [28]

Indeed, despite the immense biocomplexity, Mankind is made of chemical matter,[29] the same matter as any other matter in the universe. We might say then that the most compact "book of census"[30] of the inhabitants of this universe is the periodic table; see Figure 1.10. This table arranges the 118 chemical atoms, which were discovered by chemists or artificially created by chemists and physicists primarily over the past three centuries. Only a small part of them participate in molecules of life. When we die, our molecules decompose, and their atoms become part of other molecules in other biosystems, which may be of other humans, or of cats, snails, plants, or viruses ... *For sure, what is conserved is the atom*!

Recognizing this truism makes chemistry not only a science of making new materials, but also, and maybe primarily so, a practical form of introspection—an unconscious introspection that has taken place as Mankind journeys through matter. Chemistry brings to life Man's internal experience of change. All in all, chemistry touches the human existence, well-being, and commitment to be a benevolent master of matter – his essential constituent. All these features make chemistry a central pillar of human culture and a major driving force in its future evolution.

1.12 REFERENCES AND NOTES

[1] The material discussed herein is based on the essay: S. Shaik, *Angew. Chem. Int. Ed.* **2003**, *42*, 3208. The references already cited in the essay are not generally repeated here.

[2] B. Bensaude-Vincent, *Bull. Hist. Chem.* **1999**, *23*, 1 ["peux curieux de son commerce et n'attendant presque rien de son industrie"].

[3] For recent articles, see: (a) J. Schummer and T. I. Spector, *HYLE Int. J. Philos. Chem.* **2007**, *13*, 3; (b) J. Schummer, B. Bensaude-Vincent, and B. Van Tiggelen, *The Public Image of Chemistry,* World Scientific Publications Co. Pte. Ltd. 2007, Danvers, MA.

[4] See, for example: (a) R. Hoffmann, *The Same and Not the Same*, Columbia University Press, New York, 1995; (b) R. Hoffmann and S. Leibowitz-Schmidt, *Old Wine New Flasks*, W. H. Freeman & Co, New York, 1997; (c) R. Hoffmann and V. Torrence, *Chemistry Imagined*, Smithsonian Institute Press, Washington, DC, 1993; (d) A. L. Buchachneko, *Herald of the Russian Academy of Sciences*, **2001**, *71*, 311; (e) M. Blondel-Mégrelis, *HYLE- International Journal of Philosophy of Chemistry*, **2007**, *13*, 43; (f) P. Atkins, *What Is Chemistry?* Oxford University Press, Oxford, 2013.

[5] See a recent exciting discovery: A. R. Hariri, V. S. Mattay, A. Tessitore, B. Kolachana, F. Fera, D. Goldman, M. F. Egan, and D. R. Weinberger, *Science* **2002**, *297*, 400. For example, a recent discovery reveals that individuals with different forms (alleles) of the

gene that encode the transporter protein of the brain chemical serotonin exhibit different patterns of activity in the brain's emotional center, the *amygdala*, such that a hyperactive *amygdala* is associated with permanent anxiety expressed even in nonthreatening situations.

[6] Among the laureates of the 2013 Nobel Prize in Physiology and Medicine, T. Südhof was awarded for his discoveries of the protein molecules that are responsible for transporting the neurotransmitter molecules across the synaptic gaps of neuron cells and bringing about wiring of our brains. See: E. Pennisi and E., Underwood, *Science* **2013**, *342*, 176.

[7] (a) On the role of serotonin in sexual behavior in mice, see: Y. Liu, Y. Jiang, Y. Si, J.-Y. Kim, Z.-F. Chen, and Y. Rao, *Nature* **2011**, *472*, 95; (b) It was found that medical students at the early stage of "romantic love" exhibit changes in serotonin transporter similar to the changes reported in obsessive-compulsive disorder. See: D. Marazziti, H. S. Akiskal, A. Rossi, and G. B. Cassano, *Psych. Med.* **1999**, *29*, 741.

[8] (a) A. Walsh, *The Science of Love*, Prometheus Books, Buffalo, NY, 1991; (b) M. Leibowitz, *The Chemistry of Love,* Berkeley Book, New York, 1983; (c) M. B. Krassner, *Chemical & Engineering News*, August 29, 1983, 22; (d) T. L. Crenshaw, *The Alchemy of Love*, G. P. Putnam's Sons, New York, 1996; (e) C. B. Pert, *Molecules of Emotion,* Simon & Schuster, 1997; (f) H. E. Fisher, *Anatomy of Love*, Norton & Co., New York, 1992.

[9] S. I. Wilkinson, *Chemical & Engineering News*, June 2, 2003, 33.

[10] The PET technique in the study mentioned in Wilkinson[9] and the resulting image shown in Figure 1.2 measure the amount of the sugar called glucose that is consumed by the neurons during activity. In the following study, the authors show a good correlation between altered serotonin activity (measured with PET) and associated glucose metabolism in Alzheimer's disease patients. As the connection between glucose metabolism and serotonin was noted in this study, in patients with Alzheimer's disease, it may be speculated that the connection can be extrapolated to other individuals and situations, as is done in this chapter. See: Y. Ouchi, E. Yoshikawa, M. Futatsubashi, S. Yagi, T. Ueki, and K. Nakamura, *J. Nuc. Med* **2009**, *50*, 1260.

[11] J. LeDoux, *Synaptic Self. How Brains Become Who We Are?* Viking, New York, 2002.

[12] (a) A. J. Thompson, H. A. Lester, and S. C. R. Lummis, *Quart. Rev. Biophys.* **2010**, *43*, 449; (b) L. R. Squire and E. R. Kandel, *Memory: From Mind to Molecules,* Scientific American Library, New York, 1999; (c) M. D. Lemonick, *TIME*, September 13, 1999, 52; (d) Y. Kovalchuk, E. Hanse, K. W. Kafitz, and A. Konnerth, *Science* **2002**, *295*, 1729.

[13] The receptors of dopamine and other neurotransmitters, in Scheme 1.2, belong to a special family of G-protein-coupled receptors (GPCRs), which include seven helices. These receptors activate guanine nucleotide-binding proteins (G-proteins) in the interior of the cells. The G-proteins are signaling molecules within the cell, and they stimulate ion channels and cause the synthesis of second messenger molecules. See descriptions and discussions of G-proteins in: (a) C. A. parent and P. N. Devreotes, *Science* **1999**, *284*, 765; (b) S. P. Neves, P. T. Ram, and R. Iyengar, *Science* **2002**, *296*, 1636.

[14] Serotonin inhibits the enzyme superoxide dismutase, which protects cells against oxidative damage (by "radicals") that leads to the death of the cell due to its chemical degradation.

[15] My world-renowned neurologist colleague, the late Rami Rachamimov from the medical school in Hebrew University, used to tell this story.

[16] (a) C. S. Carter, *Psychoneuroendocrinology* **1998**, *23*, 779; (b) L. J. Young, M. M. Lim, B. Gingrich, and T. R. Insel, *Hormones Behav.* **2001**, *40*, 133.

[17] It was demonstrated that ejaculation is chemically triggered and controlled by a population of spinal neurons called LSt and located in the two segments of the lower back (so-called the lumbar region). These neurons relay sensory information from the body to the thalamus area of the brain, which in turn uses a chemical relay that results in ejaculation. See: W. A. Truitt and L. M. Coolen, *Science* **2002**, *297*, 1566.

[18] (a) See the Nobel Lecture: R. F. Furchgott, *Angew. Chem. Int. Ed.* **1999**, *38*, 1870; (b) NO binds there to another molecule called guanylate cyclase. This molecule is then stimulated to make cGMP (a key messenger molecule) from guanosine triphosphate (GTP). Increased quantities of cGMP lead to a decrease of Ca^{2+} concentration in the muscle cells, causing the muscles to relax and blood to flow into the penis, thereby causing erection.

[19] The L-glutamate receptor NMDA (*N*-methyl *D*-aspartate).

[20] K. E. Wilson, *Chemical & Engineering News*, June 29, 1998, 29.

[21] (a) M. Sher, G. Gisselmann, A. Poplawsky, J. A. Riffel, C. H. Wetzel, R. K. Zimmer, and H. Hatt, *Science*, **2003**, *299*, 2054; (b) See there the highlight by S. Wilkinson.

[22] (a) For a discussion of "taste" and "smell," see: C. L. Wilkinson, *Chemical & Engineering News*, May 1, 2000, 1; P. Zurer, *Chemical & Engineering News*, October 7, 1996, 5; (b) On memory impairment and reinforcement by small molecules, see: S. Borman, *Chemical & Engineering News*, February 20, 2012, 11 (Memory impairment is caused by the sugar molecule O-GlcNAc that glycosylates the protein CREB and affects the axon and dendrite growth in neuronal cells. Phosphorylation of CREB has an opposing effect).

[23] What rats dream about, *TIME*, February 5, 2001, 51.

[24] The neurons of the visual system, which respond to different parts of the same object, become "bound" by firing in synchronicity. This synchronicity of neural firing tends to prefer symmetric objects. See: S. H. Lee, and R. Blake, *Science* **1999**, *284*, 1165.

[25] A beautiful metaphor for chiral recognition is the "Handshakes in the Dark," described by Hoffmann on p. 40 in Hoffmann and Leibowitz[4b]. This can become very "handy" in a public talk—shake hands with your audience to illustrate chiral recognition.

[26] J. D. Watson, *The Double Helix*, Weidenfeld and Nicolson, London, 1971.

[27] For the connection of body and spirit, in the context of how the brain influences memory, see: R. Brazil, *Chemistry World*, October 1, 2014, 58.

[28] Our current language is not precise, because of the necessity to shift from the level of chemical–mechanical details (neurotransmission, brain regions, etc.) to the level of integrated experience (depression, love, motivation). The chasm between these levels can be likened to the relationship between the letters of the alphabet and a poem. The letters are certainly the only constituents of the poem, but not every pile of letters makes a poem. Bridging this language gap in a manner that draws the relation between the whole and its parts is a great challenge that lies ahead.

[29] In humans, the neocortex (the gray matter) regulates the biological information into emergent pattern recognition modes that make a mind. See: R. Kurzweil, *How to Create a Mind*. Penguin Books, New York, 2012.

[30] F. Tibika, *Molecular Consciousness*, Inner Traditions • Bear & Co., Rochester, VT, 2013, p. 17.

1.A   APPENDIX

1.A.1   Proposed Demonstrations

I am aware of the difficulties of performing demonstrations (demos) in class due to strict safety measures. Nevertheless, demos should be an essential part of most courses in chemistry, certainly in introductory courses. Let me state the goals of the proposed demos for this opening lecture.

The first goal is to demonstrate the "magic of chemistry": *one molecule disappears and another one with different properties appears.* The motto here is "fun"—the demos should be fun, and the demonstrator should be entertaining. The second goal is to "prepare the minds" of the students for the next lecture.

There are demos on the Web (e.g., at the time of the writing of the book, I found a great site called "Gorgeous chemical reactions": http://io9.com/look-in-wonder -at-the-most-gorgeous-chemical-reactions-1651407454). Others can be found in the *Journal of Chemical Education*.[1A] Some great demos can come to mind by watching the BBC series *The Elements*.[2A]

Before you perform any of the demonstrations, read carefully the hazards of the various chemicals, and the precautions in using them! Any of the students who participates in the demo or is close to you when you perform the demo must wear a lab coat, gloves, and a pair of goggles! Also, be sure to read all safety measures for laboratory work. Here are some suggestions for demos:[3A–8A]

(a) Demonstrations to set the stage for Lecture 2:
- **Mimicking the electronic shell structure of the atom**: This demo is called also the floating magnets experiment of Alfred Mayer, which inspired J.J. Thomson when he proposed his model of the atom.[3A] The experiment shows that a central force enforces ordering on small objects that are affected by it. The so-attained ordering is similar to the shell structure of the electrons around the nucleus. The demo uses magnetized needles that are inserted into corks. These corked needles are allowed to float on water in a vessel. An electric wire that is connected to a DC power supply wraps the vessel. As the power supply is turned on, the mounted needles begin to arrange in "orbits," which contain two and then eight corks, much like the shell structure of the electrons in an atom. If you practice it well, the demo is truly striking, and it breathes life into the shell structure of the atom, which otherwise remains abstract for the beginning student.
- **Demonstrating the existence of two bond types, ionic and covalent bond:** This demo is also called the buzzer experiment. It is based on the fact that some materials, like kitchen salt, have *ionic bonds*, while others, like sugar, have *covalent bonds*. Ionic bonds are made from two oppositely charged ions, like Na^+ and Cl^- in the common kitchen salt, and hence when a piece of this solid salt is connected to wires that are hooked to a battery and a buzzer, the ions move in the circuit, create current and cause the buzzer to buzz (or a radio to start playing). The sugar will not elicit any effect on the

buzzer, since the covalent bonds of the sugars are not made of charged ions. This demo can be quite entertaining, and the striking difference between the salt and the sugar prepares the students to expect two types of chemical bonds.[4A]

(b) Demonstrations of the "magic" of chemistry:

There are many experiments that show the "magic" of chemistry, and which could be found in the *Journal of Chemical Education* and on the Web. Here is one example followed by some suggestions.

- **Elephant's toothpaste**
  The experiment involves the decomposition of hydrogen peroxide (H_2O_2) in the presence of a catalyst (an added molecule that speeds the rate of a given reaction), a surfactant (a dishwashing detergent), and a colorant, carried out in a 2L plastic bottle. The catalyst, potassium iodide (KI), speeds up the decomposition reaction, and the evolved O_2 gas is captured by the surfactant, which in turn, foams like toothpaste. The foam shoots out of the bottle, and the heat released by the reaction causes the bottle to shrink.[5A] The experiment can be used to demonstrate the energy aspect of reactions, and tell about endothermic and exothermic reactions, about gases and their properties, and about Lavoisier and the other discoverers of the O_2 molecule.

- **A list of fun experiments:**
  Here are some of the many fun experiments that we tried through the years of teaching the course. The teacher is encouraged to design his/her own experiments or look for others.

- **Flame retardation, light, and relight**
  In this demonstration the combustibilities of oxygen and carbon dioxide are examined by producing each gas in a separate vessel and testing its effect on a wooden splint.[6A] This experiment is similar in many ways to those used by Scheele, Priestley, Lavoisier, and Cavendish in their discovery of the different gases that make up our atmosphere.

- **The permanganate volcano**
  This experiment demonstrates the rapid oxidation of the sweet-tasting viscous alcohol glycerin by the powerful oxidizing agent potassium permanganate. The reaction produces a great amount of heat, which leads to a dramatic eruption of smoke and fire. The presence of potassium ions (from potassium permanganate) colors the flame violet.[7A]

  When I was a child playing with chemicals, I added potassium permanganate to HCl in a test tube, and to my great surprise, the liquid started bubbling vigorously. Like a curious discoverer, I smelled the bubbling gas and almost died. The permanganate oxidized the HCl and produced the gas Cl_2, one of the gases used in chemical warfare during World War I ... I do not recommend trying this discovery ...

- **Cool experiments with liquid nitrogen**
  Liquid nitrogen, which boils at about 220°C Celsius below room temperature, freezes its surrounding environment and expands as it evaporates.

These two behaviors are exploited to make fun experiments such as freezing flowers, vegetables, etc., shrinking a balloon filled with air, smashing rubber, and more.[8A]

1.A.2   References for Appendix 1.A

[1A] See, for example: R. D. Sweedler and K. A. Jeffery, *J. Chem. Edu.* **2013**, *90*, 96.

[2A] A beautiful series on the discovery of the elements (atoms) is the BBC series *The Elements*, with J. Al-Kahlili.

[3A] A. M. Mayer, *Sci. Am.* **1878**, Supplement 5 (No. 129), 2045.

[4A] http://www.sciencelearn.org.nz/Contexts/Super-Sense/Teaching-and-Learning-Approaches/Testing-for-conductivity. Permits from University of Waikato.

[5A] L. R. Summerlin and J. L. Ealy, Jr. *Chemical Demonstrations, A Sourcebook for Teachers*, Washington, DC, **1985**, 71.

[6A] http://www.stevespanglerscience.com/lab/experiments/flame-light-relight#. Permits from Steve Spangler Science.

[7A] T. Lister, C. O'Driscoll, and N. Reed, *Classic Chemistry Demonstrations*, The Royal Society of Chemistry, London, **1995**, 69.

[8A] http://www2.physics.ox.ac.uk/sites/default/files/2012–04–27/accelerate_liquid _nitrogen_17590.pdf. Permits from University of Oxford.

1.R   RETOUCHES

1.R.1   More Drugs Looking like PEA

I mentioned that the hallucinogen mescaline looks very similar to the neurotransmitters in Scheme 1.2. There are many other drugs that have similar structures to PEA or epinephrine or norepinephrine, all containing the six-membered ring called phenyl, a variable short chain of carbon atoms, and an aminic tail. The small differences between all these molecules translate to significant differences in their function compared with that of the brain chemicals. Let me give just three examples, which one can pick up easily from the Web.[1R]

One of these is Ephedrine where the carbon chain plus aminic tail is $CH(OH)CH(CH_3)NH(CH_3)$, looking like a close "brother" of PEA. Ephedrine is a natural product extracted from a plant called *Ephedra* and used in traditional Chinese medicine. It can also be made in the laboratory. Ephedrine is administered as a stimulant, appetite suppressant, and concentration aid. It seems to work by increasing the activity of norepinephrine. It is used, for example, by professional weightlifters and athletes. Another close "brother" of PEA is pseudoephedrine, which is a chiral isomer of ephedrine, and it causes vasoconstriction (constriction of blood vessels) and relaxation of smooth muscles, thus leading to bronchial dilation. It is administered for nasal/sinus congestion and difficulty in breathing. Methamphetamine is another close brother of PEA and ephedrine, having $CH_2CH(CH_3)NH(CH_3)$ attached to the phenyl group. It is a psychostimulant (hallucinogenic) drug that was approved by the Food and Drug Administration in the United States, and is it used to increase sexual desire, to lift the mood, and to increase energy. This function is somewhat similar to PEA.

1.R.2   The Atomic Hypothesis

This hypothesis dates back to Democritus (~460–370 BC) or even earlier to Leucippus (born 475 BC). Democritus was probably the first materialist. He believed that everything, including the soul, was made of atoms … This hypothesis remained dormant for years because it lacked the "spiritual component" and was atheistic, and hence has never been endorsed (to say the least) by the Church in the western world.

In 1803, the meteorologist-turned-chemist John Dalton resurrected the hypothesis and breathed new life into it. Not only did he hypothesize that each atom had a characteristic weight, he also showed how this weight could be determined by the available technical methods of the day, by weighing. He further suggested that atoms combined to give molecules where the atomic ratio is simple, for example, 1:1, 1:2, 2:1, etc., and thereby created molecules in a simple LEGO principle. He also made some wrong assumptions, but all in all, his hypothesis gave a productive research agenda to chemists, who started to "weigh" atoms and determine molecular constitutions. This activity has ushered chemistry into its constitutional revolution. For further reading, we propose the excellent book by the historian Siegfried.[2R]

1.R.3   The Uncertainty Principle, The Exclusion Rule, and Valence

When the subatomic particles, like the electron, were discovered at the beginning of the twentieth century, they gave physicists serious headaches because they did not behave like the regular objects (so-called macroscopic bodies), like stones, balls, etc. Within a very short time, the physicists have reformulated a new theory of matter called quantum mechanics (QM), which outlines the new basic laws of elementary submicroscopic particles like the electron. A beautiful account of this epoch was written by Emilio Segrè.[3R]

The uncertainty principle: This principle (due to W. Heisenberg) states that the product of the uncertainties in the measured position and velocity of an electron is larger than a quantity, h, which is Planck's constant (very, very small, but *not* zero!). This principle excludes the possibility that the negatively charged electron will rest on the positively charge nucleus (in which case, there is no uncertainty in either its position or its velocity). Therefore, the electron diffuses in space (becomes delocalized) around the nucleus, and the position of its highest probability may be used to define the atomic radius.

In advanced chemistry courses, one learns that the electrons reside in "orbitals," which are mathematical functions that describe the probability of finding the electron around the nuclei, *in terms of little volume-occupying shapes*. The Web is full of sites

Click, like in a game of LEGO...

Chemical bond

SCHEME 1.R.1   Making a chemical bond when the spins of the two electrons are opposite.

which describe these orbitals.[4R] Therefore, the drawing of a circle in Scheme 1.4 could be replaced by a more sophisticated one that defines the "radius" of the atom by the highest probability to locate the electrons in the space around the nucleus.

The Exclusion Rule and Valence: This rule (due to W. Pauli) has become associated with the discovery that the electron has another property, so-called *spin*. This property can be visualized to arise if the electron were to behave like a topspin that revolves around its own axis. Then there would be two different spin directions, clockwise and counterclockwise, which are represented by up (↑) and down (↓) arrows. The exclusion rule states that two electrons with the same spin cannot occupy the same space (orbital). *This means that when the electrons couple to make a bond, their spins have to be opposite to one another, as shown in* Scheme 1.R.1, which is identical to Scheme 1.5, with the exception that we now show the spins.

The valence of an atom is also derived from the rules of QM, which set a limit on the number of electron pairs that an atom can have in its valence shell (the outermost electronic shell).[5R] The chemists have discovered these rules and have given them other names, which we shall be using during the course. For example, the Octet Rule (developed by the chemist G.N. Lewis) limits many atoms to have up to eight electrons in their valence shell, namely four pairs of electrons.[5R] This determines the connectivity of the atom in the bond construction game.

1.R.4   Units of Size

We mentioned the term Å (Ångström—after the name of a Swedish scientist), which is used to measure units of length in the atomic world. 1 Å is 10^{-8}th of a centimeter (cm), namely 0.00000001 cm. This is approximately the radius of an average atom. Distances between bonded atoms range between 0.75 Å and 2.5 Å. Molecules and atoms are then very small. Nevertheless, today we are able to visualize atoms and molecules!

1.R.5   References for Retouches

[1R] http://en.wikipedia.org/wiki/Ephedrine; http://en.wikipedia.org/wiki/Pseudoe phedrine; http://en.wikipedia.org/wiki/Methamohetamine

[2R] R. Siegfried, *From Elements to Atoms: A History of Chemical Composition.* American Philosophical Society, Philadelphia, PA, 2002.

[3R] E. Segrè, *From X-rays to Quarks: Modern Physicists and Their Discoveries.* Dover Classics of Science and Mathematics, Dover, 2007.

[4R] See, for example: http://en.wikipedia.org/wiki/Atomic_orbital

[5R] See, for example: http://en.wikipedia.org/wiki/Octet_rule

LECTURE 2

THE CHEMICAL BOND AND THE LEGO PRINCIPLE

2.1 CONVERSATION ON CONTENTS OF LECTURE 2

Racheli: This must be a big moment since you are actually starting the teaching material. How do you plan to proceed?

Sason: What is more natural than starting this lecture with atoms and how they bind to each other to make molecules?

Racheli: Absolutely! But I hope the teachers took your advice and performed some demos and emphasized to their students that the chemical bond is behind the chemical magic that they saw in the demos!

Sason: The "chemical magic." What a wonderful phrase, Racheli, to describe what our eyes see in any chemical demonstration: "one material disappears and a completely new one appears," with new features, be these colors, gas evolutions, gases that burn, gases that put off fire ... All these are due to changes in chemical bonds. Molecules "play a game of LEGO" with one another. They decompose and "click" ... their fragments rebind to one another and make new molecules.

Today's scientific techniques enable scientists to image bond breaking and bond making.[1] What has been imagined for 100 years is transpiring in front of our eyes. Yes, it is a chemical bond time!

Racheli: Imaging is fine, but I prefer the buzzer demo we performed; some materials (e.g., the salt NaCl) buzzed when wired to a buzzer, while others (e.g., the sugar crystal) did not. The buzz of the first types notifies us of the ionic nature of their

Chemistry as a Game of Molecular Construction: The Bond-Click Way, First Edition. Sason Shaik.

chemical bond, while the silence of the second types "speaks" of the covalent nature of their bonds.

Sason: Aren't you being a bit old fashioned, Racheli? Imaging and buzzer testing are both methods that chemists use for "listening" to the language of the molecules.

Racheli: I disagree. Imaging is intellectual, but for an introductory course, the buzzer test is ever more so revealing because it appeals to our sense of hearing. Is there a better way to make the contrasting bond types flesh out?

Sason: I agree. This experiment allows us to be guided by our senses, and conclude that the sugar molecules are made of a bond type in which the atoms share electrons and remain glued. As you already said, the bond is called *covalent* to emphasize the sharing of the bonding electrons.

Racheli: Mind you, most of the "molecules of life" are made from covalent bonds, where electrons "click" to form electron pairs that glue the atoms.

Sason: Let's start then with the covalent bond and construct some molecules …

Racheli: Wait, Wait! Aren't you going to teach the students how the atoms are organized in chemistry?

Sason: How forgetful of me, Racheli! I am definitely going to start with the periodic table. This will give me an opportunity to comment about the language of chemistry by mentioning some basic terminology and names of atoms and molecules.

2.2   THE PERIODIC TABLE: THE STOREHOUSE OF ATOMS

The known atoms are organized in the table shown in Figure 2.1, which we already saw in the previous lecture, and is called "The periodic table."[2] The organization of the atoms in rows expresses *periodicity* of the atomic properties. Let us look at periods 2–7; these numbers are marked on the left-hand side of the table. In the second period, we start on the left with the metallic atom lithium (Li), and as we proceed, the metallic property diminishes, and we end up with the gaseous element neon (Ne). The third period exhibits precisely the same variation, and so on and so forth. As we shall see, in each period, the number of bonds that a given atom can make changes along a period, and exhibits periodicity, as one moves from one period to the other.

Another result of the periodicity is that each column in the periodic table constitutes a "family" (or *group*) of elements that share similar properties. Each family has a number, and in some cases, the English letter "A" attends the number, for example, 5A. These families, which are marked by the letter A, are called main group elements (the word "main" signifies that these elements are most abundant on earth). These families have also unique names. Some of the families are called after the first element in the family, like the boron or carbon families, etc. Others have special names, which are indicated in Figure 2.1. For example, the members of the 1A family, with the exception of H, are called "Alkali Metals," referring to their historical role in soap making. Family 7A is called "Halogens," referring to their classical role as generators of salts; for example, table salt includes chlorine (Cl), which is the second member

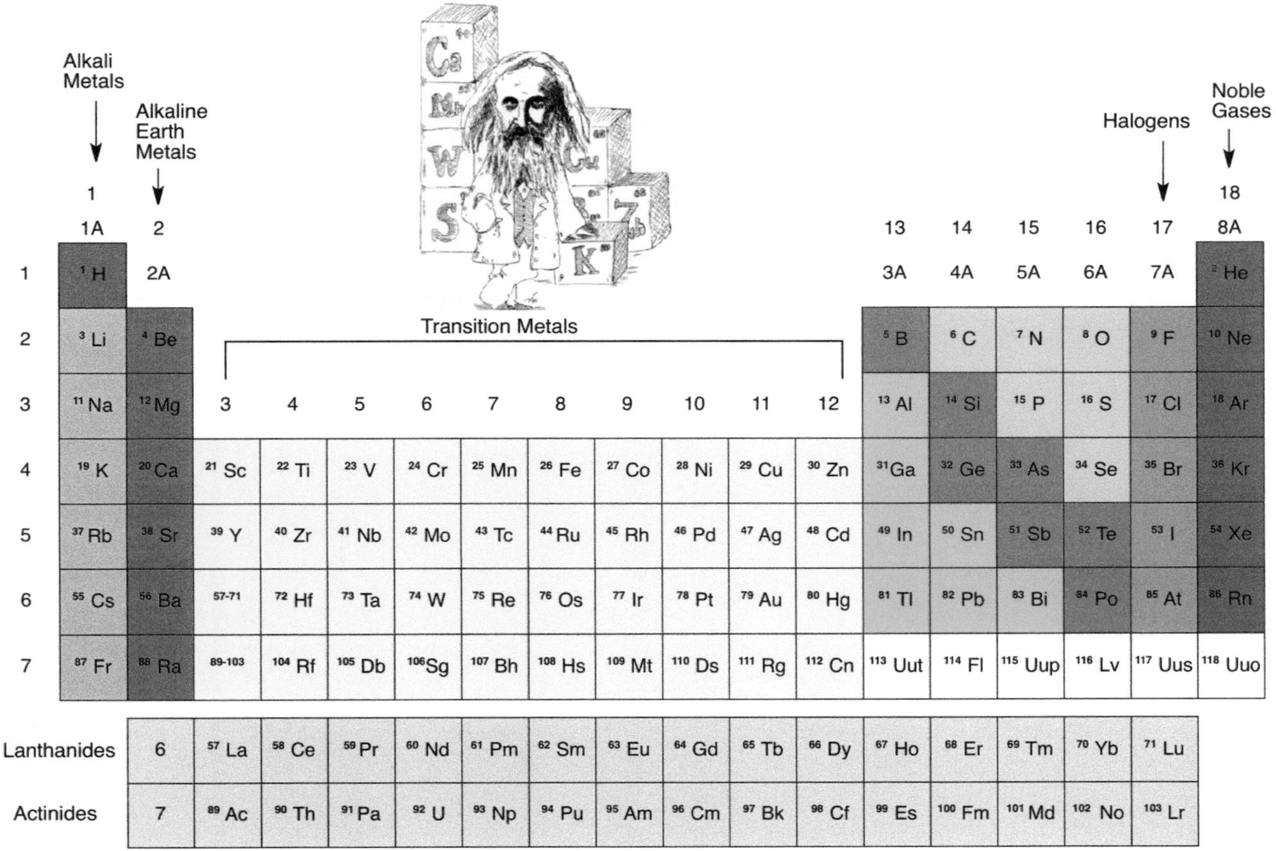

FIGURE 2.1 The periodic table of the elements and Dmitri Mendeleev's caricature (drawn by the historian W. B. Jensen and reproduced here with his permission). The numbering of the columns is done with two alternative systems. We use the one with 1A, 2A, etc.

of the halogens. Family 8A is called the "Noble Gases." This term refers to the fact that these are gases which are reluctant to participate in chemical bonding, behaving like "nobles," which do not "mix" with the "common" elements. It is a must to note here the wonderful book of Primo Levi, *The Periodic Table*,[2b] where the author likens people to chemical elements, and specifically his family members to the Noble Gases. We shall start making molecules with main group elements.

Figure 2.1 shows also blocks of other elements. The stretched block, separating the two parts of the main groups, includes metals, which have the general name "Transition Metals" to denote their location in transit between two blocks of main elements. Among the transition metals is iron (Fe), which is contained in many of the essential molecules of life, for example, hemoglobin, which captures oxygen from air during breathing, and thereby, the molecule sustains life.

The final block of elements is stretched below the table, and constitutes the two long periods of metals called "Lanthanides" and "Actinides," or together "Rare Earths." Among them are the radioactive elements Uranium (U) and Plutonium (Pu).

Chemists achieved this ingenious organization of the chemical elements in the nineteenth century. The foremost credited originator of this idea is the Russian chemist Dmitry Mendeleev, who is shown in the caricature on top of Figure 2.1. Mendeleev was facing a myriad of elements, facts, and properties that accumulated in the chemistry of his time, and he was trying to organize everything in some compact and useful manner. And suddenly, within the chaos, he saw how he could create order in the material world if he just arranged the elements in order of increasing weight, from the lighter to the heavier. This is a kind of *eureka* that happens to those fortunate scientists who suddenly "see the light"[3] and thereby reshape our conceptual worlds.

You may have noticed that we keep alternating between the terms "atoms" and "elements." Scientists are a very conservative crowd, and they generally do not like to refer to objects that are not visible or observable (this is called "positivism"). In the nineteenth century, atoms were not visible, and many chemists refrained from using the term "atoms"; they preferred the term "elements," which had a very practical meaning: this was the chemical entity that could not be decomposed to a simpler one with the available techniques at any given time (see Retouches section 2.R.1). After learning about the covalent bond, we shall see that the term "elements" includes actually some polyatomic species which are made from the same atom, and which exist in our universe. For example, hydrogen (H) exists as H_2, sulfur (S) as S_8, phosphorus (P) as P_4, etc. All the more, in our days, atoms can be imaged,[4] so we can speak conveniently about atoms without fearing the wrath of positivist scientists.

2.2.1 The Chemical Language

As we already saw in the previous section, chemistry, like any discipline of science, has its own system of naming material species. Initially, students may be intimidated by the myriad of names that do not mean much to them. But like in any new language, the feeling improves tremendously after acquiring a basic vocabulary. For our purposes, it is advisable to know the names of the first 18 elements/atoms in the periodic table. There will be no special punishment for those who would know more names! The 18 atom names are given in Appendix 2.1 at the end of the chapter. If you look at the names, you will soon realize that the elements are represented in the periodic table generally by the first letter in the name of the atom; for example, C symbolizes carbon. When a few elements have the same initial letter, we add an additional regular letter; for example, Cl denotes chlorine, and so on (some elements have three letters).

The origins of the names have historical, cultural, trivial, or functional backgrounds.[5] For example, O (oxygen) is a name given to this element by one of its discoverers, the French scientist Lavoisier, who called it by this name because he believed that this element generates the chemical materials called acids (oxy-gen = acid-generator). The symbol He for helium comes from the Greek name of the Sun God (Helios), and the fact that helium was discovered first in the solar atmosphere and only then in the earth's atmosphere. P (phosphorus) means in Greek "light bearing" because it spontaneously radiates light. The symbol Au for gold originates in the Latin word *aurea*, which describes the color "golden," or from the Greek word *oreia*, which means beautiful. Other names I would like to mention, are Po (polonium), called so by its discoverer Marie Curie, who was born in Poland, and Pm (promethium), called after the mythological figure Prometheus, who gave the fire to the human race and was severely punished for that by the Greek gods. These names reflect the human search for significance and the desire of the discoverers of the element to make connections to events/places that are dear to their hearts.

The main inventor of this chemical language by letter representation was the great Swedish chemist Jöns Jakob Berzelius, who started using capital letters with numerals that specified the number of atoms of a given kind in the molecule. Scheme 2.1 shows

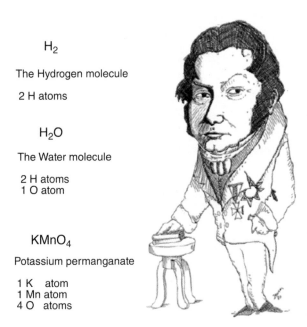

H_2

The Hydrogen molecule

2 H atoms

H_2O

The Water molecule

2 H atoms
1 O atom

$KMnO_4$

Potassium permanganate

1 K   atom
1 Mn atom
4 O   atoms

SCHEME 2.1   Some molecules and the conventional way of writing their atomic composition are shown along with a caricature of J. J. Berzelius, the inventor of this language. (The caricature is reproduced with kind permission from its artist, the historian W. B. Jensen)

Berzelius along with some molecules. It is seen that the molecular formula tells us the types of atoms it contains and their numbers. Furthermore, even a rather complex molecule with three different atoms in differing numbers, like $KMnO_4$, can be written in a compact manner. And what is more important, it is so easy to communicate with this compact language and to convert chemistry from the "art" of the isolated alchemists encapsulated in their smoky labs to a world-embracing culture that knows no boundaries. This was a great invention by Berzelius!

2.3   THE LEGO PRINCIPLE

Bonds are made between atoms, and we wish to present the process in an effective manner, reminiscent of the game LEGO we all played as kids. But mind you, the principle goes well beyond the game, and it is one of the most useful scientific principles that originated in the western world, in the 4th–5th centuries BC, by the Greek philosophers. The Greek philosophers (Empedocles, Plato, Aristotle, Democritus, etc.) were trying to understand the world around them in some functional ways, and in so doing, they invented the wonderful concept of "elements" from which the entire universe could be constructed. Democritus even invented the word "atom," which in Greek means "the indivisible"—namely, the basic building block entity of matter (see Retouches section 2.R.1). This *principle of constructing the whole from building blocks* has gained wide acceptance with the discovery of the elementary particles of matter in the twentieth century. It might have to do with the recognition that humans use "LEGO Memory,"[6] which makes us understand the universe around us in a hierarchical manner.

The principle can be illustrated using Scheme 2.2. Thus, one starts from elementary particles on the left and moves to the right. If there are cohesive forces between these elementary particles, then "click," and they combine into more complex particles! In turn, these more complex particles can combine into larger and larger entities with appropriate cohesive forces. For example, atoms combine to molecules, molecules make aggregates (e.g., molecules of water make the liquid we call "water"), and

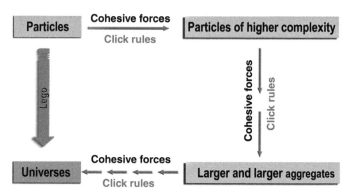

SCHEME 2.2   The principle of constructing the whole from building blocks in science.

many different molecules and aggregates can combine to give living cells, and those in turn, aggregate to give organisms, including humans. And the process goes on and on to generate universes. In this manner, entire 'material universes' can be usefully and playfully constructed from elementary entities.

Generally, the "click" combinations are not arbitrary. Not everything goes! The allowed combinations follow rules, and all other combinations are forbidden. We shall experience this construction process during this lecture and the ones that follow it, as we build molecules starting from the simplest diatomic H_2 molecule all the way to the DNA molecule that is composed of many, many thousands of atoms.

2.3.1   The Covalent Bond in H_2

The covalent bond is made from electrons, which are the elementary particles residing in the atoms and hovering around the nucleus, where the protons and neutrons are encapsulated. Any covalent molecule, even the largest possible, is made from a collection of covalent bonds, which are created by "clicking" electrons.[7]

Let us see how this principle works for the simplest covalent bond between two hydrogen atoms. The rule of bonding is that atoms combine *by pairing electrons.* Scheme 2.3 repeats this electron pairing for H atoms to establish the language we shall be using henceforth. H has a single electron, which is drawn as a heavy red circle near the atom, H•. Hence, H• is an atom *with a single connectivity*, which can therefore form only one electron-pair bond with another H• or with any atom that has a single connectivity. As such two H• atoms meet, "click," they form an *electron-pair bond* that holds the H_2 molecule. We draw this bond with a line connecting the atoms, and from this point on, such a line will symbolize for us an electron-pair bond. *When the pair is formed, the hydrogen atoms share their electrons and each atom feels two electrons in its sphere.* As we shall immediately see, it is possible to construct an infinite number of molecules by "clicking" their electrons into electron-pair bonds. Do not be shy to "click"—this will create a mental image that will help you visualize the construction of molecules.

Why does the rule allow electron pairing but not larger groupings of electrons? We sort of alluded to this selection rule in the Retouches to Lecture 1. There is in fact a highly sophisticated theory called Quantum Mechanics (QM), which explains why the pairing is the choice for covalent bonding (see Retouches section 2.R.2.). We are not going to learn this theory in the course. And in all truth, this idea of electron

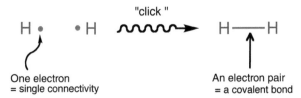

SCHEME 2.3   The covalent bond in H_2. The red line connecting the two H atoms on the right symbolizes *an electron-pair bond, also known as a covalent bond.*

FIGURE 2.2   The caricature of Gilbert Newton Lewis pointing to the Cl_2 molecule with an electron pair between the atoms (the caricature is reproduced with kind permission from its artist, the historian W. B. Jensen). Note that the Lewis representation for the bond, Cl:Cl, is different than the one we use, Cl–Cl. His Cl:Cl representation tells precisely that the bond is made from an electron pair. But this representation is less convenient for large molecules. Just imagine having to dot 100 different bonds ... Therefore, the line "binding" the two atoms replaced this electron dot representation.

pairing came to a great chemist called Gilbert Newton Lewis,[8] not from any theory, but simply from inspection of facts and great intuitive powers.

When Lewis (Figure 2.2) wrote his famous 1916 paper,[8] the electron was already discovered, and the British scientist Moseley found a technique that allowed him "to count electrons and protons" in atoms.[9] Moseley found that the atom helium (He) has two protons and two electrons. Since helium refrains from making molecules by covalent bonds, Lewis postulated *that an electron pair creates a bond, and this is the fundamental building block of all molecules.* He then showed in his paper how powerful this new idea was for understanding many, many chemical facts. Twelve years later, the emerging new QM theory showed that Lewis was right in defying the laws of classical physics and assuming pairs of electrons (despite their repulsion due to their negative charges).[10] According to QM theory, *electron pairing is favorable in the small subatomic world.* We shall therefore proceed to construct molecules by pairing electrons, and not worry anymore about justifying this "bond" since QM theory has amply justified it. Our interest is to learn how to make molecules and how to understand their shapes and properties.

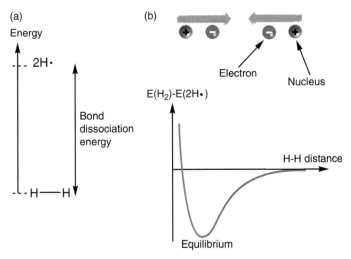

SCHEME 2.4 (a) The separated 2 H• atoms and the H–H molecule placed on an energy scale (the arrow points in the direction of increasing energy). The molecule is "more stable" (lower in energy) than the atoms. To break the bond, we need to invest the bond dissociation energy, which is the energy difference between the energy of the molecule, $E(H_2)$, and the energy of the separate atoms, $E(2H•)$. (b) During the H–H bond-making event, the two positive nuclei are attracted (see the blue arrows above the positively charged nuclei) to the electron pair residing in between them, and thereby the atoms form a stable bond. As shown by the red curve, the energy is initially lowered as the atoms approach one another from far away and then, past a certain point, the energy rises because of the repulsion between the nuclei. The bond reaches "equilibrium" at the minimum of the energy curve.

When a bond is generated, this is attended by another change, which chemists refer to as *energy stabilization*. As depicted in Scheme 2.4a, the bond formation lowers the *energy* of the molecule compared with the energy of the atoms, and this lowering is the *stabilization*—the molecule H_2 is more "stable" than its two constituent atoms. Energy is one of the key driving forces in Nature—take any object, hold it in the air, and let it loose. It will fall down spontaneously because this will lower its energy. If you want to sense this energy release, imagine your head in the way of the falling object. To raise the object to the original height, you will have to invest this energy by working your muscles. The same applies to the H–H bond. If we wish to break this bond, we must invest the "bond dissociation energy" in Scheme 2.4a.

To gain some feeling for the root cause of bond formation, consider the drawing in Scheme 2.4b. Each hydrogen atom is made of a positively charged proton and a negatively charged electron. When electron pairing occurs, the two positive nuclei feel attraction from the two paired electrons (instead of only one electron when the H• is not bonded), and they approach one another and become bonded. As such, the energy of the molecule is lowered, and it gets stabilized relative to its atoms. The approach of the nuclei to the electron pair is arrested at some distance between the atoms by the repulsion between the two positive charges of the nuclei. In this manner, the bond reaches "equilibrium" at some distance between the atoms where the attraction of the nuclei to the electron pair equals the repulsion between the nuclei. Later in the course, we shall learn how to gauge the bond distances in a molecule.

2.4   THE BONDING CAPABILITY OF ATOMS AND THE LAW OF NIRVANA FOR MAIN GROUP ELEMENTS

If we have the connectivity values for all atoms, we can make bonds with great facility. It turns out that we can achieve this knowledge by inspecting the periodic table and by using rules that limit the atomic connectivity to certain magic numbers.

Scheme 2.5 uses carbon (C) as an example. This atom is the sixth one in the periodic table. Inspecting the respective box, where this element resides, one can see a number 6 on the top of the box. This superscript is called "the atomic number"; it counts the number of protons in the nucleus and is therefore, also, *the total number of electrons that the atom possesses*. Additionally, carbon belongs to the family 4A,

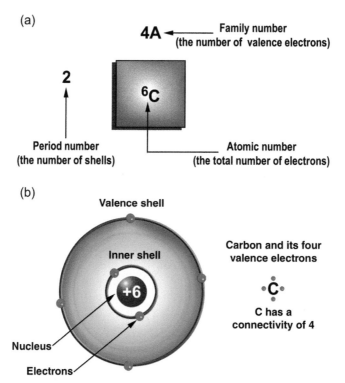

SCHEME 2.5   (a) The atom C (carbon) in its periodic table box. The superscript 6 is the total number of electrons. The period number (2) refers to the number of electron shells. The family number, 4A, signifies that four of the six electrons owned by C are in the valence shell. (b) A schematic representation of the atom C, depicting its electronic shells around the nucleus that has a charge 6+ of six protons. The electrons are represented by the red spheres: two electrons reside in the inner shell, and four reside in the valence shell. On the right, we show C with its connectivity of four.

and this numeral counts the number of the *valence electrons*—namely these are the electrons that will participate in bonding. But, how do these valence electrons actually group together? And what do the other two electrons of carbon do?

As elementary particles, electrons are very obedient, and they follow without coercion certain laws (the rules of QM theory, which we mentioned already). It turns out that the electrons in the atoms are organized around the nucleus *in shells*. The number of shells possessed by the atom *is given by the number of the period in the periodic table*. Carbon is in the second period, and hence it has two electronic shells. The first electronic shell can accommodate a maximum of two electrons, and the rest of the electrons must then reside in higher-energy shells—the highest of which is the *valence shell*. In the case of C, in Scheme 2.5b, the remaining four electrons reside in the second shell, which is the valence shell. Hence, carbon has two electrons in the inner shell and four in the valence shell. Only this group of four electrons that are relatively far from the nucleus (compared with those in the inner shell) can participate in the bonding to other atoms. *This valence shell can contain maximum of eight electrons*, and therefore, the connectivity of C is four, as specified in Scheme 2.5c. Thus, carbon will form four covalent bonds with other atoms.

2.4.1 The Valence Shell and Connectivity in a Family

The effectiveness of this electron counting method can be illustrated by looking at Scheme 2.6, which shows carbon and two of its other 4A family members, silicon (Si) and germanium (Ge). As can be seen, their total electron counts are very different (6, 14, and 32), but nevertheless, all the three atoms have four electrons in their valence shells. Since the valence shell in main group elements can contain a maximum of eight electrons, *the connectivity of all these atoms will be four*.

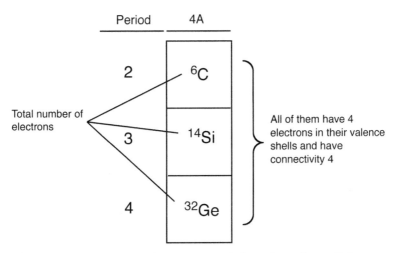

SCHEME 2.6 Carbon, silicon, and germanium; their total numbers of electrons are indicated by the superscripts. All these atoms, which belong to family 4A, have the same number of electrons in their valence shell, four electrons, and all share the same connectivity of four.

2.4.2   The Octet and Duet Rules: The Law of Nirvana

Thus, by simply reading the superscripted number near each atom, the period number where the atom is in, and its family number, we can count the total number of electrons, the number of electron shells, and the number of valence electrons. We need, however, to further clarify and generalize the issue of the maximum number of electrons that can be accommodated in the valence shell and how this number determines the atomic connectivity. The Noble gases at the end of each period give us the clue. As we said already, Noble gases are chemically inert and are not inclined to generate covalent electron-pair bonds (read Primo Levi's description of his family members[2b]).

Consider first helium (He), which belongs to the first period (Figure 2.1), and hence has only one electronic shell, which is also its valence shell. Being first, this shell can contain only two electrons, and since He has already two electrons (look at the superscript near He in the Periodic table), then this atom is electronically "satisfied", and need not form any covalent bonds; *its connectivity is then zero (0).* The first atom in this period is H, which has one valence electron. By making a bond with another H, as in Scheme 2.3, both hydrogen atoms now feel in their valence shells two electrons. Both reached the same degree of electronic "satisfaction" or saturation, as in helium, and will not make any further bonds. Thus, H• has a single connectivity, and will form only one bond with any other element, thereby reaching two electrons in its valence shell. In the language of chemistry, the bonding of H follows the "Duet Rule" (2-electron rule).

Look now at the next noble gas, neon (Ne). Its place in family 8A means that its valence shell has already eight electrons.[11] Ne is electronically "satisfied" and will therefore not form covalent electron-pair bonds. The other atoms in the same period will form electron-pair bonds until they fill eight electrons around them—chemists use the term "octet" for the figure 8. Thus, the "Octet Rule"[8] states that the connectivity of such an atom will be the number of electrons it lacks to reach an octet in its valence shell. The same applies to all other periods of main group elements; *the connectivity of the atoms in these periods will be equal to the number of electrons they lack to reach octet in their valence shells.*

In order to use only a single term to describe this attainment of electronic "satis-faction," we can call the Duet and Octet Rules the "Law of Nirvana"—Nirvana in the case of the atom would mean for us "reaching the state of no more need to bind."[12] Thus, by attaining duet (for the first period) or octet (for all other periods), the atom reaches "Nirvana."

Let us use Scheme 2.7 to illustrate how selective is the Law of Nirvana, by making a molecule from an atom of fluorine (F) and as many H atoms as needed. Since the inner shell electrons do not participate in bonding, we show here and in all examples to follow, only the valence electrons. F resides in family 7A, and hence it has seven electrons in its valence shell, while H has one. Let us suppose we were tempted to think that F utilizes all of its valence electrons to make electron-pair bonds; then, as shown in Scheme 2.7a, F will need to surround itself with seven H• species, and "click," it will form the molecule H_7F. Since F has seven electron-pair bonds in this

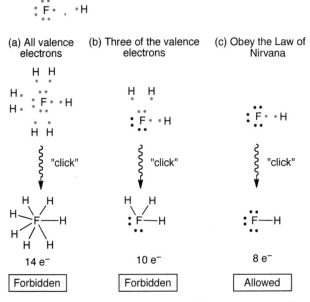

SCHEME 2.7 Making hypothetical molecules and the existing one from H and F atoms (starting H and F atoms are shown at the top), using different numbers of valence electrons of fluorine: (a) Using all the valence electrons to make H_7F surrounds F with 14 electrons ($14e^-$). This is forbidden. (b) Using three valence electrons of F to make H_3F surrounds F with 10 electrons ($10e^-$). This is forbidden. (c) Using only one valence electron on F to make H–F surrounds fluorine with eight electrons, which is precisely the number required to reach Nirvana. H–F is the only allowed molecule.

imagined molecule, it will be surrounded with 14 electrons. The Law of Nirvana tells us that *this is forbidden*, and indeed, such a molecule does not exist.

We could have thought that there is some freedom in utilizing the valence electrons for bonding, and as such we could have considered F to have a connectivity of three, by utilizing just three of its valence electrons, as in Scheme 2.7b. Since H• has a connectivity of unity, we would need three H's, and "click," we could form the molecule H_3F. Inspecting the F in this imagined molecule, we find it is surrounded by five electron pairs, three bonds, and two "idle" pairs that chemists call *lone pairs*. The name *lone pair* signifies a pair of electrons which exists in the molecule but does not participate in bonding. In total, therefore, F is surrounded in H_3F by 10 valence electrons. Hence, H_3F violates the rule, and it does not exist.

Finally, if we simply obey the Law of Nirvana, then F will have a connectivity of one, that is, F•, and hence it can bind only one H•, and "click," they make the molecule H–F, in Scheme 2.7c. Here the F atom is surrounded by one bond pair and three lone pairs; together eight electrons; at the same time, H is surrounded by two electrons. Both atoms reached Nirvana, and this is the only molecule that is allowed and is observed.

To summarize this exercise, we show in Scheme 2.8 the correct molecule H–F that can exist under the Law of Nirvana. It has an electron-pair bond symbolized by a line

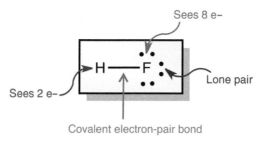

SCHEME 2.8   A summary of the electronic structure of H–F, showing that F has one bond with H and three lone pairs, and as such, both atoms are in a state of Nirvana.

connecting the two atoms, and there are three lone pairs on F. Taken together, F sees an Octet and H sees a Duet; both obey the Law of Nirvana.

The H–F molecule is a poisoning gas. Upon dissolution in water, it becomes an "acid," which is a class of a materials we shall learn more about during the course (acids tend to liberate a proton H^+ to other molecules called "bases," which accept the protons). H–F has many uses in pharmaceutical and petroleum industries. One of its uses is also in art, for glass etching. The corrosive H–F can dissolve the glass by a chemical reaction. If the glass were covered initially with wax, on which we drew for example, a flower and removed the wax from the lines of the drawing, and then applied H–F, the result would be a glass with an etched flower.

2.5 MAKING MOLECULES USING THE AVAILABLE ATOM CONNECTIVITY AND THE LAW OF NIRVANA

The example of H–F illustrates the power of the Law of Nirvana, and we can now generalize it and proceed to make some new molecules. Figure 2.3 is a connectivity-based mini periodic table that summarizes the maximum connectivity of main group elements based on the available number of valence electrons and the Law of Nirvana. The number of valence electrons each atom possesses is identical to the numerical indicator of the chemical family. In order to emphasize the connectivity, we label these electrons as red spheres, while the other valence electrons that will form lone pairs are indicated by black dots.

It is seen that the connectivity in each period (from the second on) initially rises from 1 to 4, and then falls gradually to zero at the Noble gas atom. This connectivity pattern is periodic, and it repeats itself in every period.

Another piece of information in Figure 2.3 is the division of the atoms to three types. The Noble gases have already achieved Nirvana. The atoms of the families 4A–7A will reach Nirvana by bonding, while the atoms in families 1A–3A cannot achieve the desired Nirvana by utilizing their connectivity and will form *electron-deficient molecules.*

To illustrate electron-deficient molecules, consider in Scheme 2.9 the molecules that can be constructed from beryllium (Be) and boron (B) with as many hydrogen

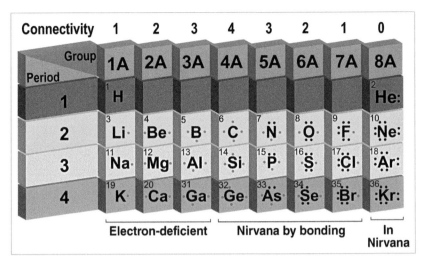

FIGURE 2.3 The mini periodic table of connectivity of main group atoms, shown down to the fourth period. The electrons participating in connectivity are highlighted in red. Valence electrons that do not participate in bonding are indicated by the double black dots. As indicated beneath the brackets, atoms in groups 1A–3A generate electron-deficient molecules, and atoms in groups 4A–7A generate molecules that achieve Nirvana, while atoms in group 8A are already in Nirvana.

atoms as required. Since the Be atom has a connectivity of two, it can bind only two H• particles, "click," and BeH_2 with two Be–H bonds is generated. Counting electrons will show immediately that in BeH_2, Be is surrounded by only four electrons, and it is hence electron deficient. Similarly, since B has a connectivity of three, it will "click" with three H• particles to make the BH_3 molecule. With three B–H bonds, B is surrounded by only six electrons and it is hence electron deficient.

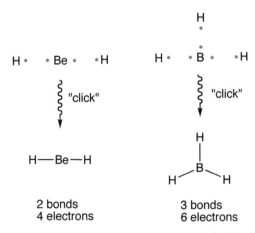

SCHEME 2.9 Two electron-deficient molecules, BeH_2 and BH_3. The numbers of bonds and electrons surrounding the central atoms are indicated beneath the drawings.

Note that both atoms use *all* their valence electrons as connectivities to make bonds with the appropriate number of H's. But since the number of available electrons for the atoms Be and B is too low, even after bonding with the H's, these atoms do not reach Nirvana. Consequently, these molecules will be chemically reactive (high propensity to react with other molecules) and will have a propensity to attain Nirvana by utilizing electrons from other molecules (e.g., lone pairs of other molecules). We shall work out later an example. Starting from carbon on, the molecules have sufficient electrons and can attain Nirvana (consult Figure 2.3) by using their valence electrons.

2.5.1   Using the Table of Connectivity to Make Molecules That Attain Nirvana

To illustrate the utility of the connectivity table, in Figure 2.3, we proceed with examples of a few important molecules in our universe, which are shown in Scheme 2.10.

Water: Let us start in Scheme 2.10a by constructing a molecule made from an oxygen (O) atom and as many hydrogen atoms as required for Nirvana. Since oxygen has a connectivity of two, it would need two H• particles to bind to. Bringing the

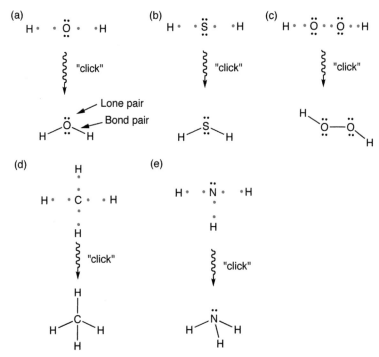

SCHEME 2.10   Constructing molecules by clicking atoms with appropriate connectivities into electron-pair bonds. The electrons participating in connectivity are highlighted in red, while valence electrons that do not participate in bonding are marked as black dots. (a) Water (H_2O). (b) Hydrogen sulfide (H_2S). (c) Hydrogen peroxide (H_2O_2). (d) Methane (CH_4). (e) Ammonia (NH_3).

H• atoms and clicking the connectivities generates the molecule H_2O, which is a molecule of water.

Let us count electrons now and ascertain that the atoms in the water molecule reached Nirvana. Thus, it is seen that O is surrounded by four electron pairs, two lone pairs, and two bond pairs, and hence it attains octet within the molecule. Each hydrogen atom has a duet, so all the atoms in the molecule achieved Nirvana, and the molecule is stable. Note that we draw the molecule with a bent angle. The water molecule is indeed bent, and this is very important for its ability to aggregate with many other molecules and make the liquid water we all know well. But at this stage, we still do not have the know-how to predict that this would be the shape of the molecule. We will! Until then, it is also OK if you draw your water as linear H–O–H.

Hydrogen Sulfide: Let us turn now to Scheme 2.10b to construct a molecule from one atom of sulfur (S) and hydrogen atoms as required. This is so simple; the connectivity is conserved down the family column, and hence S has connectivity 2. As before, two H• atoms will be needed to make a molecule in Nirvana. Bringing them and clicking the connectivities we obtain H_2S (hydrogen sulfide).

Unlike water molecules that aggregate to form a liquid state at ambient temperatures, H_2S is a gas, where the individual molecules keep apart. And what stinking gas it is! This is the gas that emanates from rotten eggs (and other rotten materials of living creatures) and from sewage. As kids and amateur chemists, we used to prepare this gas (it is very easy to do so) in class near the teacher's desk. Kids tend to be cruel to their teachers, for the sake of having some fun … Odd as it may sound, despite its stench, H_2S is also beneficial. It turns out that it is formed in living cells and used to regulate blood pressure. Janus-like is the chemical matter!

Hydrogen Peroxide: Scheme 2.10c illustrates how a connectivity of two can lead to a variety of molecules. Let us now take two oxygen atoms and add hydrogen atoms as needed to construct a molecule. Since H• is a single connector, while •O• is a double connector, we cannot connect the two H• particles to a single •O•, because this will leave the other •O• particle unbound. The way to satisfy the connectivities is to link to each •O• with one H•, and interconnect the •O• atoms to each other. So, we bring all the four atoms, and "click," the connectivities to three bonds, thereby generating the molecule hydrogen peroxide (H_2O_2).

The most common daily use of H_2O_2 is to convert dark hair to blonde, and during hair dyeing to any desired color. This is achieved by a chemical reaction of H_2O_2 with the melanin pigment, which is responsible for dark hairs.[13] H_2O_2 is also used for disinfection of wounds to avoid bacterial damage. Our immune system generates H_2O_2 as means of fighting foreign materials that enter the living body, and in signaling pathways. Recent studies show[14] that H_2O_2 is also generated in the brain and causes stem cells to become neurons in aging people! It is really so marvelous to realize how almost every molecule plays two, three, and more roles, like the two-faced Roman God Janus in Picture 2.1, only that molecules are many-faced little creatures catering to many needs.

Methane: In Scheme 2.10d, we construct a molecule from an atom of carbon (C) and hydrogen atoms as needed. Since C is a quadruple connector and H• is a single

PICTURE 2.1 The two-faced Roman God, Janus. This photograph was obtained from http://en.wikipedia.org/wiki/File:Janus1.JPG.

connector, we are going to need four H• atoms. Assembling all the atoms and clicking their connectivities leads to the molecule CH_4, which is called methane.

The molecule is also called "swamp gas," because it is produced in swamps by decaying plant matter. In high concentrations, it is claimed to spontaneously ignite, and it may be responsible for some UFO (unidentified flying object) reports. In fact, the gas methane is one of the greatest carbon-based energy reserves on earth. As we shall see later, by reacting with the O_2 molecule, methane and other hydrocarbons (made from C and H) burn and give energy. But methane is also one of the gases causing the greenhouse effect on our earth. The molecular Janus again …

Ammonia: Finally, let us construct a molecule from an atom of nitrogen (N) and H atoms as needed. As shown in Scheme 2.10e, since N is a triple connector, it will require three H• particles. Clicking the connectivities leads to the molecule NH_3, which is called ammonia, after the name of the old Egyptian city No-Ammon (where the palace of Karnak and Luxor are located), from where a mineral that breathes out ammonia used to be mined. Ammonia is a gas with a very sharp pungent odor, so pungent "that it can revive the fainted" and in large quantities, it is very poisonous. At the same time, this molecule is the main source of artificial fertilizers and a starting material for making explosives. The molecular Janus again …

Finally, you all noticed that we drew both methane and ammonia not with right angles. This is because these molecules are three-dimensional (3D) objects, as shown in Figure 2.4. We shall learn during the course about the 3D shapes of molecules. Right now, we are simply flagging this fact by drawing the molecules as we did, but it is still fine if you draw them with right angles.

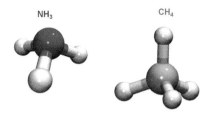

NH₃   CH₄

FIGURE 2.4   3D shapes of NH_3 and CH_4, using ball-and-stick models, where a ball represents an atom and a stick represents a bond. Color coding: N (blue), C (green), and H (gray).

2.5.2 Bonding in Atoms with Multiple Connectivity

Multiple connectivity of an atom endows it with great many possibilities of making a variety of molecules. Let us see some of these molecular constructions in Scheme 2.11.

Carbon Dioxide: In Scheme 2.11a, we wish to make a molecule from a single atom of C and as many O atoms as needed. Since C has quadruple connectivity and O is a double connector, we need two O atoms. Assembling these atoms and clicking their connectivity leads to the molecule CO_2, also called carbon dioxide. It is seen that in CO_2, the carbon atom has two double C=O bonds with the oxygen atoms. A double bond is, of course, stronger and *more stable in energy* than a single bond. CO_2 is one of the gases in the earth's atmosphere, and it serves as a warming mantle of our earth. However, as its concentration in the atmosphere has been steadily increasing, it is now considered a major cause of global warming (the so-called "greenhouse effect"). If your teacher presented the CO_2 demo during the last lecture, you will recall that this gas is also a fire retardant.

Dinitrogen: Let us construct now the minimal molecule made from nitrogen (N) atoms. As shown in Scheme 2.11b, since N is a triple connector, two atoms can click and form an N_2 molecule with a triple bond. A triple bond is even stronger than a double bond. In fact, the N≡N bond is one of the two strongest bonds in nature.

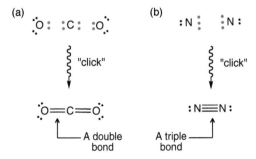

SCHEME 2.11   Constructing molecules from atoms with multiple connectivities: (a) Carbon dioxide (CO_2) with double bonds. (b) Dinitrogen (N_2) with a triple bond.

As we just mentioned, Mankind needs a lot of ammonia (NH_3) to be used for making fertilizers. Since N_2 is the major gas in the earth's atmosphere, it would be wonderful to be able to make ammonia from the abundant N_2 by reacting it with H_2. However, this is not easy since the N≡N bond is very strong. Nevertheless, Nature has found an efficient way here to achieve this conversion (which is also called "nitrogen fixation") at ambient temperatures in bacterial cells that use *nitrogenase*, which is an *enzyme* that facilitates the process (see Retouches section 2.R.3).

Fertilizers used to be available to the western world by exporting the mineral *Guano*, from the Chincha Islands, 21 km off the southwest coast of Peru, where it was deposited in 10,000 tons a year by the native sea birds. Alexander von Humboldt, who explored Latin America, brought it for the first time to Europe in the late eighteenth century because he learned from the Indians that Guano possessed the power of aiding plants to grow. The French chemists Fourcroy and Vanquelin analyzed the mineral and identified it to consist of bird excrement and to contain nitrogen. By 1863, Peru's economy boomed due to the export of Guano, which was used in Europe and the United States as fertilizer. Then Guano was found also in Chile along with nitrate salts, which were used as explosives and fertilizers. After the Chincha Islands War in which Peru, Bolivia, Chile, and Spain participated in an attempt to control the Guano commerce, Chile emerged as the major exporter of these materials to Europe and the United States.

During World War I, Britain and its allies blocked the commerce of Germany with Chile. Therefore, Germany could not anymore export the Guano and nitrates needed for agriculture and explosives. As it usually happens, necessity is the "mother of invention." This acute need of Germany led to the development of the Haber–Bosch process, which generates ammonia from N_2 and H_2 by applying huge pressure and high temperatures. Figure 2.5 shows Fritz Haber[15] and Carl Bosch, and the ammonia-making process.

Fritz Haber was one of the great German chemists pre- and during World War I.[15] He came up with a way of conducting the reaction using additives called "catalysts," which facilitate the reaction (see Retouches section 2.R.3). Carl Bosch was an employee of the German company BASF (*Badische Anilin- und Sodafabrik*), and he undertook to conduct the Fritz Haber process on an industrial scale. He succeeded, and this is the process, which is still used to this day to make ammonia. Both Fritz Haber and Carl Bosch became very rich, and both received the Nobel Prize for their invented process. If you ever visit Heidelberg, go to see Villa Bosch, where Carl Bosch used to live in a kingly style. By the way, the facility of ammonia making in the twentieth century is considered to be a major cause of the global population growth in this century. BASF, which is centered in Ludwigshafen (Germany), is a huge plant the size of a city with 35,000 employees (more than 110,000 globally). It still makes ammonia using the Haber–Bosch process.

Before moving on, it is important to recognize that N_2 is important in its own right. Being the major gas in the earth's atmosphere (78.1%), it coexists with the other major component, O_2 (20.9%). And as such, it dilutes O_2 and prevents thereby violent spread of fire. Breathing of O_2-rich gases is also very risky to humans, leading to the buildup of fluids in the lungs and to strokes. We are lucky to have N_2 surrounding us!

Fritz Haber   Carl Bosch

$$N_2 + 3H_2 \rightleftharpoons 2NH_3$$

High pressure
High temperature
Catalyst

FIGURE 2.5   Fritz Haber's caricature and Carl Bosch's photo, both shown above the ammonia-producing reaction. (Fritz Haber's caricature was drawn by W. B. Jensen and used with his permission, and Bosch's photograph is used from www.nndb.com/people/405/000100105)

White Phosphorus: Finally, let us proceed to Scheme 2.12 and construct a molecule made only from phosphorus (P) atoms, but one that does not possess multiple bonds. Of course, without this qualification, we would have used an analogy to N_2 and built a P≡P molecule. However, since we asserted that we cannot have any multiple bonding, we must use the triple connectivity of P to connect to three other P atoms, thus having together four P atoms, each connected to the other three. So, we bring the 4P's together and click their connectivities to make the P_4 molecule shown in Scheme 2.12, which is called *white phosphorus* because of its white color.

In Figure 2.6, we show the 3D model of white phosphorus (P_4). It is a beautiful cage molecule! White phosphorus is the most common structure of the element in nature. Since P is a triple connector, it can make many other molecules which contain only P atoms. Thus, we have also red and black phosphorus, which are much more complex molecules than P_4.

From red phosphorus, we make matches and also explosives. As kids, we used to mix red phosphorus with an oxygen-rich compound and some other ingredients,

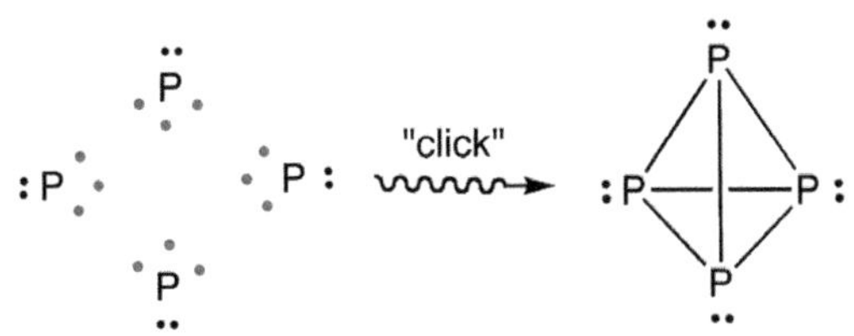

SCHEME 2.12   Constructing white phosphorus (P_4).

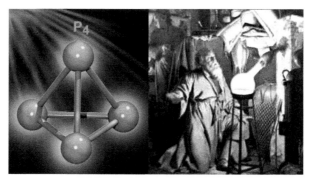

FIGURE 2.6 The 3D shape of white phosphorus (P_4), shown alongside its discoverer Hennig Brand. (This painting, by Joseph Wright of Derby, was obtained from http://en .wikipedia.org/wiki/File:Henning_brand.jpg)

and then we inserted everything into a small capsule. If you stepped on it, it would explode and make a terrible noise, which scared the neighbors who did not let us play soccer near their house! It is also highly dangerous to make these explosive mixtures, but as kids, we were not aware of that. Do not try it!

White phosphorus was discovered by the German alchemist Hennig Brand in the seventeenth century. Brand was convinced that he could extract gold from urine. Why urine? I guess that is because people in the seventeenth century did not drink much water, and their urine was "golden." But your guess for Brand's decision may be as good as mine. Anyway, Brand collected gallons of urine in his basement and started treating the urine with alchemical techniques (which were usually aggressive heating and distillation). After some such treatments, he was left with a white paste that was glowing in the dark. He called this paste *Phosphorus Mirabilis*, which means "the glowing magic." Figure 2.6 shows near the 3D shape of the molecule the painting by Joseph Wright of Derby of Hennig Brand discovering *Phosphorus Mirabilis* while distilling urine. The picture is imbued with the fascination with chemistry.

As I mentioned already, the red form of phosphorus can be used for production of matches and gunpowder. As a result, producing phosphorus garnered a lot of interest by states and by private industry. For years, enlisted soldiers in Europe had to give urine as part of their service. You can imagine the stench of these phosphorus products! The scientist who ridded the world from this stench was the great Swedish chemist Carl Wilhelm Scheele, who developed a method to extract phosphorus from bone ashes and plants. Sweden became a major exporter of matches in the nineteenth century.

2.6 THE PRINCIPLE OF CONSERVATION OF THE NUMBER OF ATOMS IN CHEMICAL REACTIONS

We keep mentioning the term "chemical reaction," which is the process behind the magic of chemistry—molecules react and change into other molecules. Let us look now at the ammonia-producing reaction in Figure 2.5. All chemical reactions are written in this manner. This is part of the chemical language: there are starting molecules (on the left-hand side), called *reactants*, which react and generate new molecules, called *products* (on the right-hand side). The reactants and products are separated by an arrow showing the direction of the reaction from reactants to products[16]:

$$N_2 + 3H_2 \rightarrow 2NH_3 \tag{2.1}$$

We can also see that for each molecule of N_2, one needs three molecules of H_2, which together lead to the formation of two NH_3 molecules. If you now count the number of atoms of each kind on both sides of the arrow, you will find two N and 6 H on either side. As such, the equation is said to be "balanced." This balancing simply reflects a very clear principle: the "conservation of the atom"; the same number of atoms we started with in the reactants will appear in the products. No atom is lost, no atom is gained! *All chemical reactions have to be balanced.*

2.7 SUMMARY

Some Principles of Chemical Bonding: We have learned in this lecture to use the periodic table (Figure 2.1) to provide us information about the electronic structure of the atom, with focus on main group atoms, so we can construct molecules. Based on that, we derived the atomic connectivity in a mini periodic table in Figure 2.3 for the main group elements.

The full periodic table in Figure 2.1 has seven periods numbered as 1–7, and eight families of main elements, numbered by the numerals 1A–8A. Each element has a superscripted number that marks its sequential place in the table. The table provides the following information about an atom:

- Its total number of electrons is identical to the number of the element in the table. For example, the sixth element, C, has six electrons.
- The number of the period where the atom is found tells us the number of electronic shells it possesses. For example, C is in the second period, and it has two electronic shells.
- The inner shell, which is the first electronic shell closest to the nucleus, can contain a maximum of two electrons. For example, in the first period, which contains only H and He, the atoms have only one shell. He, the second element in the table, has two electrons in this shell, while the first element H has only one electron, which makes H a single connector.
- The outer shell of the atom is called the "valence shell," and the number of electrons it contains is identical to the numeral indicator of the family to which the atom belongs. For example, C belongs to group 4A, and hence, it has four

electrons in the valence shell. Eight is the maximum number of electrons in the valence shell of a main element atom from period 2 onward.

- The number of valence electrons is conserved as we move down the family column. For example, Si and Ge in group 4A have four valence electrons, as does C.

Figure 2.3 provides the following information about the connectivity of atoms:

- The connectivity of the atom is the number of covalent electron-pair bonds it can make by utilizing all or part of its valence electrons. Atoms make bonds by "clicking" their connectivities into electron pairs.
- When the number of valence electrons of a certain atom is larger than its connectivity, the molecule will have *bonds* and *lone pairs*. For example, F has a connectivity of one and seven valence electrons. It will therefore make one bond, for example, as in H–F, and the rest of the six valence electrons will reside on the F as lone pairs.
- The atoms of group 8A have completely filled valence shells, which is a state of atomic Nirvana. For example, He has two electrons in the valence shell, while Ne, Ar, and Kr all have eight electrons. These atoms have a connectivity of zero and will not form covalent electron-pair bonds with other atoms.
- The atoms in families 4A–7A will reach a state of Nirvana by covalent bonding as dictated by their connectivities (which are the number of electrons needed to complete the valence shell to 8). For example, C will form four bonds, and N only three bonds, etc.
- The atoms in families 1A–3A will generate electron-deficient molecules. For example, Be has only two valence electrons, and hence it can form only two bonds with other atoms. Thus, it does not reach the state of Nirvana. Electron-deficient compounds are very active chemically and will try to achieve Nirvana by binding to molecules, which possess lone pairs (we will study this in detail in Lecture 3).

The Energy Due to Chemical Bonding: We have also learned that by making bonds, the energy of the molecular species is lowered (Scheme 2.4a). The molecule is more "stable" than the atoms. The chemical bond is extremely ubiquitous in our universe because Nature prefers objects of low energy.

The Atom Conservation Principle: In a chemical reaction, the bonds of the initial molecules (called "reactants") are broken and remade to yield the final molecules, called the "products." The chemical reaction is governed by the principle of *conservation of the atom*, which requires all chemical reactions to be "balanced" and have identical numbers of each atom on both sides of the equation arrow, as in Figure 2.5.

We used these principles to construct a few molecules. We told stories about them because these stories reflect the intimate and intricate connection of man and matter,[2] and the fact that science is a part of human culture. The principles we learned in this lecture will enable us to generate entire molecular worlds in the next lecture.

2.8 REFERENCES

[1] See a summary in: F. L. Giessibl, *Science*, **2013**, *340*, 1417.

[2] (a) S. Kean, *The Disappearing Spoon*, Little, Brown & Company, 2012, New York; (b) P. Levi, *The Periodic Table*. G. Einaudi Ed., Penguin Books, Torino (English translation, 1984).

[3] See the wonderful BBC series *The Elements*, with J. Al-Kahlili.

[4] (a) See, imaging of a single molecule and its atoms: R. Zhang, Y. Zhang, Z. C. Dong, S. Jiang, C. Zhang, L. G. Chen, L. Zhang, Y. Liao, J. Aizurua, Y. Luo, and J. L. Yang, *Nature* **2013**, *498*, 82; (b) See also, imaged H and C atoms, *Chemical & Engineering News*, July 21, 2008, p. 9.

[5] D. W. Ball, *J. Chem. Educ.* **1985**, *62*, 787.

[6] R. Kurzweil, *How to Create a Mind*, Penguin Books, London, 2012, p. 126.

[7] Note that there is a very popular term, "click chemistry," invented by B. Sharpless, to describe chemical reactions that occur with facility, as in a click. For a review, see: M. V. Gill, M. J. Arévalo, and O. Lòpez, *Synthesis* **2007**, 1589.

[8] G. N. Lewis, *J. Am. Chem. Soc.* **1916**, *38*, 762.

[9] E. Segrè, *From X-rays to Quarks: Modern Physicists and Their Discoveries*. Dover Classics of Science and Mathematics, Dover, 2007.

[10] S. Shaik, *J. Comput. Chem.* **2007**, *28*, 51.

[11] The place of helium in family 8A may appear a bit awkward since it has only two electrons in its valence shell. However, because this valence shell is full, helium is a Noble gas, and its place above Ne is natural.

[12] Nirvana is an ancient Sanskrit term, which describes the situation of profound peace of mind upon liberation from the need to achieve. It is used here as a metaphor for the state of atoms after achievement of the number of electrons to resemble the Noble gases.

[13] http://en.wikipedia.org/wiki/Melanin

[14] C. Drahl, *Chemical & Engineering News*, January 10, 2011, pp. 32.

[15] On Fritz Haber, see: F. Stern, *Angew. Chem. Int. Ed.* **2012**, *51*, 50.

[16] On history of arrow usage in chemistry, see: S. Alvarez, *Angew. Chem. Int. Ed.* **2012**, *51*, 590.

2.A APPENDIX

Table 2.A.1 Names[a] of the First 18 Elements

H (Hydrogen)							He (Helium)
Li (Lithium)	Be (Beryllium)	B (Boron)	C (Carbon)	N (Nitrogen)	O (Oxygen)	F (Fluorine)	Ne (Neon)
Na (Sodium)	Mg (Magnesium)	Al (Aluminum)	Si (Silicon)	P (Phosphorus)	S (Sulfur)	Cl (Chlorine)	Ar (Argon)

[a]In parentheses is the element's name.

2.R RETOUCHES

2.R.1 Elements versus Atoms

Figure 2.R.1 provides a quick journey from the Greek philosophers to the nineteenth-century chemistry of atoms and molecules.

The idea that matter was made of four or five elements appears in all the great cultures of the ancient world. As we told in the Retouches for Lecture 1, this idea seems to have been originated by Greek philosophers,[1R] who postulated the existence of the four elements. These were *Earth*, *Water*, *Air*, and *Fire*, which were supposed to transform into each other, and by blending, to give rise to all the different forms of matter in the universe. As opposed to these abstract notions about matter, the philosopher Democritus thought that all matter was made from indivisible material particles, which he called "atoms." His idea was not accepted, but resurfaced again in the seventeenth to nineteenth centuries and ultimately remained as the central paradigm in chemistry.

In the western world, the theory of four elements was dominant almost till the eighteenth century. However, in the seventeenth to eighteenth centuries, chemistry also became a materialistic science (as opposed to being abstract, as it was in the Greeks' times when theories were conceived without experimentation), with different materials that had different properties. With the increasing sophistication of techniques for analysis and synthesis,[2R] the language also changed. The "element" was given a practical definition; it was called "a simple body." The great French chemist Antoine Laurent Lavoisier suggested calling "a simple body" every material that could not be further decomposed using the available techniques at a given time. He also warned against the chase after atoms, which he felt were merely philosophical elements, "… it is extremely probable we know nothing at all about them [because they cannot be observed]."[2R]

Aristotle Democritus Lavoisier Dalton Avogadro Cannizzaro

FIGURE 2.R.1 From ancient Greek philosophy to materialistic principles in the nineteenth century: Aristotle (four elements), Democritus (atoms), Lavoisier (simple bodies), Dalton (atoms with characteristic weights), Avogadro (molecules), and Cannizzaro (atoms vs. molecules). Aristotle's image by Francesco Hayez (with credits to The Yorck Project), Democritus' portrait by Antoine Coypel (credits to the Los Angeles County Museum of Art), Lavoisier's portrait by Jacques-Louis David, Dalton's photo by Charles Turner (credits to the Library of Congress), Avogadro's drawing by C. Sentier (credits to the Edgar Fahs Smith Memorial Collection), and Cannizzaro's photo (credits to the supplement of *Nature* magazine). All images are in the public domain.

In the nineteenth century, all the gases (hydrogen, oxygen, nitrogen, carbon dioxide, etc.) were already known, and it was agreed that some of them (e.g., hydrogen, oxygen, nitrogen) were "simple bodies." Their reactions were also known and could be followed by weighing the amounts needed to complete the reactions, for example, the required weights of hydrogen and oxygen that formed water without leaving any trace of the "simple bodies." John Dalton, who was investigating gases, advanced *the atom as the basic constituent of matter.* He showed that based on this concept and on the assumption that atoms combine in ratios of simple integer numbers (e.g., 1:1, 2:1, etc.), it was possible to determine the relative weights of different atoms and to form a scale of relative atomic weights, giving the lightest atom hydrogen a weight of unity. Thus, since he found that O and H combined in a weight ratio of 8 (O)/1 (H), the weight of oxygen became 8, assuming that H and O combined to give a simple compound HO. His idea shows the power of simple postulates, but also their eventual weakness, which brings about their change into better ones. This is how science progresses.

In Dalton's days, connectivity was not yet conceived, and hence, the Dalton approach ran quickly into controversies since different atomic weights were determined for the same atoms from different compounds. The concept of the molecule was missing. A bold assumption by Amedeo Avogadro that identical atoms can combined to molecules, for example, H forms H_2, etc., seemed to solve all the controversies, but it was initially ignored and vehemently objected to, especially by Dalton. In the fall of 1860, in the congress of organic chemistry in Karlsruhe (in Germany), another Italian Chemist, Stanislao Cannizzaro, gave a lecture in which he utilized Avogadro's hypothesis and showed that it removes all the controversies and leads to a fixed scale of relative atomic weights. *The atom thus became the constituent of all chemical matter!*[2R]

2.R.2   Electron Pairing

As we discussed in the Retouches section for Lecture 1, in addition to its negative electric charge, the electron has a property called spin. Thus, the electron is in fact also *a small magnet.* When electrons are paired, this happens only when their spins are opposite.[3R] As such, *while a single electron is a magnet, the electron pair is devoid of any magnetism.* Another change that occurs with electron pairing is that *the energy of the pair is lowered*, and this provides a driving force for pairing electrons during bonding.

2.R.3   Enzymes and Catalysis

Enzymes are pretty large molecules that exist in all living organisms and plants, and their role is to enhance the rate by which reactions occur.[4R] It turns out that chemical reactions do not occur spontaneously because they face *energy barriers.* Because of the need to break bonds during the chemical reaction, the energy must first go up, and only then it can go down, thus forming a barrier. Barriers make many of the

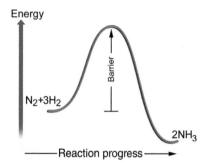

FIGURE 2.R.2   A representation of the energy change during the chemical reaction: $N_2 + 3H_2 \rightarrow 2NH_3$. Note that, as the reaction progresses, there is an energy barrier that impedes the production of NH_3.

molecules persistent with long lifetimes. Just imagine a world without barriers. In such a world, there would be no molecules, only *a soup of atoms*!

The barrier is illustrated in Figure 2.R.2 for the reaction of N_2 with H_2. It is very high, and without an enhancer, the reaction will take eternity. The enzyme nitrogenase[5R] is composed of a large molecule made of iron (Fe) and molybdenum (Mo) atom centers embedded in a protein, and what it does is to facilitate the conversion of N_2 and H_2 to ammonia (so-called "nitrogen fixation") by lowering the barrier and allowing the reaction to proceed at ambient temperatures.

Catalyst: This is a generic name for a molecule that when added to a reaction causes rate enhancement. Enzymes are natural catalysts. There are also catalysts that have been invented by chemists; for example, Fritz Haber used metal-based catalysts like osmium (Os) and iron (Fe).

2.R.4   Alchemy

In the pre-chemistry era, there was alchemy (it is believed that this originates in the Arabic name of Egypt, Al-Khem). Alchemists believed in the theory of four elements and in the principle "as above so is below," which basically meant that they were trying to emulate the heavenly practices on earth. Alchemists believed that the ultimate fate of everything is to improve and reach higher degrees of perfection, and that this could be achieved by following the principles of alchemy. This philosophy included the belief that through alchemy one could convert base metals such as lead to gold.[6R] At some point, many of the alchemists, especially those in Europe, focused all their attention on making gold, and this has caused the science to degenerate. Nevertheless, alchemists achieved many great technical achievements,[7R] which are still with us. Even the great Newton was an avid alchemist. There is in fact a modern movement that tests the procedures of alchemists to find ways for them to work (not to convert lead to gold though!). A recent paper in the *Science* magazine calls this movement "The Alchemical Revolution."[8R]

2.R.5   References for Retouches

[1R] (a) I. Asimov, *A Short History of Chemistry*, Heinemann Educational Books Ltd., London, 1973, Chapters 1–3. (b) W. H. Brock, *The Norton History of Chemistry*, W.W. Norton & Co., New York, 1992, Chapter 1. (c) E. Heilbronner and F. A. Miller, *A Philatelic Ramble through Chemistry*, Wiley-VCH, Weinheim, 1997, Chapter 1.

[2R] R. Siegfried, *From Elements to Atoms: A History of Chemical Composition*, American Philosophical Society, Philadelphia, PA, 2002.

[3R] S. S. Zumdahl and S. A. Zumdahl, *Chemistry*, 9th Ed., Brooks College, Cengage Learning, 2010, Chapters 7 and 8. Available at: www.cengage.com/global

[4R] Reference 3R, pp. 583–589.

[5R] S. J. Lippard and J. M. Berg, *Principles of Bioinorganic Chemistry*, University Science Book, Mill Valley, CA, 1994, pp. 134–135.

[6R] Reference 1Ra, Chapter 2.

[7R] See, for example, the story of Maria Hebraea (Maria the Jewess), who left many tools which are in use today in all chemistry laboratories: R. Patai, *The Jewish Alchemists*, Princeton University Press, Princeton, NJ, 1994.

[8R] S. Reardon, *Science*, **2011**, *332*, 914.

2.P   PROBLEM SET

2.1 What do the rows and the columns represent in the periodic table?

2.2 Answer the following questions:
- **(a)** What is the atomic number of phosphorus (P)?
- **(b)** How many electrons does phosphorus have?
- **(c)** What is the number of valence electrons of phosphorus?
- **(d)** How many electrons are needed for phosphorus in order to reach the state of Nirvana?
- **(e)** What is the connectivity of phosphorus?

2.3 Repeat exercise 2.2 for boron (B), and answer also the following questions:
- **(f)** Will boron achieve Nirvana?
- **(g)** What kind of molecules will it form?

2.4 Construct the simplest molecules from the following atoms according to the Law of Nirvana, and answer the following questions: How many electrons does each atom contribute for bonding? How many electrons surround each atom? How many nonbonding electrons does each atom have, if any?
- **(a)** O.
- **(b)** He.
- **(c)** H, C, and N.

(d) One beryllium atom and as many chlorine atoms as needed.

(e) Two nitrogen atoms and as many hydrogen atoms as needed.

2.5 The burning (combustion) reaction of methane (CH_4, the main component of "natural gas" fuel) with oxygen molecules (O_2) produces carbon dioxide (CO_2) and water.

(a) Construct the molecules that react and are produced in the combustion reaction.

(b) Write the balanced equation that describes the combustion reaction. Which principle requires balancing chemical equations?

(c) From your own daily experience, does this reaction consume energy or does it release energy? Based on this answer, specify which bonds are stronger and which ones are weaker.

LECTURE 3

ELECTRON-DEFICIENT MOLECULES, GIANT MOLECULES, AND CONNECTIVITY OF LARGE FRAGMENTS

3.1 CONVERSATION ON CONTENTS OF LECTURE 3

Racheli: Sason, isn't it the time to stop keeping all of us in suspense?

Sason: Please? Explain.

Racheli: You promised to tell us how electron-deficient molecules force their way to Nirvana. Well, we are all waiting …

Sason: Racheli, I am going to show an ingenious solution these molecules find for achieving Octet, or Nirvana as we refer to it, by utilizing lone-pair electrons of other molecules. Parasites, aren't they?

Racheli: Just talking about these parasite molecules would be a short lecture. What then?

Sason: Hey, give me a chance to complete the thread of my thoughts …

I will subsequently construct a giant covalent molecule from silicon (Si) and oxygen (O) using the click bonding idea. It will show that *chemistry is endless*. It will also demonstrate that constructing large molecules from individual atoms is tedious. We are going to need more creative ways for constructing large molecules from large modular parts. So, we will have a full lecture. Won't we?

Racheli: Aaha … I remember that when we were kids, at some point we got LEGO kits with larger building blocks, called DUPLO, and we were able to build beautifully complex objects. Is this what you have in mind?

Chemistry as a Game of Molecular Construction: The Bond-Click Way, First Edition. Sason Shaik.
© 2016 John Wiley & Sons, Inc. Published 2016 by John Wiley & Sons, Inc.

Sason: You read my mind. Maybe you should do the teaching, and I will do the kibitzing?

Racheli: No, thanks. I feel comfortable where I am and am just happy to leave you the hard work.

Sason: OK then, I will show that we can generate large modular fragments with connectivities of 1–4 and build from them new and large molecules. I will also take advantage of the molecular construction stories and will introduce some more terms chemists use, in particular the term *stoichiometry*, which was invented by a nineteenth-century chemist named Richter.

Racheli: By the way, *stoicheion* in Greek means element, and this makes the literal meaning of stoichiometry "the measurement of elements."

Sason: You know, Racheli, chemists are very funny about inventing terms. When they want to make a term sound important, they resort to Greek or Latin. This is a cultural habit that was (maybe still is) common in the western world, presumably because of the veneration to the contribution of the Greeks and the Romans to the western culture. But how many of your friends know Latin these days?

Racheli: Not many …

But, Sason, as usual, you are digressing and getting carried away. Please do not forget, while you build your giant covalent molecule, that it is advisable to demonstrate this process in "real life." It is actually the art of making glass from *water glass.*

Sason: Yes. This is how we get these glasses with ravishing colors.

The "chemistry" of glass making is in fact ancient, and one can couple this demo with a reading of a 2,700-year-old recipe for making colored glass, and with the story of the great seventeenth-century alchemist Jan Baptista van Helmont, who used *water glass* to prove that the entire material world is made from the Greek element "Water."[1]

Racheli: Stories are wonderful, Sason! Just do not digress too much …

Sason: Wait, wait! I have planned one more thing …

Racheli: OK …

Sason: I am going to try and describe the wonderful O_2 molecule, which does not adhere to simple "click" bonding. This will be done in the Retouches section (**3.R**). I think we should mention also exceptions to the rules. Rules are made, sometimes to be broken, because this is how we test our worldview.

Racheli: As you know, O_2 is a magnetic molecule. There are also demos on the Web that show the magnetism of O_2.

Sason: This is your job … Mine is to start now with the lecture.

3.2   ELECTRON-DEFICIENT MOLECULES

As we saw in the preceding lecture, the atoms in groups 1A–3A have small numbers of valence electrons, and hence their intrinsic connectivity creates electron-deficient

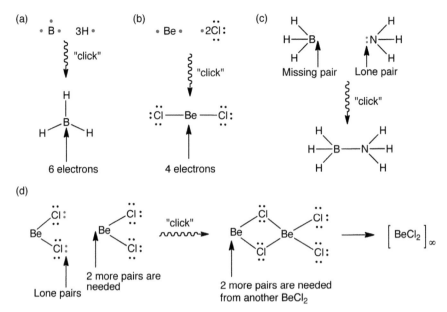

SCHEME 3.1   The electron-deficient molecules (a) BH_3 and (b) $BeCl_2$, and their modes of achieving Nirvana: (c) BH_3 combines with NH_3, and (d) $BeCl_2$ combines with another $BeCl_2$ molecule. The process in (d) can occur an infinite number of times such that the final molecule is $[BeCl_2]_\infty$ (the horizontal numeral 8 represents infinity). The electrons participating in connectivity are highlighted in red.

molecules. The generation of BH_3 and $BeCl_2$ by clicking the valence electrons to covalent electron-pair bonds is shown in Schemes 3.1a and b. In both cases, the central atom remains electron-deficient even after its bonding to the other atoms. Electron deficiency means basically that the corresponding atom has a "yet unsatisfied" connectivity.

Free lone-pairs on molecules are also connectivity centers, since a lone pair can be shared with an electron-deficient center. Thus, as shown in Scheme 3.1c, to achieve Nirvana, BH_3 requires an additional electron pair in the valence shell of B. What comes to the rescue is a molecule of ammonia (NH_3) that possesses a lone pair on the nitrogen atom and can use it for making a bond to the electron-deficient B. "Click" and there is a new bond pair between B and N. Now, all atoms are in a state of Nirvana (see Retouches section 3.R.1).

Similarly, to attain Nirvana, $BeCl_2$ requires two additional electron pairs around Be. As seen in Scheme 3.1d, since the Cl atoms have lone pairs, two molecules of $BeCl_2$ can cooperate, and "click," the right-hand beryllium atom attains four electron pairs around it, and it reaches Nirvana. The left-hand beryllium atom is still lacking two electron pairs, and these will be provided by a third $BeCl_2$ molecule, which will create two new Cl–Be bonds with the left-hand Be atom. This "click" process can continue indefinitely to create an infinite structure of $BeCl_2$ units linked to one another. We may label this molecule as $[BeCl_2]_\infty$, where the horizontal numeral 8 is the sign of infinity. We shall later see another "infinite" molecule with a unit that repeats itself, in principle, an infinite number of times.

The above examples show the power of the Law of Nirvana. Molecules get stabilized by reaching Nirvana, and hence they become also less reactive. Molecules that lack Nirvana will always strive to look for a source that can provide the missing electrons. In the above examples, the missing electrons came from lone pairs. One can imagine many other examples, and some are given in the problem set following this lecture.

3.2.1 Electron-Deficient Free Radicals

Our universe is full of electron-deficient species, which are called "free radicals" and are created when bonds are broken, such as in high-energy processes. For example, methane (CH_4) molecules exist in our atmosphere due to volcanic activity, coal, manure, rice farms, bacteria that form methane (methanogens), farm animals (ruminant animals, e.g., cows, sheep, pigs), wetlands, huge quantities of the gas being trapped in oceans, etc. In the upper levels of the atmosphere, there is a high level of radiation from ultraviolet rays (so-called UV radiation) coming from the sun. As shown in Scheme 3.2, the radiation provides sufficient energy to break a C–H bond in the CH_4 molecule. Breakage of the C–H bond generates the species H• and CH_3•, each having a single unpaired electron, and both are representatives of the free radical species. There are many more free radicals, which swarm around our atmosphere (e.g., NO•, OH•, Cl•, etc.) and form new free radicals by abstracting (chemists' term for grabbing) atoms from other molecules. Our atmosphere is not benign, since free radicals are major causes of aging, and the hole in the ozone layer allows cancer-causing radiation to penetrate the atmosphere, etc. However, like in other cases, here too free radicals have opposing effects on human life. Too much methane is a major source of global warming, and hence destruction of methane by producing free radicals is beneficial. On the other hand, free radicals cause damage to the biological material like DNA and proteins (e.g., our skin, hair, etc.), because this is how the radicals achieve their Nirvana. Thus, free radicals can abstract an atom from another molecule they happen to encounter, as CH_3• might do by abstracting an H atom from a DNA molecule. We shall come back to free radicals at the end of the book.

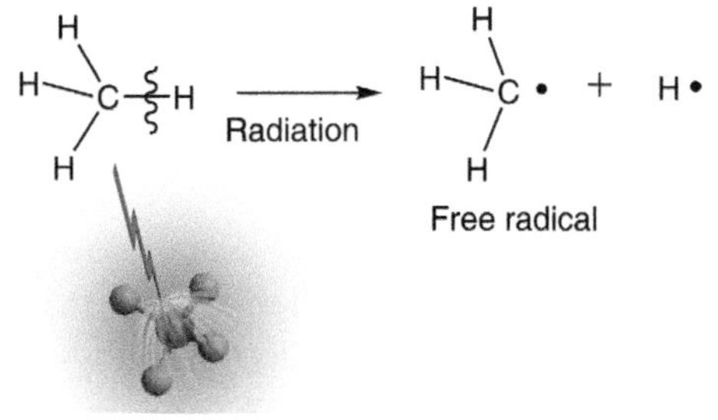

SCHEME 3.2   Radiation will cause breaking of a C–H bond in CH_4 and will form two free radicals.

FIGURE 3.1 From left to right: sand, a quartz crystal, and glass. The golden sand photograph was obtained from http://commons.wikimedia.org/wiki/File:Fulong_Beach_golden_sands.jpg, and the quartz photograph was obtained from http://commons.wikimedia.org/wiki/File:Quartz_crystals_Macro_1.JPG. The glass image, which is in the public domain, was obtained from http://commons.wikimedia.org/wiki/File:VibratingGlassBeam.jpg?fastcci_from=20353.

3.3 THE POWER OF MULTIPLE CONNECTIVITY: SiO_2—A GIANT MOLECULE

In the previous lecture, we saw that when the atomic connectivity exceeds unity, the atom is capable of forming quite a few different molecules (recall P_2 vs. P_4, etc.). I want to show you here a spectacular example, of building silicon dioxide, which is the essential constituent of the sand in our earth, of quartz, and of the glass we make from it (Figure 3.1). Making glass can also be demonstrated in class (see demos at the end of this chapter).

3.3.1 SiO_2—A Giant Molecule

Let us make a molecule from a silicon atom (Si) and as many oxygen (O) atoms as are required to satisfy the connectivities of the two atoms. We have only one qualification: we do not allow any multiple bonds; we require a molecule in a state of Nirvana and one that forms a solid material. To demonstrate the many, many possibilities of making molecules from Si and O, let us turn first to Scheme 3.3 to see two of the molecules that do not qualify as the sand of the earth.

SCHEME 3.3 Two molecules made from Si and O atoms: (a) SiO_2, which is an analogue of CO_2, and (b) SiO_4. The electrons participating in connectivity are highlighted in red.

In Scheme 3.3a, we use the quadruple connectivity of Si and the double connectivity of O, and "click," we form SiO_2, which is an analog of the CO_2 molecule. We are not interested in this molecule, both since it has Si=O bonds (see Retouches section 3.R.2) and since, like CO_2, it will be gaseous. Small molecules tend to be gases, liquids, or, at best, very soft solids, at ambient temperatures (about 20°C). While we shall learn later why this is so, it is clear now that this is not our target molecule. Had this been the constituent of the sand on our earth, we would have been "walking on clouds" rather than on solid earth.

Scheme 3.3b shows another possibility of clicking the connectivities of Si and two oxygen atoms, to make SiO_4 with Si—O and O—O bonds. This is a nice-looking molecule, which most likely exists somewhere or can be made in the laboratory. However, in all likelihood, it will not be a hard solid. Moreover, for reasons addressed in the Retouches section (3.R.3), the O—O bond is quite weak, and molecules having many such bonds tend to explode easily. This is not the sand of the earth!

The molecular constituent of sand (quartz, glass) must be a giant molecule, which cannot contain weak O—O bonds and must possess only Si—O bonds, which are very strong. As such, we might expect to form a hard solid fitting to be the sand of the earth. So, let us proceed with constructing such a molecule following the steps in Scheme 3.4.

In step (a), we start by bringing in one silicon (Si) atom and four oxygen atoms. By clicking their connectivities and without creating O—O bonds, we form the species I, where the four oxygen atoms remain with a dangling single connectivity. For simplicity, we do not draw anymore the lone pairs of the oxygen and just mark the dangling connectivities. In step (b), we take I and bring now four Si atoms, which have quadruple connectivity. We "click" and form species II, wherein the four terminal silicon atoms each remain with a dangling triple connectivity. In step (c), we bring three oxygen atoms for each dangling triple connectivity on a silicon atom. By "clicking" the connectivities, we form species III, which now has 12 terminal oxygen atoms, each with a dangling single connectivity. It is very clear what the next step should be if we wish to build a giant molecule. Thus, we shall have to bring 12 Si atoms and "click" the connectivities to form 12 terminal silicon atoms, each with a dangling triple connectivity, and so on.

We can continue this process indefinitely. However, we can already provide a molecular formula of this infinite structure, by focusing on species III, as done in Scheme 3.5a. Here one can clearly see that the giant molecule is made of SiO_4 units that share oxygen atoms in the Si—O—Si linkages and at the same time repeat infinite times in all directions. We can write a general chemical formula for this giant molecule, as done in Scheme 3.5b, where the SiO_4 motif is infinitely repeated, as indicated by the infinity sign to the right of the square brackets. Note, however, that each one of these four oxygen atoms is bonded not to one but to two silicon atoms, and therefore the correct formula is $[SiO_{4/2}]_\infty$, as shown in Scheme 3.5c. Further abbreviation of the formula may be achieved by writing the molecule as being made up of an infinite number of SiO_2 units $[SiO_2]_\infty$. Now, since chemists are well aware of the infinite structure, they drop the infinity sign and use the simplest formula, SiO_2.

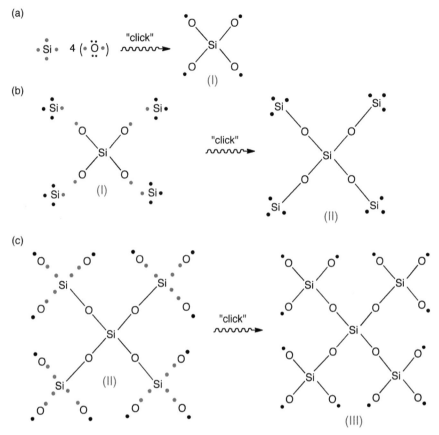

SCHEME 3.4 Initial steps of constructing a giant molecule made from Si and O in three steps (a–c). The electrons participating in connectivity are highlighted in red (lone pairs are not shown).

3.3.2 Definitions of Terms That Follow from the SiO₂ Story: Stoichiometry and Polymers

The use of the SiO₂ formula by chemists to describe this giant molecule may seem like a bit of laziness, but actually it is not. Explaining this point gives me an opportunity to digress and define some more terms that chemists use to describe molecules. One is *stoichiometry* (from Greek, meaning measurement of elements), which concerns the ratios of atomic combination in a molecule. The other is *polymer*, which is the word chemists use to describe giant molecules which have a repeating motif.

   The Power of Connectivity—Stoichiometry: The last two formulae of sand, $[SiO_2]_\infty$ or simply SiO_2, show that Si and O combine in an atomic ratio of 1:2. Thus, even though the molecule may have zillions of atoms, *still the basic ratio of the two constituents remains 1:2*. Ever since Lavoisier showed the power of quantitative analysis in chemistry and defined the notion of an element as a simple body, chemists tried to find quantitative relationships between the combined weights

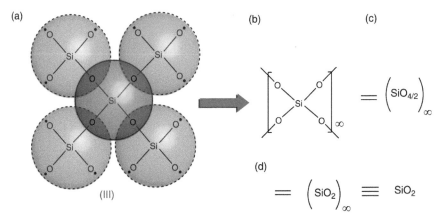

SCHEME 3.5 (a) The central SiO_4 unit of species III from Scheme 3.4 is circled, and it is apparent that this is a repeating motif of the molecular structure. (b) A condensed formula that shows the SiO_4 units repeating to infinity. (c) Since each O is bonded to two Si atoms, then the formula is in fact $[SiO_{4/2}]_\infty$. (d) Shorthand notations of writing the molecular formula: $[SiO_2]_\infty$ or simply SiO_2.

of elements. Richter (Figure 3.2), who was a student of the philosopher Kant, and liked mathematics very much, defined the term *stoichiometry* as the laws by which elements combine to form chemical compounds.

After Dalton revived the atomic hypothesis, *stoichiometry* became the smallest integer ratio of atomic combination in a molecule. Chemists are able to determine these ratios using a rather simple experiment, and so we know that the stoichiometric ratio of our compound, which is made from Si and O, is 1:2, respectively. Once the electron was discovered and Lewis defined the covalent bond as an electron-pair

FIGURE 3.2 Jeremias Benjamin Richter (1762–1807), the originator of the term *stoichiometry*. This image is in the public domain, and it was obtained from http://commons.wikimedia.org/wiki/File:Jeremias_Benjamin_Richter.jpeg.

bond, it became possible to reason the experimentally derived *stoichiometry* from the number of electrons the atoms have in their valence shell. Thus, by using the formula SiO_2, the chemist actually describes the stoichiometric ratio of the elements in the sand molecule. The construction of this molecule, as we just did, demonstrates the power of connectivity and the Law of Nirvana, which we presented in the previous lecture.

Polymers: In the description of SiO_2, we kept using the term "infinity," although it is a philosophical term rather than a practical one. What we actually wanted to say is that the molecule is made from a motif that repeats itself many, many times. Thus, instead of using the sign of infinity, we can write instead the formula as $[SiO_2]_n$. The subscript n, in this formula, counts the number of times the motif repeats itself, which can be a lot, such as $n = 1$ million, but not infinity. The $[SiO_2]_n$ molecule is an example of molecules that are giant molecules, which are called by the generic name *polymers*, where *poly* means many, and *mers* means units; so together, this term refers to a molecule made from many units. We shall see many more *polymers* in the next lectures.

3.4   SiO₂ AND GLASS MAKING

Sand is an old chemical, which has existed on the earth ever since it was formed. The formation of glass from sand was probably known very early in human history, as early as Mankind discovered fire. Glass making has become a human trade and was industrialized very early on in history. We know this because glass artifacts were found in all the ancient cultures, in places such as Egypt, Mesopotamia (the Middle East), China, and the Indus Valley. The ancient chemists (the protochemists) used to make glass by cooking sand with "washing soda" (Na_2CO_3) or limestone ($CaCO_3$) and then molding the material into shapes while it was still hot and soft. They also knew how to make the glass more lucid and how to incorporate all kinds of additives to make special glasses (e.g., adding lead oxide, PbO, creates flint glass) and to color glass (e.g., adding Fe^{2+} (iron) ions makes a green glass, adding selenium or sulfur makes a red or golden glass, respectively, etc.). The protochemists believed that the chemical process succeeded by the grace of the Gods,[2] and therefore glass making involved a ritual, sometimes with a human sacrifice or stillborn embryos. Here, for example, is a recipe for glass coloring, from the times of the Assyrian King Asurbanipal (668–626 BC), translated into Old English:

"When thou settest out the plan of a furnace for minerals thou shalt choose a favourable day in a fortunate month … While they are making the furnace … thou shalt offer the mineral into the furnace. Thou shall bring in embryos; another … shall not enter, nor shall one that is unclean; the day thou puttest down the mineral into the furnace thou shalt make a sacrifice before the embryos; thou shalt set a censer of pine incense."[3]

3.5 GLASS MAKING FROM *WATER GLASS*

One can dissolve sand by adding the ionic material Na^+OH^-, which is called also a *base* (a type of material we shall discuss in detail in Lecture 8). In this manner, one can prepare a clear solution called *water glass*. By adding an *acid* like HCl to the *water glass*, the solid SiO_2 forms and precipitates. This is perfect demonstration of the difference between the solid glass polymer and the *water glass*, wherein the SiO_4^{2-} species exist as isolated units in the solution of the *water glass*. We highly recommend doing this demonstration in class (see demonstrations at the end of this chapter). If one adds a colorant to the *water glass* and then an acid, the colorant will be trapped in the solid glass and generate a colored glass. This is also an entertaining demonstration. Chemistry can be a lot of fun!

But chemistry is also useful, and *water glass* has become now the basis of the sol–gel technology[4] that manufactures glasses with a variety of molecules trapped in to perform different functions (e.g., an entrapped catalyst or enzyme). As often happens in chemistry, the practical knowledge of preparing *water glass* is ancient. As we already said, the Greek philosophy of the "four elements" (*Air, Earth, Fire,* and *Water*) dominated scientific thought up until the eigtheenth century. This theory claimed that the entire material world could be constructed from the four elements. Furthermore, the Greek philosopher Aristotle argued that there is in fact a fundamental element, which he called *Protyle*, from which the four elements, and in turn all matter, were made.

Jan Baptista van Helmont, the greatest alchemist of the seventeenth century, believed that "Water" was the *Protyle* and that all matter was made from it. van Helmont, shown in Figure 3.3, designed many beautiful experiments to prove his point.[1] In one of these experiments, he used *water glass*. Thus, van Helmont took sand (for him, this was the element "Earth") and by adding the base ingredient, all the sand dissolved and became "Water." Then by adding an acid, all the "Water" solidified and gave back the solid material ("Earth"). By weighing the sand before the dissolution and after the solidification, and finding the exact same weights, van

FIGURE 3.3   A portrait of Jan Baptista van Helmont (painted by Mary Beale in 1674). This image, which is in the public domain, was obtained from http://commons.wikimedia.org /wiki/File:Jan_Baptist_van_Helmont_portrait.jpg.

Helmont "proved" that he could make *Earth* from *Water* and convert the *Earth* back to *Water* without losing anything. For his time, this was a pretty good proof that "everything was made from Water" (even though this is a circular argument). Now, a few centuries later, we see farther than our ancestors and can tell students that both *water glass* and sand are made from molecular species containing Si and O.

3.6 MUST WE WORK SO HARD TO CONSTRUCT LARGE MOLECULES?

I constructed the SiO_2 molecule with two goals in mind. One was to show you that whenever the atomic connectivity is multiple, there exist plenty of ways of making molecules using two types of atoms. In fact, already in step 3 in Scheme 3.4, we could have used a different number of oxygen atoms and generated another polymeric molecule with the same fundamental (stoichiometric) formula but a different arrangement of the atoms. *This is the beauty of chemistry—it is endless*! The second goal I had in mind was to impress upon you that using atoms and clicking their connectivities can become tedious for large molecules. In the next section, I present a more compact and productive molecular construction process.

3.6.1 Creating Larger Modular Building Blocks

We can use the molecules we already made to enrich our construction kit with larger modular fragments. This can be done by "unclicking" the bonds and creating dangling connectivities on the central atoms. This is demonstrated in Scheme 3.6 using the molecules CH_4, NH_3, and H_2O.

Inspection of Scheme 3.6a illustrates the process for CH_4. Note that when we "unclick" a C—H bond, we keep only the larger fragment, while discarding the H• fragment. Thus, by unclicking one, two, and three C—H bonds, we create modular fragments made from C and an appropriate number of hydrogens which have connectivities of 1–3. Scheme 3.6b shows this process for NH_3, and Scheme 3.6c shows the procedure for H_2O, where we create fragments with single and double (in the case of NH_3) connectivities.

We can now add these building blocks to the corresponding atomic fragments and generate a new connectivity table, as shown in Figure 3.4. These modular fragments will appear in many chemical families of compounds, and if you use them, this will endow you with the ability to recognize patterns in the molecular world.

3.6.2 Making New Molecules from the New Modular Fragments

To provide a glimpse of what can be done with the new modular fragments, let us use the connectivity table Figure 3.4 and build some of the molecules that can be made from it, while restricting ourselves to combinations of two fragments. Other molecules will be made when you solve the problem set at the end of the lecture.

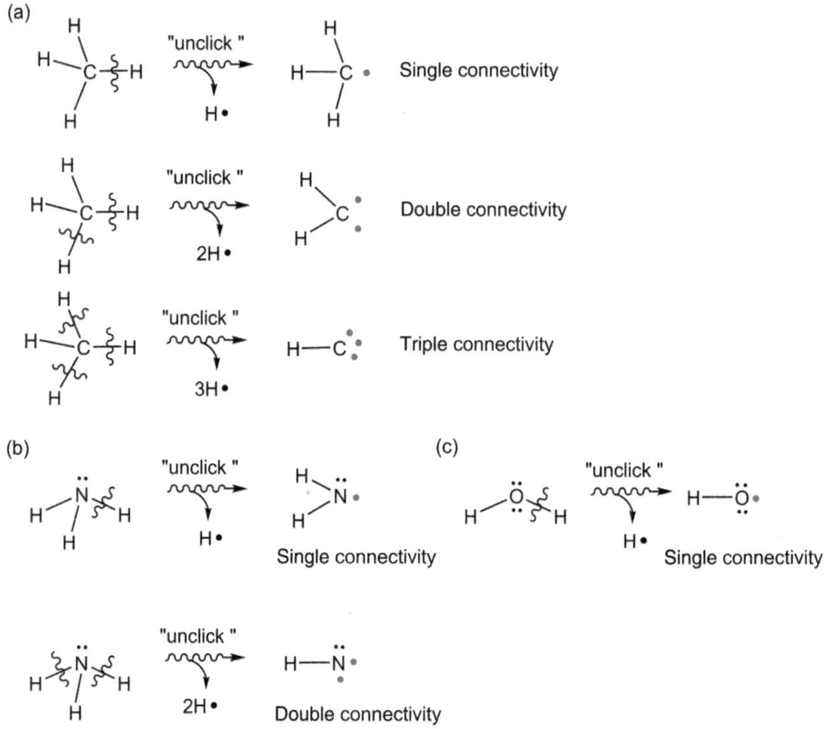

(a)

"unclick" H·   Single connectivity

"unclick" $2H$·   Double connectivity

"unclick" $3H$·   Triple connectivity

(b)

"unclick" H·   Single connectivity

"unclick" $2H$·   Double connectivity

(c)

"unclick" H·   Single connectivity

SCHEME 3.6   Creating modular fragments with variable connectivities by "unclicking" bonds: (a) Fragments of CH_4. (b) Fragments of NH_3. (c) Fragments of H_2O. Connectivity is highlighted by red circles.

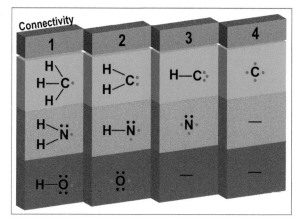

FIGURE 3.4   Table of fragments of CH_4, NH_3, and H_2O, depicting their corresponding connectivities. Electrons participating in connectivity are highlighted in red. Lone pairs are shown as black dots, and bonds are shown as lines. Fragments are derived from molecules shown in Scheme 3.6 (as well as other images).

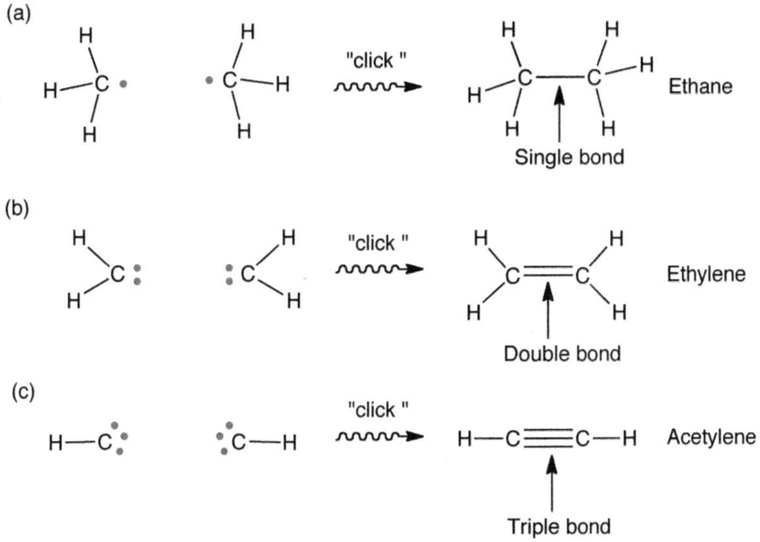

SCHEME 3.7   Molecules made from two (in this case identical) modular fragments: (a) H_3C, (b) H_2C, and (c) HC. For each molecule, we show also its name and the multiplicity of its C—C bond.

Scheme 3.7 shows the molecules that can be constructed by combining two identical fragments, which were derived from methane (CH_4). We are interested only in molecules that are in states of Nirvana. Let us make then a molecule, using only CH_3 fragments. Since this fragment has a single connectivity, we can combine only two of them, and as shown in Scheme 3.7a, "click" and we make a C—C bond in the molecule H_3C—CH_3 (C_2H_6). Already at this point, it is possible to see the great simplification of using the molecular fragments. Just imagine how many "clicks" you would have had to go through to construct even the small C_2H_6 molecule, starting from the constituent atoms.

The C_2H_6 molecule is called *ethane*, and like methane, it is also a natural gas, which can be used as fuel. As we shall see in the next lecture, methane and ethane are members of a vast family of fuels, which includes an infinite number of members.

Let us construct now a molecule that is made from the minimal number of H_2C fragments. As shown in Scheme 3.7b, since H_2C is a double connector, the minimal molecule will be constructed from two such fragments. "Click" and we make a C=C bond in the molecule H_2C=CH_2. This molecule is called *ethylene*, and like methane, it is also a gaseous material. From ethylene, we make the *polymer* polyethylene (those plastic bags that are so convenient to have but that pollute the earth). Ethylene is also a member of a family of molecules that possess C=C bonds, which are called by the common name *alkene*.

The final molecule in Scheme 3.7c is made from HC fragments. If we were asked to construct the smallest possible molecule made from these fragments, this would require just two fragments, and "click," we would generate the molecule HC≡CH, with a triple bond between the two carbon atoms. This molecule is called *acetylene*.

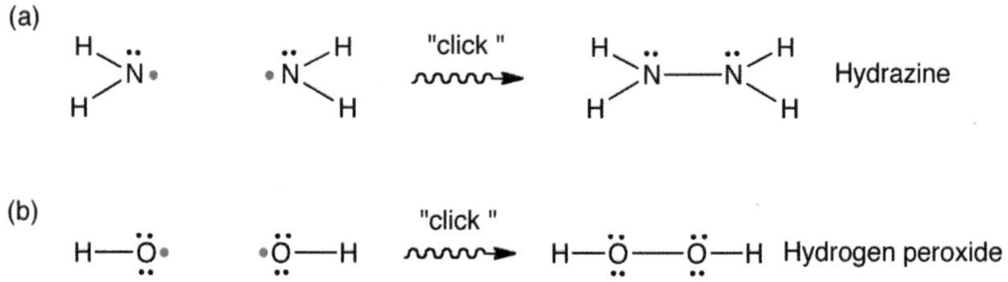

(a)

(b)

SCHEME 3.8   Molecules made from the fragments (a) H_2N and (b) HO, and their names.

It is a gas, which is kept in cylinders and used for soldering iron. This shows that the flame arising from the burning of acetylene is very, very hot. From acetylene, we make the *polymer* polyacetylene (it is now used in batteries for electric cars). Acetylene is also an example of molecules containing triple C≡C bonds, which, as a group, are called *alkynes*.

At this point, some of you may wonder "Why should I bother to memorize these names if I can find them in the book or on the Web?" Of course, this complaint is understandable. But try looking at it differently. Just imagine that you were living with a family but decided not to learn the names of your parents and siblings, because you could find these names in a family document. You would be missing something, wouldn't you? It is the same with the names of molecules; these names belong to a language, and if you want to speak the language properly and communicate comfortably with others who know chemistry as well as you do, some knowledge of the language is going to be necessary.

If you are convinced by my argument, we may continue this tour of new molecules by inspecting Scheme 3.8. Thus, suppose we are requested to construct a molecule made only from $H_2N\bullet$ fragments. Since the fragment has a single connectivity, we need two such fragments, and "click," we generate the molecule $H_2N–NH_2$, having an N–N bond. It is called hydrazine. Hydrazine is a gas with the odor of ammonia. Like the many-faced Janus, hydrazine has many uses, and two of them are as a component of rocket fuels (along with H_2O_2; this is quite an explosive mixture that has caused many accidents) and in airbags in cars. It is also poisonous and a source of creating NO free radicals in the atmosphere.

In Scheme 3.8b, we generate a molecule known to us, but now we make it from HO• fragments. Since the fragment has a single connectivity, only two fragments can be used, and "click," we generate the molecule we already made before, HO–OH, the so-called hydrogen peroxide.

Scheme 3.9 shows three molecules made from two different fragments. Thus, in Scheme 3.9a, we make a molecule from $H_3C\bullet$ and $\bullet NH_2$ fragments. Because the fragments each have a single connectivity, we can use only one fragment of each type. When we "click" their connectivities, we obtain the molecule $H_3C–NH_2$ (CH_5N), which is called *methylamine*. There is a simple logic in the name. The initial part of the name, *methyl*, is the name of the H_3C fragment, which derives from methane (CH_4), and the latter part of the name, *amine*, refers to the NH_2 "tail." This molecule is a member of a family called *amines*. The family members are typified by having

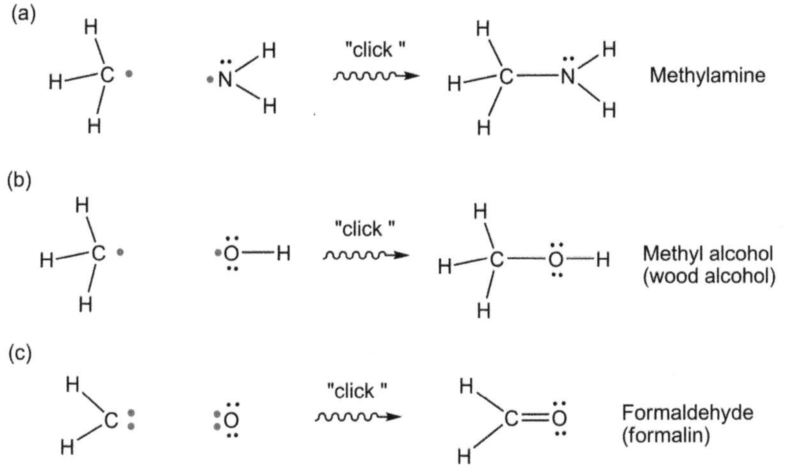

(a)

(b)

(c)

SCHEME 3.9 Molecules made from the fragments H_3C, H_2N, HO, and O, and their names: (a) $H_3C–NH_2$. (b) $H_3C–OH$. (c) $H_2C=O$ (the name *formalin* refers to the state of formaldehyde in water solutions).

N–C bonds. This is a very important family, and you may recall that all the important neurotransmitters (Lecture 1) are *amines*.

In Scheme 3.9b, we make a molecule from the single connectors $H_3C\bullet$ and $\bullet OH$. By clicking their connectivities, we generate the molecule $H_3C–OH$, which is called *methyl alcohol* (methanol) or "wood alcohol." It is a part of a large family of *alcohols*, in which all the members have the bond C–OH. This particular alcohol, which we just constructed, exists in wood. As an amateur chemist in my childhood, I used to produce this alcohol by heating a few pieces of wood (taken from used matches) in a test tube that was equipped with a cork into which we inserted a narrower glass tube (called a pipette); then, by warming the glass gently over a fire, we bent the tube and stretched its tip to narrow its opening. As we warmed the test tube, the wood started "sweating," and some liquid began to drip from the pipette. Methyl alcohol is a key component of this liquid that is stored in the wood.

You cannot drink methyl alcohol, because it causes blindness, among other problems. During the Prohibition, when the "Dry Law" was introduced in the United States, many illegal distilleries made methyl alcohol and sold it to customers. The results were disastrous for those who drank this alcohol. We shall see the alcohol present in wine in one of the next lectures.

Finally, in Scheme 3.9c, we are asked to make the minimal molecule from the fragments H_2C and O. Since both are double connectors, we need only one of each, and "click," we form $H_2C=O$, with a C=O bond. The molecule is called *formaldehyde*, and its solution in water is known by the name *formalin*, which, due to its antibacterial action, is used to preserve tissues of animals and humans. Formaldehyde is also the smallest member of the family of molecules called *aldehydes*, wherein the C=O bond is connected to at least one H fragment.

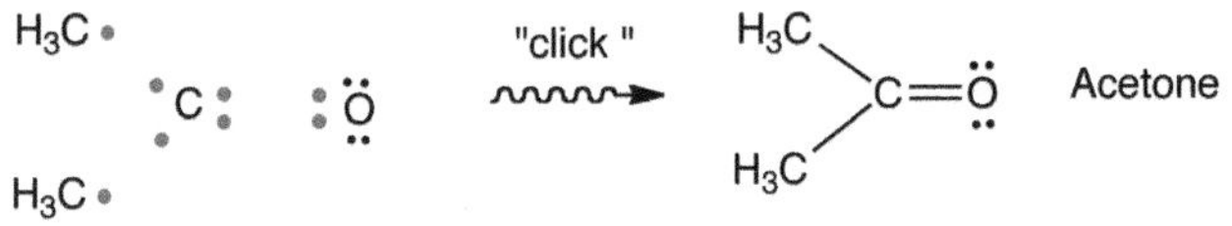

SCHEME 3.10 Construction of acetone from four building block fragments.

There are so many other molecules we could make by using building blocks from the connectivity table in Figure 3.4. For example, as shown in Scheme 3.10, suppose we are required to construct a molecule made from single C and O fragments and as many $H_3C\bullet$ fragments as needed. It is clear that we need two $H_3C\bullet$ fragments to satisfy the available connectivities, and "click," we form the molecule $(CH_3)_2C{=}O$, which is called *acetone*, and about half of us use it regularly to remove varnish from our nails. However, it has many other uses as well. Acetone is a great solvent for synthetic polymeric materials. Do not try to use it to clean the stains from your upholstered TV chair, like I once did to mine and had to buy a new one! Acetone is also a member of the family of molecules called *ketones*, which are all typified by a C=O fragment bonded to two other carbon fragments.

We now have the way to generate many molecular worlds by using the large molecular building blocks we formed in Figure 3.4. This will be done in the next lecture.

3.7 SUMMARY

Electron-Deficient Molecules: We learned in this lecture how electron-deficient molecules, made from atoms in the groups 2A and 3A, manage to attain Nirvana despite being initially electron-deficient. "Adopting" a lone pair on another molecule to create an additional bond achieves the desired Nirvana (this may occur more than once, as in the case of $BeCl_2$). The take-home lesson is that *lone pairs are connectivity elements* for binding to electron-deficient molecular fragments.

Free Radicals: Free radicals belong to a special class of electron-deficient molecules. These species have one unpaired electron, like $H_3C\bullet$. They can attain Nirvana by abstracting an atom along with its electron from another molecule. In this manner, they cause damage to biological material (among other systems), such as our DNA.

SiO$_2$—The Power of Multiple Connectivity of Atoms: We started with Si and O, which are quadruple and double connectors, respectively, and we proceeded to construct from them molecules. We showed that the multiple connectivities of the two atoms form a basis for generating many, many possible molecules. Initially, we made the simplest O=Si=O molecule and then gradually proceeded to the giant covalent molecule that makes the sand of the earth and the raw material for glass making.

Stoichiometry: This exercise gave us an opportunity to define the term *stoichiometry*, which tells us the ratio in which the atoms are combined in a given molecule. So, even though the giant sand molecule contains many thousands of atoms of Si and O, still the stoichiometric ratio of Si to O is 1:2. It obeys the relative connectivities of the atoms (4 for Si and 2 for O; the ratio is 2:1).

Polymers: Constructing SiO_2 gave us an opportunity to introduce the term *polymer* (many units), which describes giant molecules with a molecular motif that repeats itself many times. Thus, the molecule of sand can be written as $[SiO_2]_n$, where n is a large number. In this manner, the formula tells us about the stoichiometric ratio of the atoms and about the fact that this molecule is made from many repeated units. Another way to write it is $[SiO_{4/2}]_n$, which adds some structural information, namely that the repeated unit is made from one Si atom connected to four O atoms and that each O is shared by two Si atoms.

Creating Large Modular Building Blocks: The SiO_2 exercise drove us to seek a more compact and productive way of constructing molecules, by generating larger modular fragments. Thus, by "unclicking" bonds in molecules we already made (CH_4, NH_3, H_2O), we generated new fragments. The new construction kit is shown in Figure 3.4, along with the connectivities of the fragments. These modular fragments were used to generate example molecules that represent important chemical families. We therefore learned a very important principle of pattern recognition in molecules: every molecule forms a basis for new modular fragments for creating new chemical worlds. Chemistry is beautifully endless!

3.8 REFERENCES

[1] P. Ball, *Elegant Solutions. Ten Beautiful Experiments in Chemistry*. RSC Publications, Cambridge, 2005, pp. 11–21.

[2] M. Eliade, *The Forge and the Crucible*, University of Chicago Press, 1978.

[3] C. J. S. Thompson, *The Lure and Romance of Alchemy*, Bell Publishing Company, New York, 1990, p. 29.

[4] D. Avnir, *Acc. Chem. Res.* **1995**, *28*, 328.

3.A APPENDIX

This lecture was designed to be short to enable teachers to split 1 hour between demos and tutoring. It is advisable to tutor the construction of molecules using the connectivity table of atoms and of larger modular fragments. The problem set here and in Lecture 2 may be helpful toward this end.

3.A.1 Proposed Demonstrations

We highly recommend demonstrating the *water glass* experiment. If possible, the O_2 magnetism demo should be done too. Both are really stunning. In case the O_2 demonstration cannot be presented, please direct the students to watch web videos for this experiment.

- **The *Water Glass* Experiment**

This experiment shows how one starts with a nice clear solution, which contains monomers, and by adding drops of acid, this solution becomes increasingly viscous until it completely solidifies, thus becoming a *polymer*.

Another aspect of this demonstration is glass coloration, which is associated with the "magic of chemistry," with the old art of glass making, and with modern sol–gel technologies.

In the experiment, a concentrated solution of sodium silicate is diluted (1:1) with deionized water, and an indicator colorant such as bromocresol purple is added, just to make the demo colorful. The purple solution is stirred on a plate equipped with a magnetic stirrer. Then, diluted hydrochloric acid is added to the mixed solution (4:1 HCl 1M to concentrated sodium silicate). As the polymerization reaction progresses, the color of the solution changes from purple to pink (the indicator is sensitive to the change from basic to acidic solution), and the solution becomes more and more viscous until the stirrer suddenly stops! A captivating moment in the experiment is when the vessel is turned upside down unexpectedly (by the students), and yet the liquid does not spill. It has turned into a gel of a colored glass. R. Ben Knaz home-designed this experiment.

• The O_2 Magnetism Experiment

O_2 is a hugely important molecule. As shown in Scheme 3.R.3, it violates the pairing rules and exists as a magnetic molecule. The demonstration of this magnetic property is simple to do and striking to watch.[1A]

Liquid oxygen is prepared by either of the two methods described in the literature.[1A] It is stored in a Dewar flask filled with liquid nitrogen. The blue color of liquid oxygen can be observed and discussed when it is poured into a test tube. The demonstration begins by pouring liquid oxygen between the magnetic poles of a strong magnet. A small amount of the liquid should get trapped and held by the magnetic field, showing the magnetism of oxygen. For comparison, replacing liquid oxygen with liquid nitrogen does not lead to any accumulation of liquid between the poles, thus demonstrating that nitrogen is not magnetic.

3.A.2   References for Appendix 3.A

[1A] B. Z. Shakhashiri, G. E. Dirreen, and L. G. Williams, *J. Chem. Educ.* **1980**, *57*, 373.

3.R   RETOUCHES

3.R.1   Formal Charges

When an electron-deficient molecule adopts a lone pair from another one, as was described in Scheme 3.1c, chemists tend to depict the molecule with *formal charges*, as shown in Scheme 3.R.1. Thus, since B accepted formally an excess electron (one

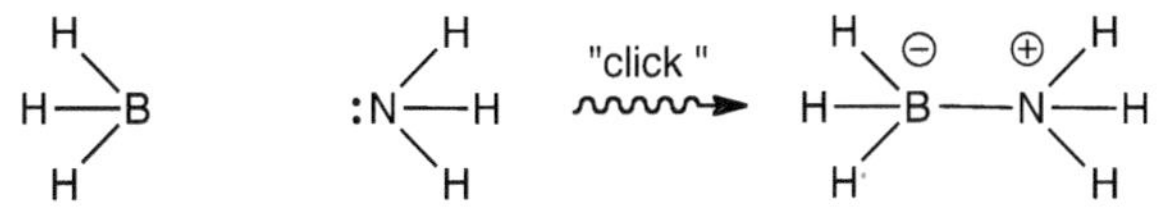

SCHEME 3.R.1 Formation of a bond between H_3B and NH_3, along with *formal charges* in the so-formed molecule.

electron from the new bond pair), it "gained" a negative charge. At the same time, N in NH_3 had the lone pair all to itself, but during the new bond formation process, it "lost" formally one of the electrons, and therefore it acquired a positive charge.

As expressed by the term *formal charges*, these are not real charges. They are just formal ways of keeping track of the electrons. In this case, the *formal charges* tell us that the electron pair that initially "belonged" to NH_3 is now shared with BH_3. Even though, *formal charges* appear in advanced textbooks and can become sometimes useful, I decided to waive them since they are not a fundamental feature of the molecule, and they do not add any additional explanatory power in this course.

3.R.2 Multiple Bonds to Silicon

Carbon and nitrogen form multiple bonds alongside single bonds. However, as we step down the family columns, the atoms have lesser propensity to form multiple bonds. Thus, multiple bonding to silicon is rare, and such kinds of molecules are made with difficulty and require a great deal of chemical ingenuity.[1R] One of the main reasons is that the multiple bonds of silicon are weaker than the sum of the single bonds. For example, the Si=O double bond is considerably weaker than two Si—O single bonds. The same applies to phosphorus and to other elements in the 4A and 5A families.

3.R.3 The Lone-Pair Bond Weakening Effect

I stated in the text that O—O bonds are weak and molecules containing them (so-called peroxides) tend to explode. Scheme 3.R.2a shows a molecule called methyl peroxide with an O—O bond. Inspection of the bond shows that the oxygen atoms have each two lone pairs. Since electrons are negatively charged, the lone pairs on the adjacent atoms will repel one another and will raise the energy of the molecule. When we break the bond to two $CH_3CO\bullet$ fragments, this repulsion disappears, and therefore, as shown in the energy diagram in Scheme 3.R.2b, the O—O bond will have a small bond dissociation energy. Hence, O—O is said to be "a weak bond," unlike C—O, which does not bear lone-pair repulsion and is "a strong bond." This effect is not limited to O—O bonds. The F—F bond, with three lone pairs on each F, is also very weak, and the N—N bond, with one lone pair on each N, is not a very strong one. This is known as the "lone-pair bond weakening effect."[2R]

Generally, it is relatively easy to break the O—O bond in peroxides and replace it by much stronger bonds, which have a lower energy. The difference in the bond energies is released as energy (e.g., heat), as shown in Scheme 3.R.2c. The energy

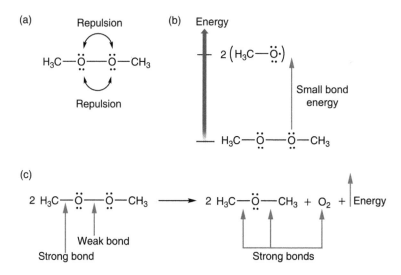

SCHEME 3.R.2   Dimethyl peroxide has two types of bonds: O—O and C—O: (a) The strong repulsion between the lone pairs on the two oxygen atoms raises the energy of the molecule. (b) Since the molecular energy is high, the resulting energy required to break this O—O bond is relatively small. (c) Peroxides cleave their relatively weak O—O bonds and replace them with much stronger bonds. Here we exemplify such a putative decomposition reaction where the O—O bond is replaced by an O—C bond and a doubly bonded O_2. The energy released in this process may cause explosions and may also cause a fire.

released causes heating of the apparatus, and in case we have a small volume or large quantities of peroxides, this results in a fire and/or explosion.

Try googling "peroxide explosion", and you will see some videos that describe explosions of peroxides. Do not perform such an experiment! It can be fatal. I recall such an incident, when I spent a leave of absence in a research institute. On one of the holidays, I went to the lab to do some work (I am a theoretician, so the riskiest part of my lab work was working with the heavy desktop computers which could fall on my foot …). The institute was almost deserted. I suddenly heard the sound of an explosion. I ran out of the lab looking for the source of the noise, but like most chemistry departments, there too, each floor had a long corridor with many small extensions. So it took me some time to find out from where the noise was coming. Lying on the floor, there was a young person bleeding from his neck. A piece of glass that flew during the explosion (he experimented with a peroxide) cut his artery. Luckily, there was an older PhD student in the floor, who happened to be also an MD and was getting his PhD in chemistry. He stopped the bleeding and saved the young student's life. Sometimes, these accidents end up differently …

3.R.4   The O_2 Molecule and Its Magnetism

The dioxygen molecule, O_2, constitutes 20.95% of all the gasses in our atmosphere, and it makes life possible. It may seem puzzling that we have not yet constructed

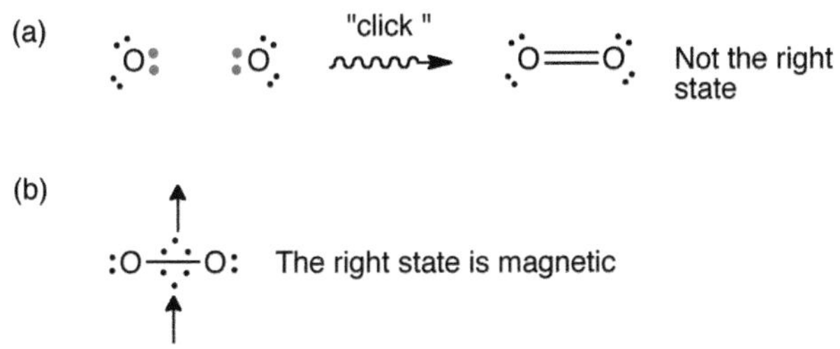

(a)   :O: :O: **"click"**   :O=O: Not the right state

(b)   :O⋮O: The right state is magnetic

SCHEME 3.R.3 Constructing O_2 from two O atoms: (a) The straightforward click bonding approach generates O=O, which is not the right state of the molecule. (b) The actual state of bonding in O_2, showing two unpaired electrons with their spin magnets pointing in the same direction.

O_2 in the text from its O fragments, because as shown in Scheme 3.R.3a, this should be very easy in principle. Since O is a double connector, we need only two oxygen atoms to form a molecule in Nirvana. We bring them together, and "click," we form the O=O molecule. Sounds perfect! The only problem is that this is not the state of O_2 in our atmosphere. The atmospheric O_2 is a magnetic molecule. It turns out that O_2 is an exception that does not adhere to the "click" electron-pair bonding, and it prefers to bind by using a different bond type, called *an odd-electron bond*. It is not simple to explain this exception in our course because we must use quantum mechanics for such an explanation.[3R] I shall try to describe the bonding in this molecule without quantum mechanics.

The bonding in O_2 is illustrated in Scheme 3.R.3b. It is seen that the two O atoms form one electron-pair bond (the connecting line in O—O) by clicking one of the connectivities of each O. The remaining connectivity on each O engages a lone pair on the other O and forms an odd-electron bond (three-electron bond). An odd-electron bond is considered to be equivalent to half of an electron-pair bond. Thus, we have a "normal" electron-pair bond and two odd-electron bonds, and this is equivalent to having two electron-pair bonds.

The two odd-electron bonds retain the two unpaired electrons. We drew near those unpaired electrons arrows pointing in the same direction. As we explained in the Retouches in the previous lectures, an unpaired electron is a little magnet due to its spin. The spin has a direction up or down, which is indicated by the direction of the arrow. During the covalent bond pairing, these little magnets are arranged so that they cancel each other. However, when the electrons remain unpaired, like in Scheme 3.R.3b, the spin is not cancelled, and the molecule attains a magnetic property.

The beauty of science is that when we say something, we can test it by doing an experiment. Indeed, there are beautiful experiments on the Web showing how a test tube of liquid O_2 (at very low temperature) is attracted to the poles of a magnet (see demonstrations). Note that the O=O formulation is available to O_2, but it belongs to a higher-energy state.

3.R.5   References for Retouches

[1R] R. West and M. J. Fink, *Science* **1981**, *214*, 1343.

[2R] R. T. Sanderson, *Polar Covalence*, Academic Press, New York, 1983, pp. 20–22.

[3R] S. S. Zumdahl and S. A. Zumdahl, *Chemistry*, 9th Ed., Brooks College, Cengage Learning, 2010, Chapter 9. Available at: www.cengage.com/global

3.P   PROBLEM SET

3.1 What would you expect to occur if we add F^- ions to BF_3? Show and explain the reason.

3.2 One of the products formed by burning hydrazine with hydrogen peroxide is NO. Construct this molecule and characterize it in terms used in this lecture.

3.3 Construct a molecule made from a single NH fragment and as many CH_3 fragments as needed. What type of molecule did you construct?

3.4 Construct a molecule made from a single CH_2 fragment with (a) a single NH fragment, and (b) two NH fragments.

3.5 Construct the minimal molecule made from CH and N fragments.

3.6 Generate a connectivity table (analogous to the one in Figure 3.4) using the molecules CF_4, NF_3, and F_2O. Construct from these fragments one molecule with a double bond and one with a triple bond.

3.7 Generate a connectivity table analogous to Figure 3.4 but with replacement of C, N, and O by the next atoms in the respective families.

3.8 BH_3 forms B_2H_6. Try to describe it (a bonus problem).

CONSTRUCTING MOLECULAR WORLDS OF CARBON–HYDROGEN FROM LARGE LEGO FRAGMENTS

4.1 CONVERSATION ON CONTENTS OF LECTURE 4

Racheli: By now you have a full construction kit of fragments. How do you plan to proceed?

Sason: This is the time for the great construction task! I want to show the teachers/students that this kit enables them to construct many chemical worlds.

   I will start with the fragments derived from methane (CH_4), and will build chains, rings, cages, necklaces, and polymers made from rings and cages, and whatnot…

Racheli: Aren't you getting carried away a bit? How can you achieve all this without having taught about isomers?

Sason: We'll learn on the fly, Racheli! As we construct molecules, we are going to encounter cases where the molecules have different structures but the same stoichiometry. These are *isomers*.

Racheli: Most of these molecules are beautiful because of their three-dimensional (3D) structures.

Sason: I always bring to class the 3D model set.

Racheli: Shouldn't you be concerned that the 3D details would confuse your students if they did not yet learn about 3D shapes?

Sason: Not even a bit! These objects are beautiful, and beauty is not about details. At some point later, I am going to teach how to predict the 3D shape of a molecule.

Chemistry as a Game of Molecular Construction: The Bond-Click Way, First Edition. Sason Shaik.
© 2016 John Wiley & Sons, Inc. Published 2016 by John Wiley & Sons, Inc.

At this moment, the beauty and understanding will merge into something more beautiful!

Racheli: You mean understanding will come on the fly?

Sason: Absolutely! Beauty first, followed by understanding on the fly, will enable me to demonstrate that the chemical world is endless, and the chemical matter is malleable. And that through chemistry, mankind acts as a *demiurge*, a semi-god creature that reshapes the world by the sheer guidance of imagination.

Racheli: It reminds me that you used to read in class from the book of Bruno Schulz, "*A Treatise on Mannequins or, The Next Book of Genesis.*" He also speaks about *demiurges*, and when he refers to matter, he writes the following:

"Matter is prone to endless fecundity, an inexhaustible vital force, and a beguiling power of temptation that enriches us to become creators on our own right… All matter flows out from endless possibilities, which run through it in sickly shudders… Anyone may knead and shape it [matter]; it submits to all. All arrangements of matter are impermanent and loose, liable to retardation and dissolution."

Sason: I like the word *demiurge* and its sense of creativity, which fits chemistry perfectly. In the ancient world, those who possessed the secrets of reshaping matter were revered as though they were graced by heavenly powers.[1] Just read the biblical description of *Bezal'el*, who was in charge of constructing the shrine for the wandering Israelites in the desert. Bezal'el was a metallurgist, a protochemist. And the Bible says (Exodus, Chapter 35, verses 31–32 in the King James version):

"And he hath filled him with the spirit of God, in wisdom, in understanding, and in knowledge, and in all manner of workmanship: And *to devise curious works, to work in gold, and in silver, and in brass …*"

The protochemist is graced with a demiurgical creative ability to reshape matter. The knowledge of chemistry makes mankind a creator of new matter.

Racheli: OK. Now that we are both done being a little poetic, what else are you going to teach?

Sason: My plan is to teach a mixture of things, both practical and creative. Molecules, especially those made from C and H, are very practical and serve as sources of energy. We shall learn what happens when fuel burns.

But making new molecules is also a creative endeavor of the synthetic chemist. I shall mention the beautiful cage molecule *dodecahedrane*, $(CH)_{20}$ and the race to make this molecular mimic of one of the classical Platonic bodies. Our myths of beauty and symmetry come to life in chemistry because of its creative element.

Racheli: This is very challenging. How would you draw these molecules on the board?

Sason: Not easy. A model and a slide presentation are going to be necessary. And once seen, the construction of these molecules from the modular fragments will generate understanding of their structural principles.

Racheli: You mentioned fuel. Burning it generates CO_2. This is air pollution. How would you teach the students about the quantity of pollution generated by, let's say, one liter of liquid fuel?

Sason: Roald Hoffmann suggested to me to estimate the yearly CO_2 production of a car.

Racheli: But then you will have to introduce the term *mole*, which chemists use for estimating the quantities of matter in a chemical reaction.

Sason: I am going to introduce the mole concept in the Retouches section. Then, I will proceed with a small calculation that demonstrates the *huge* amount of CO_2 that is produced by an average car each year.

Racheli: You mean my little old car puffs copious amounts of CO_2? I find it hard to believe…

Sason: Wait and you shall see…

4.2 MOLECULAR CHAINS INVOLVING ONLY C AND H

Scheme 4.1 depicts the LEGO fragments made of only C and H atoms. The connectivities of these fragments range from 1 to 4. We shall now proceed to construct from these fragments a variety of molecular worlds, which chemists call families or series.

4.2.1 Extended Chains

In the previous lecture, we built the molecule ethane (C_2H_6) from two $H_3C\bullet$ fragments. To refresh our memories, I repeat this construction in Scheme 4.2a, where the two single-connector fragments click to form a C—C bond between the $H_3C\bullet$ fragments. Now we may proceed to a more challenging construction process, making molecules from $H_3C\bullet$ and as many $H_2C:$ fragments as needed.

Schemes 4.2b–e show that since $H_3C\bullet$ is a single-connector fragment, if we wish to click it with the double-connector $H_2C:$ fragments and still produce chains, we must utilize two $H_3C\bullet$ fragments and any number of $H_2C:$ fragments that are going to be located in between the single connectors. Scheme 4.2b utilizes one $H_2C:$ and two $H_3C\bullet$ fragments; "click" and we form a molecule called *propane*, involving a short chain of two C—C bonds. Much like methane and ethane, propane is also a gas that can be used as fuel. In fact, there are cars that run on propane. Scheme 4.2c shows the construction of a molecule with an additional $H_2C:$ fragment. This generates *butane*, with three C—C bonds. Butane is a gas too, and it is used in gas burners for home cooking. Scheme 4.2d shows the construction of *octane* using six $H_2C:$ fragments, which click to form seven C—C bonds. Octane is liquid. When you drive to a gas station to fill the tank of your car, you get octane and its isomers (see more later).

Fragment :	$\bullet\,CH_3$	$:CH_2$	$\bullet\overset{\bullet}{C}H$	$\bullet\overset{\bullet}{\underset{\bullet}{C}}\bullet$
Connectivity:	1	2	3	4

SCHEME 4.1 Connectivities of fragments made from C and various numbers of H atoms. The electrons participating in connectivity are highlighted in red.

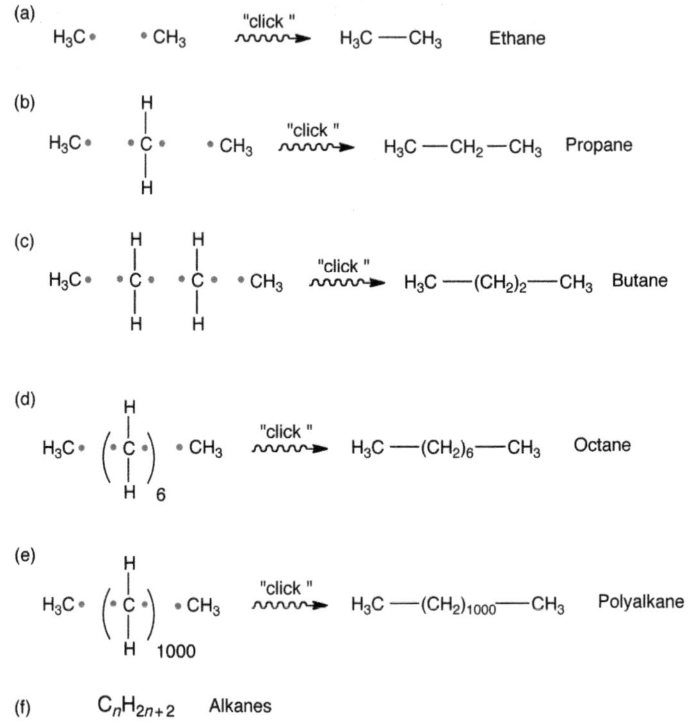

(a) $H_3C\bullet$ $\bullet CH_3$ $\xrightarrow{\text{"click "}}$ $H_3C\text{—}CH_3$ Ethane

(b) $H_3C\bullet$ $\bullet\overset{\overset{\displaystyle H}{|}}{\underset{\underset{\displaystyle H}{|}}{C}}\bullet$ $\bullet CH_3$ $\xrightarrow{\text{"click "}}$ $H_3C\text{—}CH_2\text{—}CH_3$ Propane

(c) $H_3C\bullet$ $\bullet\overset{\overset{\displaystyle H}{|}}{\underset{\underset{\displaystyle H}{|}}{C}}\bullet$ $\bullet\overset{\overset{\displaystyle H}{|}}{\underset{\underset{\displaystyle H}{|}}{C}}\bullet$ $\bullet CH_3$ $\xrightarrow{\text{"click "}}$ $H_3C\text{—}(CH_2)_2\text{—}CH_3$ Butane

(d) $H_3C\bullet$ $\left(\bullet\overset{\overset{\displaystyle H}{|}}{\underset{\underset{\displaystyle H}{|}}{C}}\bullet\right)_6$ $\bullet CH_3$ $\xrightarrow{\text{"click "}}$ $H_3C\text{—}(CH_2)_6\text{—}CH_3$ Octane

(e) $H_3C\bullet$ $\left(\bullet\overset{\overset{\displaystyle H}{|}}{\underset{\underset{\displaystyle H}{|}}{C}}\bullet\right)_{1000}$ $\bullet CH_3$ $\xrightarrow{\text{"click "}}$ $H_3C\text{—}(CH_2)_{1000}\text{—}CH_3$ Polyalkane

(f) C_nH_{2n+2} Alkanes

SCHEME 4.2   Construction of alkanes from $H_3C\bullet$ and $H_2C\colon$ fragments: (a) Ethane from two $H_3C\bullet$ fragments. (b) Propane from two $H_3C\bullet$ fragments and one $H_2C\colon$ fragment. (c) Butane from two $H_3C\bullet$ and two $H_2C\colon$ fragments. (d) Octane from two $H_3C\bullet$ and six $H_2C\colon$ fragments. (e) A long chain from two $H_3C\bullet$ and 1,000 $H_2C\colon$ fragments. (f) The general formula of the alkanes, C_nH_{2n+2}.

What is probably obvious at this point is that we can generate a whole series of molecules with increasing numbers of $H_2C\colon$ fragments and two capping $H_3C\bullet$ fragments. For illustration, we use in Scheme 4.2e 1,000 $H_2C\colon$ fragments, and we form a giant molecule (a polymer) with 1,002 carbon atoms and 2,006 hydrogen atoms. The molecule has a chain of 1,001 C–C bonds with 2,006 hydrogen atoms bonded via C–H bonds to the carbon chain.

It is clear that we can engage indefinitely in this game of construction, and in so doing, we shall generate a family of related molecules, all having chains of C–C bonds, a running number of $H_2C\colon$ units in the middle, and two $H_3C\bullet$ fragments capping the ends of the chain. As shown in Scheme 4.2f, all the family members can be summarized in a single general formula, C_nH_{2n+2}, where n is a number that in principle runs from 1 to infinity. The members of this family are called *alkanes* (a name that indicates the constitution of these molecules from C–C and C–H bonds that form chains). In practice, there is no infinity, but chain lengths can reach large numbers.

Three-dimensional (3D) Shapes and Dynamics of Some Alkanes: Before proceeding, it is worthwhile to appreciate the pretty 3D shapes of some alkanes. Figures 4.1a–c show ethane, propane, and butane using a *ball-and-stick representation*,

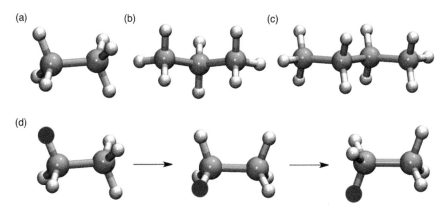

FIGURE 4.1 3D shapes of some alkanes using balls (for atoms) and sticks (for bonds) to depict: (a) ethane, (b) propane, and (c) butane. (d) The dynamics of ethane as it rotates around the C—C bond and changes the relative orientations of the C—H bonds. The variable location of the distinctly colored H emphasizes the rotation.

where balls represent the atoms and the sticks are the bonds. First, one can see (Figure 4.1a–c) that the C—C chain zigzags and the C—H bonds are arranged alternately on the front and back of the chain. The second interesting feature is illustrated only for ethane, where we show in Figure 4.1d two forms that arise when the CH_3 groups rotate relative to one another around the C—C bond and pass via a third form where the C—H bonds of the two CH_3 groups are parallel to one another. Such a rotation around C—C bonds is virtually free and occurs extremely fast, over 10 billion times a second, in all the molecules in Figure 4.1. This is only one of the possible motions which ethane performs. Thus, not only do molecules have shape, but they are also in constant internal motion, dancing and writhing as if they are breakdancing. We still do not have the know-how to predict the 3D shapes in Figure 4.1, but we shall soon learn to do so!

4.2.2 Branched Chains and Isomerism

So far we have made alkane molecules having extended chains (Scheme 4.2) by using $H_3C\bullet$ and $H_2C:$ fragments. We can, however, easily imagine how we could have formed branched alkanes by using also the triple-connector HC fragment and the quadruple-connector fragment, C. Let us provide exemples of this in Scheme 4.3.

Consider first a molecule that contains single HC and H_2C fragments and as many H_3C fragments as needed, which does not possess multiple C—C bonds. Scheme 4.3a shows the construction of this molecule in two steps. Initially, we click the connectivities of HC and H_2C, and we obtain the fragment drawn in the box with three dangling connectivities. In the second step, we satisfy the dangling connectivities with three $H_3C\bullet$ fragments, and "click," we form the molecule C_5H_{12}, which has a main chain of four C—C bonds with a single branch.

Let us now make a molecule that has four $H_3C\bullet$ fragments and as many C fragments as required. As shown in Scheme 4.3b, the four single-connector H_3C fragments

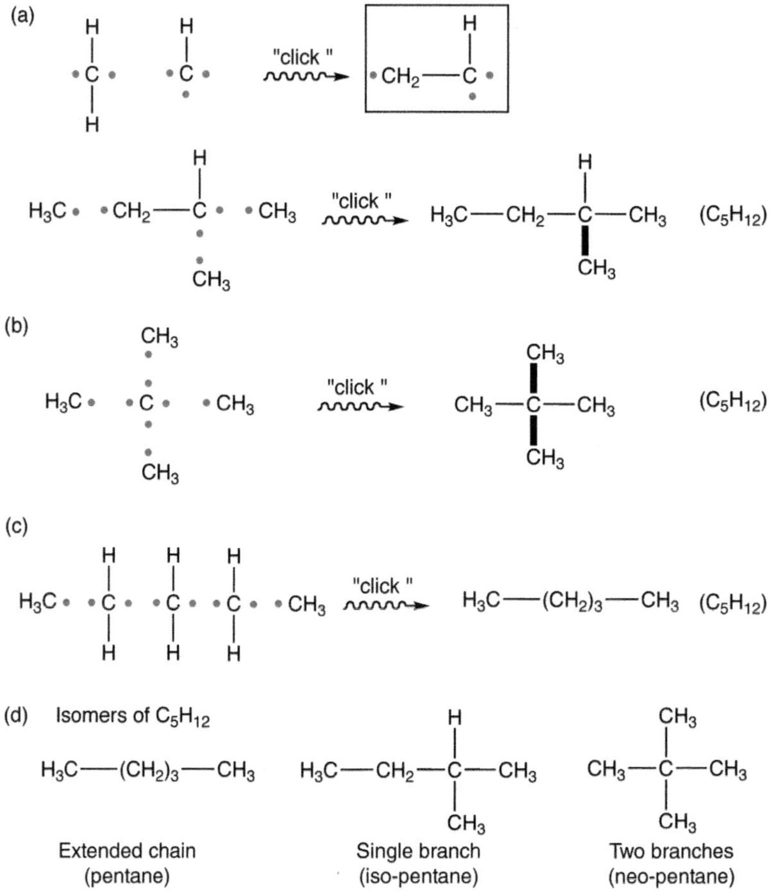

SCHEME 4.3 Construction of C_5H_{12} isomers with and without branching. The branching is depicted using thicker C–C bonds. (a) One CH fragment in the molecule creates a chain with one branch. This branching is depicted using two sequential steps. (b) One C fragment creates two branches. (c) Using only CH_2 and CH_3 fragments creates an extended chain. (d) The three C_5H_{12} isomers shown earlier in this figure and their names.

require only one C fragment. Click, and we form another C_5H_{12} molecule, with a C–C–C chain and two branches. To compare these molecules with one having an extended chain, we construct in Scheme 4.3c the molecule that is made from three H_2C fragments and two H_3C fragments capping the ends. This is the third molecule that has the same formula, C_5H_{12}.

Thus, as summarized in Scheme 4.3d, we generated three molecules with the same *stoichiometry* (the same molecular formula, C_5H_{12}) but with different structures due to the different connectivities of the constituent fragments. Chemists refer to these distinct molecules as *isomers*, which is a Greek term that means *equal parts* (*iso* = equal, *meros* = parts). In the case of C_5H_{12}, the three molecules have the same number of parts (C and H atoms), but they differ in their structures due to different arrangements of their bonds. Scheme 4.3d also lists the names of these isomers.

The name *pentane* derives from the Greek *pente*, meaning five, indicating that the molecule has five carbon atoms. The other names still use the term *pentane*, with a prefix *iso* or *neo*, trying to convey the fact that these are isomeric molecules. Naming the isomers can become cumbersome when the number of isomers increases. For this reason, chemists have created a standard naming system.[2]

The principles of creating alkane isomers are clear by now and can be summarized as follows:

- ☞ Using only CH_2 fragments with two CH_3 caps creates extended chains.
- ☞ Adding a **CH** fragment creates a chain with a single branch.
- ☞ Adding a **C** fragment to the construction process creates two branches.
- ☞ Depending on the total number of carbons, one can use a few branching fragments to create more isomers.
- ☞ The branches can be short (CH_3 caps) or longer (e.g., CH_2CH_3).

4.2.3   Isomers of Octane (C_8H_{18})

We mentioned that octane is the fuel you get in the gas station to fill the tank of your car. But the fuel in the gas station is not really pure octane. It is rather a mixture of isomers of octane with some content of C_7H_{16} (heptane). This is the reason I wish to show some of these isomers here.

As the number of carbon atoms in the alkane grows, the number of isomers increases very steeply, such as 355 for $C_{12}H_{26}$ and a phenomenally large number for $C_{32}H_{66}$, more than three times the world's population![3] C_8H_{18} has 18 isomers, and writing down all of them is already tedious. We have no intention of doing this, because there are systematic algorithms that can do it for you if you need to generate all the isomers of a given C_nH_{2n+2} molecule. We specify these numbers to simply underscore the awesome richness of the alkane world. Our goal here is to generate a few of the isomers of octane using the LEGO fragments in Scheme 4.1 and the above principles.

Scheme 4.4 shows four isomers of octane. The first (a) is made from six CH_2 fragments and two CH_3 caps, leading to the extended chain isomer of C_8H_{18}. The second one (b) involves a single CH fragment, which creates a chain of seven carbon atoms and a single branch. One should note that the CH fragment could be placed in different positions along the chain, thus creating more singly branched isomers.

The third isomer in Scheme 4.4c uses a C fragment, thus creating a six-carbon chain with two branches. Again, the C fragment can be placed in different locations along the chain, and one can thereby create more isomers.

Finally, in Scheme 4.4d, we show the isomer that would be generated by utilizing the fragment C, which forms two branches, and the fragment CH, which forms one branch. The so-obtained isomer has the name *iso*-octane.

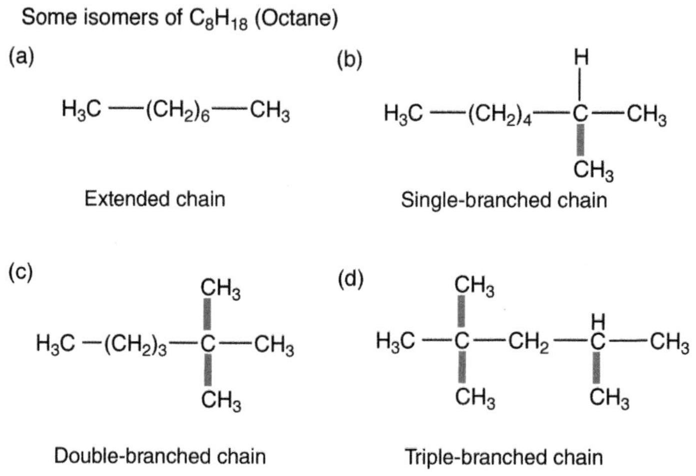

Some isomers of C_8H_{18} (Octane)

(a)

H_3C —$(CH_2)_6$—CH_3

Extended chain

(b)

$$H_3C —(CH_2)_4—\overset{\overset{\displaystyle H}{|}}{\underset{\underset{\displaystyle CH_3}{|}}{C}}—CH_3$$

Single-branched chain

(c)

$$H_3C —(CH_2)_3—\overset{\overset{\displaystyle CH_3}{|}}{\underset{\underset{\displaystyle CH_3}{|}}{C}}—CH_3$$

Double-branched chain

(d)

$$H_3C —\overset{\overset{\displaystyle CH_3}{|}}{\underset{\underset{\displaystyle CH_3}{|}}{C}}—CH_2—\overset{\overset{\displaystyle H}{|}}{\underset{\underset{\displaystyle CH_3}{|}}{C}}—CH_3$$

Triple-branched chain

SCHEME 4.4   Construction of four C_8H_{18} isomers with an extended chain (a), a singly branched chain (b), a doubly branched chain (c), and a triply branched chain (d). Thick bonds mark the branching. The isomer in (d) is named *iso*-octane.

Other isomers arise from using two branching fragments (CH and/or C), or three CH fragments, and from having longer branches (e.g., –CH_2CH_3 instead of CH_3 as in Scheme 4.4). When you drive into a gas station, your tank will be filled with a mixture of *iso*-octane and other isomers, including some C_7H_{16}. The octane number measures the percentage of *iso*-octane in this mixture. The higher it is, the better the gasoline is, as it allows the car engine to work smoothly (without knocking).

4.2.4   Some Applications of Alkanes

Scheme 4.5 summarizes the main uses of alkanes. The most important sources for alkanes are natural gas and oil. Oil drilling exposes pools of crude oil, which has subsequently to be refined and separated into liquids and solids.

The small alkanes, CH_4 to C_4H_{10}, are called together "natural gas," and they are found in deposits collected in porous rocks and deep in the oceans. Methane (CH_4), the main component of natural gas, is produced in marshes and wetlands by

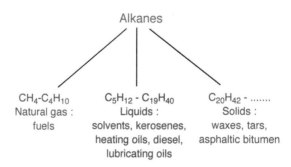

SCHEME 4.5   Alkanes and their uses.

microorganisms, and from sediments of decaying animals and plants. It is constantly puffed into the atmosphere by living animals (e.g., cows, sheep). Other planets (e.g., Jupiter and its moons) contain methane and ethane (C_2H_6) lakes. Mankind uses these gaseous materials as fuels for heating and cooking, and also in engines. Methane trapped in the bottoms of the oceans is considered to be the major reserve of C/H-based energy.

The next group of alkanes, ranging from C_5H_{12} to $C_{19}H_{40}$, constitutes of liquids or soft solids. The smaller ones like C_5H_{12} and C_6H_{14} are used as solvents, such as for paint or for dry cleaning. Then comes the gasoline group, C_7H_{16} to $C_{11}H_{24}$ and $C_{12}H_{26}$ to $C_{15}H_{32}$, which are called collectively kerosene, followed by diesel, and lubricating oils (like in car engines).

The solid alkanes from $C_{20}H_{42}$ onward are wax, tar, and asphaltic bitumen. Bitumen is used for the construction of asphalt roads. One of the uses of waxes, such as $C_{23}H_{48}$, $C_{25}H_{52}$, and $C_{27}H_{56}$, in nature is ecological. These waxes are produced by the female bees and used as pheromones (sex attractants) to attract male bees. The orchid (called the spider orchid) also produces this mixture of waxes, attracts the male bees, and makes the bees its agent of pollination. All these materials have immense importance to our well-being and, of course, to the female bees and the orchid!

Alkanes as a Source of Energy: An important use of alkanes is as a source of energy for cooking our food, warming us in freezing winters, making our engines run, and lighting our electric bulbs in the dark. It may feel to you somewhat uneasy to think that we are using our own decay products as a source of energy to keep us alive and well!

Producing energy from alkanes involves a chemical reaction of the alkane with molecular oxygen. Scheme 4.6 shows the chemical equations for these reactions in the specific case of methane (a) and for a general alkane (b). It is seen that the products of this reaction are CO_2 and water. Another "product" is *energy*, which is released as heat during the reaction (consider how hot your car engine gets when it runs).

As we discussed in previous lectures, the release of energy in a chemical reaction arises whenever the bonds in the product molecules are stronger than those in the reactant molecules. In our case, the bonds of the CO_2 and H_2O molecules are stronger than the C—C/C—H bonds in the alkane and the O—O bond in O_2, and hence the excess energy is released. And if we can harness it, we can use it to work for us, in an engine, or in an electricity wire, or simply in a heater. It is customary to feel that these

(a)
$$CH_4 \ + \ 2O_2 \ \longrightarrow \ CO_2 \ + \ 2H_2O \ + \ \uparrow Energy$$

(b)
$$C_nH_{2n+2} \ + \ \frac{(3n+1)}{2}O_2 \ \longrightarrow \ nCO_2 \ + \ (n+1)H_2O \ + \ \uparrow Energy$$

SCHEME 4.6 Alkanes as sources of energy: (a) in the reaction of CH_4 with O_2 and (b) in the reaction of a general alkane with O_2. The *vertical* arrows mean that energy is released during these reactions.

carbon-based fuels are harmful to the environment. This is true, and we have to find greener sources of energy. But until then, just imagine how you would feel in the winter without fuel!

Another important aspect of these chemical equations is that they follow *the principle of the conservation of the atom*, which I introduced in Chapter 2. To follow the principle, we must balance the numbers of each atom type on both sides of the reaction. By counting the O, C, and H atoms on both sides of the equations in Scheme 4.6, you can instantly verify that they are identical. Thus, in Scheme 4.6a, one molecule of CH_4 requires two molecules of O_2 to produce one molecule of CO_2 and two molecules of water. In fact, each carbon atom in the alkane will generate a molecule of CO_2, such that in a general alkane with n carbon atoms, we are going to produce n molecules of CO_2. So if we burn an alkane with 100 carbon atoms, each molecule of the fuel will produce 100 molecules of the undesired CO_2! A good exercise is to calculate the weight of CO_2 that is produced by an average car in a year. The quantity is hugely large, as shown in the Retouches section 4.R.4 of this lecture.

Alkanes are therefore very useful. We are so accustomed to use these molecules as energy sources, solvents for paint, fuels for our cars, raw material for our roads, etc. that all this seems mundane and rather boring. However, let us not forget that under the mundane usage, there lie so many exciting aspects of the submicroscopic constituents of our daily materials. And let us not forget the ingenious work of chemists to produce all these molecules from the ugly black oil! Our goal in this section was to project the beauty of this infinitely rich molecular world.

4.3 MOLECULAR RINGS AND CAGES MADE FROM CH_2 AND CH FRAGMENTS

In the preceding lecture, we constructed ethylene, $H_2C=CH_2$, and acetylene, which are made from two fragments of H_2C and HC, respectively. Here we wish to build rings and cages that possess no multiple bonds.

4.3.1 Molecular Rings Made of CH_2 Fragments

Let us construct now the minimal molecule that is made of CH_2 fragments and does not contain multiple bonds. As shown in Scheme 4.7a, since $H_2C:$ is a double connector, this minimal molecule will have to contain three such fragments. Clicking their connectivities, we obtain the cyclic molecule $(CH_2)_3$, called *cyclopropane*. It is easy to see that if we take four such fragments, we shall construct the four-membered ring, $(CH_2)_4$, called *cyclobutane*, in Scheme 4.7b. In fact, we can generate an entire *molecular world of cycloalkanes*, with the general formula $(CH_2)_n$, where n is a number that runs from 3 in principle to infinity (Scheme 4.7c). In practice, however, rings with more than $10\,CH_2$ groups are not common. One of the reasons is that the probability of the two ends to meeting and tying up into a ring decreases as the number of carbons increases.

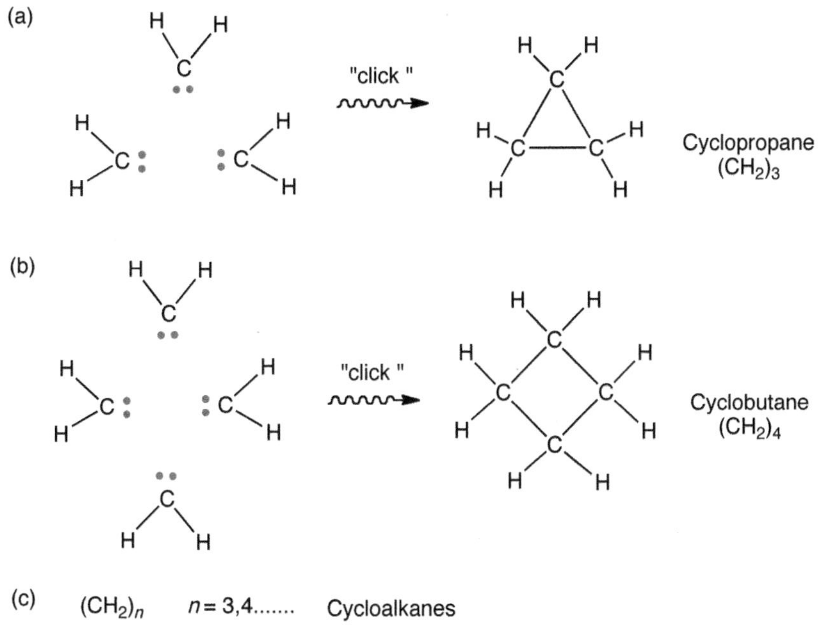

SCHEME 4.7 Making rings from H_2C: fragments: (a) Cyclopropane, $(CH_2)_3$. (b) Cyclobutane, $(CH_2)_4$. (c) A general cycloalkane formula, $(CH_2)_n$.

Cycloalkanes have similar properties to those of alkanes, and they can also be used as sources of energy by burning them with O_2.

4.3.2 Molecular Cages Made of CH Fragments

Molecular cages made from CH fragments are the outcomes of the fascination of chemists with unusual molecules and with the classical Platonic solids. Let us see some of these objects and learn how we can construct them in the "click" way. Consider first the smallest molecule which is made of HC fragments and which does not involve multiple bonds. It is clear that since HC is a triple connector, then each HC will require three other fragments, and hence the smallest such molecule will be $(CH)_4$. These fragments are shown in Scheme 4.8, to click into a cage molecule, in which each carbon is bonded to the other three. This molecule was made for the first time in 1978 by the German chemist Günther Maier, who described it in a very short paper.[4] The 3D shape of this molecule is shown in Figure 4.2. One can see that it has a nice symmetric shape of a tetrahedron—an object with four faces made from equilateral triangles. This is the reason why Maier called this molecule *tetrahedrane*; he drew a link between the molecule and the abstract tetrahedron structure.

The tetrahedron is one of the perfect Platonic bodies, which are shown in Figure 4.3 and are known by the names tetrahedron, octahedron, cube, dodecahedron, and icosahedron. These bodies were thought by the Greeks to symbolize the four elements and a fifth one called *Quintessence*. The tetrahedron symbolized *Fire*, the octahedron *Air*, the cube *Earth*, and the icosahedron *Water*. According to this philosophy, the

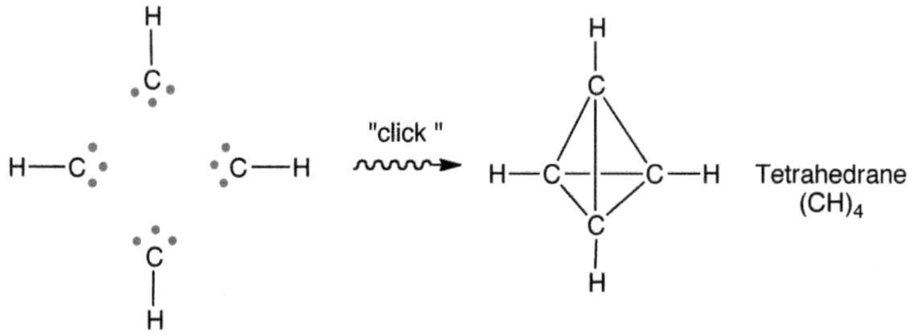

SCHEME 4.8   Formation of the cage molecule tetrahedrane, $(CH)_4$, made from four HC fragments.

dodecahedron symbolized the heavenly element, called the *Ether* (the glow) and/or *Quintessence*.[5] As we mentioned before, the Greek philosophy of Plato and Aristotle dominated western thought up to the eighteenth century. And as such, most scientists and nonscientists who have had a classical education harbor a fascination with these Platonic objects. Maier was one such individual: he made the molecule *tetrahedrane* to mimic the Platonic tetrahedron. Our myths have lives of their own, and they are incarnated into the material worlds we create.

A very special Platonic molecule is *dodecahedrane*, $(CH)_{20}$, which is shown in Figure 4.4b and c, alongside the abstract 3D dodecahedron in Figure 4.4a. It is seen that the molecule (Figure 4.4b), like its abstract object (Figure 4.4a), contains 12 pentagonal faces and 20 vertices. The only difference is that the vertices of the molecular object are CH fragments. It is not easy to draw the molecule, but it is very instructive to build a model of this molecule. If you do so, inspect the model, rotate it, and enjoy its beauty.

Dodecahedrane was finally made in the chemistry laboratory of the American chemist Leo Paquette,[6] after an odyssey of 30 years, with many attempts by the greatest chemists. Many PhD students have spent their academic years trying to

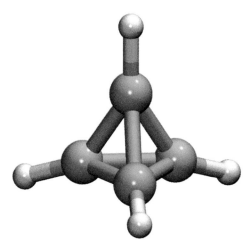

FIGURE 4.2   The 3D shape of tetrahedrane.

FIGURE 4.3   The perfect Platonic bodies represented by an artist, from right to left: a tetrahedron, an octahedron, a cube, a dodecahedron, and an icosahedron, located in Steinfurter Bagno Park (This image is in the public domain; it was obtained from http://commons .wikimedia.org/wiki/File:Platonische_Koerper_im_Bagno.jpg, and it is used with credit to Zumthie).

achieve this goal. Why go through such an enormous effort? The first reason is the fascination with the object, which is clear from the title Paquette used in his article, "Dodecahedrane—The chemical transliteration of Plato's universe." Second, making such a molecule in the laboratory is a great challenge. Challenges call for heroes, like Hillary and Tenzing, and others who followed them in climbing Everest. And it is also a sport activity—who gets there first? Chemists are not different from other humans; they want to challenge their skills and to be the first to achieve a goal. And finally, the 30 years of chasing dodecahedrane have taught chemists a great deal

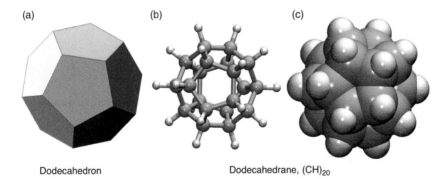

FIGURE 4.4   The abstract dodecahedron body (a Platonic body) in (a) and the 3D representations of the *dodecahedrane* molecule, $(CH)_{20}$, in (b) and (c). The representation in (b) shows all the C—C and C—H bonds as sticks, while the one in (c) shows the packing of the atomic spheres, C (blue) and H (gray). (The dodecahedron image on the left-hand side of the figure was obtained from http://commons.wikimedia.org/wiki/File:Zeroth_stellation_of_ dodecahedron.png.)

about strategies for chemical synthesis (making molecules). Thus, in the end, the art of chemical synthesis has benefited greatly from this 30-year odyssey. Every such conquest improves the ability of chemists to devise more efficient and more beautiful syntheses of yet unknown molecules.

One can think of other cages, such as the cubane $(CH)_8$ molecule, etc. You will find some in the problem set at the end of the chapter.

4.4 MOLECULAR PLANES AND CAGES MADE FROM C FRAGMENTS

The four-way-connector fragment C can give rise to molecular planes and cages made exclusively from carbon atoms. These kinds of molecules are becoming highly important in the emerging nanotechnology field. They also form very beautiful and aesthetically pleasing architectures. Therefore, I would like *at least* to describe them, though their construction process is not straightforward and will require all of you to stretch your imaginations.

We have seen three of these molecules already in Lecture 1, namely graphite, diamond, and the soccer ball-like molecule, C_{60}. Figure 4.5 shows these molecules again. In part (a), we see the solid called graphite, which is used in our pencils. The

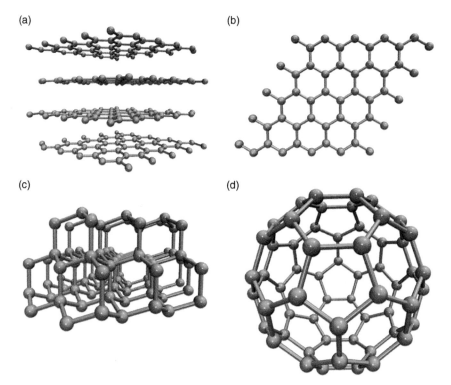

FIGURE 4.5 Structural ball-and-stick models of (a) stacked graphite layers, (b) a graphene layer, (c) diamond, and (d) C_{60}.

molecule is built from layers of hexagons in a honeycomb pattern. The layers are stacked, but they are not bonded by click bonds, and therefore it is quite easy to peel the layers of graphite and form the single-layered structure called *graphene*, which is shown in Figure 4.5b. In 2010, the Nobel Committee awarded the prize in physics to Andrei Geim and Konstantin Novoselov from the University of Manchester, for producing graphene by peeling it from graphite using Scotch tape and showing the unusual properties of this molecular plane.[7]

The graphite layers, and hence also the graphene, are molecular planes which contain in principle an infinite number (or a very large number) of carbon atoms and may be called two-dimensional polymers. It turns out that, once peeled from graphite, the graphene sheet tends to curl and generate *nanotubes*, which constitute a family of materials that are finding growing use in nanotechnology, with applications in solar cells, transistors, screens for plasma televisions and cellphones, camera sensors, gas sensors, material strengthening, and so on. Graphene is the strongest material in the world, and it is said that even an elephant would not be able to burst through a sheet of graphene with the thickness of plastic wrap. What an amazing molecule!

Figure 4.5c shows the hardest and one of the most precious materials in the world, diamond. It is seen that the honeycomb pattern of carbon hexagons still exists, but now the structure contains carbon cages within a 3D molecular structure. As we noted before, the general shape pattern of diamond constitutes in principle a small change compared with graphite, but what a huge difference this small change makes!

Figure 4.5d shows C_{60}, called *buckyball* or *fullerene*, after the name of the famous architect Buckminster Fuller, who built geodesic domes looking like C_{60}, with hexagons and pentagons. Robert F. Curl Jr., Sir Harold W. Kroto, and Richard E. Smalley found this carbon cage molecule in the process of vaporizing graphite and shared the 1996 Nobel Prize in chemistry for this discovery.[8] C_{60} has plenty of applications, ranging from catalysis all the way to drug delivery in cancer therapy. C_{60} is also a member of a family of fullerenes, which are increasingly used in nanotechnology along with the nanotubes mentioned above. All these fullerenes are carbon cages of different shapes; for example, C_{72} is a sausage-like cage. The shape and size of the fullerenes depend on the number of pentagonal rings in the molecule.

It is not a simple matter to click carbon fragments and fully generate these beautifully complex molecular structures. Nevertheless, we can have a glimpse into these molecules by inspecting their fundamental structural elements in Schemes 4.9 and 4.10. This will teach us quite a bit about these molecules.

Thus, as shown in Scheme 4.9, taking six C fragments which are four-way connectors and clicking their connectivities without exceeding C=C bonding creates a C_6 hexagon involving alternating C=C and C–C bonds, with six dangling connectivities total on the six carbon atoms. If we now bring additional C fragments (a total of 18 additional Cs) and click their connectivities to the fundamental hexagon, while restricting the construction process to generate only fused hexagons, in the end, we are going to generate a C_{24} cluster that retains the alternating C=C/C–C bonds and possesses 12 dangling connectivities. We can continue this process indefinitely by adding more carbon fragments, requiring these to form only hexagonal units and not exceed C=C bonding. At some point, we are going to generate a carbon sheet with

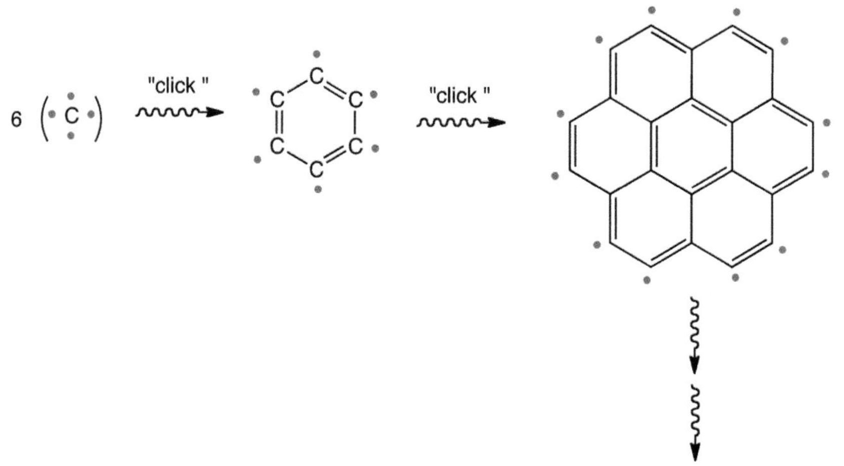

A molecular plane with a
honeycomb structure

SCHEME 4.9   The fundamental hexagonal C_6 unit and its evolution into graphene in several sequential steps. Graphite is a stack of graphene planes.

a honeycomb pattern, as found in graphite and graphene. Of course, since we still do not know how to predict the geometry of molecules, we cannot actually tell with any certitude that the honeycomb structure with alternating C=C/C–C bonds would form a *molecular plane*. But it is still good to know that it does. This will prepare us for a forthcoming lecture on the geometry and 3D structure of molecules.

The diamond structure is significantly more complex since it has a 3D structure constructed from cage units that in turn form tunnels. Let us nevertheless try to see how the elemental cage could be formed from the fundamental hexagon. Scheme 4.10a shows six C fragments that click to form a C_6 hexagon having only single C–C bonds. All the carbon atoms in the so-formed hexagon possess two dangling connectivities. Since we allow only for single C–C bonds, then to avoid clashing with each other, these two connectivities will further click in two different directions in space (actually, as we shall see later in Lecture 7, this avoided clashing is the driving force for going 3D). As we show in Scheme 4.10b, the fundamental hexagon adopts a shape of a reclining chair, where each carbon atom has two connectivities, one in a vertical direction (up or down) and the other is horizontal (right or left). The lines that guide from the C atom to the dangling connectivity, drawn by a filled circle, indicate these directions. The angle between the vertical and the horizontal directions is 109°, and we shall learn about its origins in Lecture 7.

If we now bring four C fragments connected as a $C(C_3)$ unit, we can cap the three upward-pointing vertical connectivities of the fundamental hexagon and generate the elemental diamondoid cage structure shown in (c). Count how many dangling connectivities you have in structure (c) and anticipate the molecule that will be formed if you saturated these connectivities with H fragments.[9] We can subsequently cap also the three downward-pointing connectivities of the hexagon, and we get

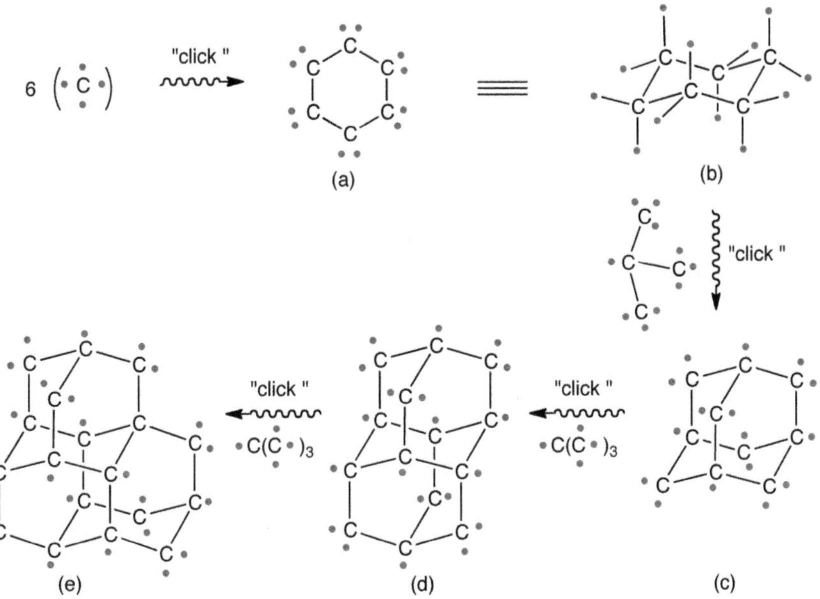

SCHEME 4.10 (a) Formation of the fundamental hexagonal C_6 unit in the diamond molecule, having 12 dangling connectivities. (b) The 3D shape of this C_6 unit, with its directed connectivities shown by lines guiding from the C to the red connectivity circle representing an electron available for bonding. (c–e) Successive capping of the fundamental hexagon in (b) with $C(C)_3$ units to generate diamondoid cages *en route* to the completion of the structure of diamond.

the double-cage structure shown in (d). Subsequently, we can cap the horizontal connectivities and generate the triple-cage structure as shown in (e). Since most of the carbon atoms in (e) still have one or two dangling connectivities, we can in fact continue indefinitely until the formation of the diamond molecule is achieved.

Understanding the structure of C_{60} may be facilitated by first inspecting a model of the molecule (if a model is not available, have a look at a soccer ball). In so doing, you will note that the molecule has fused hexagons and pentagons, which curve to form a cage molecule. To construct this cage, we start again by clicking six C fragments to form a C_6 unit with alternating C=C/C–C bonds and six dangling connectivities, as shown in Scheme 4.11a. Clicking one fragment consisting of four carbon atoms to fuse another hexagon to the C_6 unit, we obtain a C_{10} unit in Scheme 4.11b, with eight dangling connectivities. At this point, instead of fusing another hexagon, we add one fragment consisting of two carbons and fuse it to form a pentagon in the concave areas of the C_{10} unit. Repeating this once more, we get a C_{14} unit in Scheme 4.11c. Further fusion, first of four C_3 units to the dangling connectivities of C_{14}, and then of single C fragments to the new bay areas of C_{26} and so on generates finally in Scheme 4.11d the C_{60} buckyball, which contains 12 pentagons and 20 hexagons, fused by means of alternating C=C/C–C bonds, such that each pentagon is fused to five hexagons (Scheme 4.11e) and each hexagon is fused to three pentagons plus three hexagons (Scheme 4.11f).

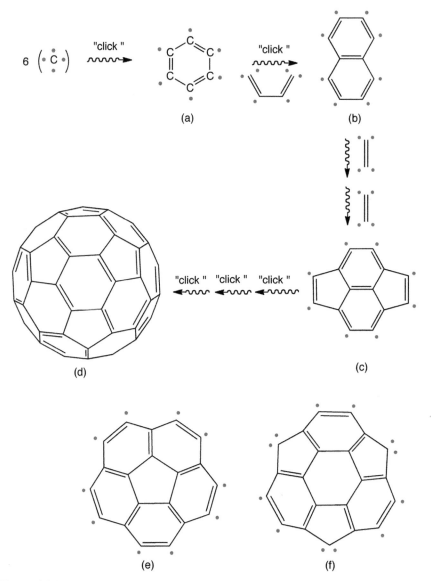

SCHEME 4.11   Construction of C_{60}: (a) Formation of a C_6 unit with six dangling connectivities. (b) Adding one fragment consisting of four carbon atoms generates a C_{10} unit with eight dangling connectivities. (c) Adding a fragment, which contains two carbon atoms, twice, to the concave regions in (b) generates a C_{14} unit with eight dangling connectivities. (d) Successive additions of C fragments eventually generate C_{60} (d). In C_{60}, one can find that each pentagon is surrounded by five fused hexagons in a C_{20} unit (e), and each hexagon is surrounded by three pentagons and three hexagons in a C_{21} unit (f). Model building shows that both molecular fragments (e and f) are bowl-shaped.

We have no simple means of predicting the spherical shape of the C_{60} molecule. However, when one builds a model of the molecule from plastic molecular models, one witnesses that when the C_{20} and C_{21} molecular motifs in Schemes 4.11e and 4.11f are constructed, *they begin to curve and assume the shape of a bowl.* Once we add more carbon fragments, the curvature increases, and the structure closes eventually into a sphere. I highly recommend model building as part of this lecture or in a tutoring session.

4.5   ISOMERS OF RINGS AND CAGES

As we already stressed, multiple connectivity gives rise to several possibilities of clicking the connectivities into isomeric molecules. Let us take, for example, six HC fragments, which have triple connectivity, and construct the possible isomers. If we allow them to click in all possible manners, as long as they all combine into a single molecule, they will give rise to the three isomeric molecules, as shown in Scheme 4.12. In (a), the HC fragments click to form a ring with alternating C=C/C—C bonds, which is a familiar motif already (note that we can exchange the location of C=C and C—C bonds; see Retouches sections 4.R.5 and 4.R.6). The resulting molecule is called *benzene.* The great British scientist Michael Faraday discovered

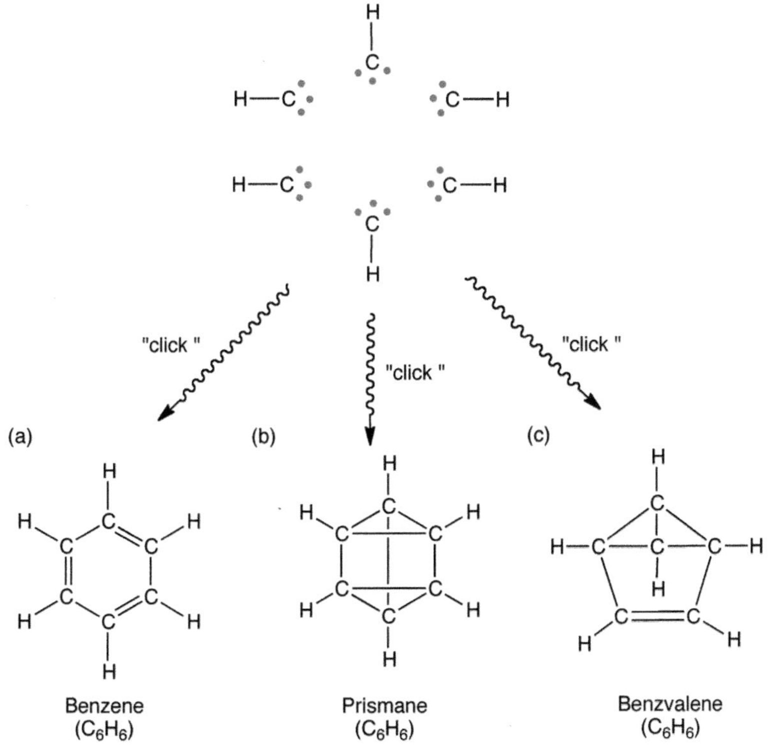

SCHEME 4.12   Construction of three isomers that are all made from six CH fragments.

benzene in candle soot.[10] This molecule is a chemical icon, which has played a key role in the development of chemistry and in the understanding of the chemical bond. Like with many other molecules discussed before, benzene is a "Janus" molecule— along with benefits, it also brings negative features. After years of careless handling of benzene by many chemists, it turns out that the molecule is highly carcinogenic and must be handled with extreme caution.

Another (CH)$_6$ isomer is *prismane* in Scheme 4.12b. Its name comes from the fact that it looks like a prism. The third isomer, in Scheme 4.12c, contains one C=C bond, and the remainder of the bonds are C–C bonds (its name is marked, but it is not too intuitive like prismane).

Thus, the triple connectivity of HC gives rise to different ways of clicking these connectivities, and hence generating molecular isomers. Let me recommend that you generate isomers in cases with four and eight HC fragments.

4.6   INFINITY OF MOLECULAR WORLDS MADE FROM C AND H

We have constructed so far many different molecules and molecular worlds, of chains, rings, cages, and mixed motifs. Yet, we did not even begin to scratch the surface of the *molecular infinity* to which our modular fragments lead. Any molecule forms a basis for new molecular worlds. This is demonstrated in Schemes 4.13 and 4.14, where we form new fragments from molecules we already constructed in this lecture and in previous ones.

Scheme 4.13 shows the modular fragments which can be generated from cyclopropane, by unclicking successively its C–H bonds. Thus, unclicking one of the C–H bonds generates a fragment with a single connectivity. Unclicking two C–H bonds can be done in two ways, which lead to two different double-connector fragments, one with the dangling connectivities on the same carbon atom and the other with the dangling connectivities on adjacent atoms. Continuing to unclick C–H bonds will generate a few triple connectors, then four-way connectors, five-way connectors, and all the way to a six-way connector.

Scheme 4.14 shows more fragments generated from ethylene, tetrahedrane, and benzene, and the respective connectivities of these fragments.

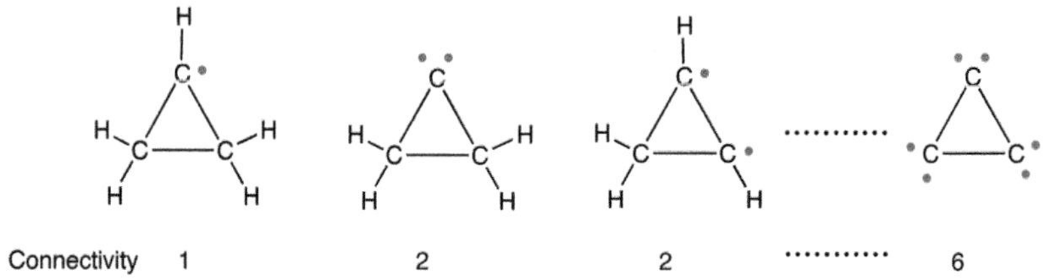

| Connectivity | 1 | 2 | 2 | ········· | 6 |

SCHEME 4.13   Modular fragments made from cyclopropane with the electrons participating in connectivity highlighted in red.

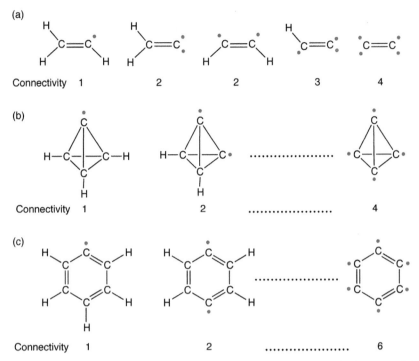

SCHEME 4.14 Modular fragments made from (a) ethylene, (b) tetrahedrane, and (c) benzene. The electrons participating in connectivity are highlighted in red.

Using these fragments, we can construct new molecular worlds. We can make chains from double-connector fragments, cages from triple connectors and quadruple connectors, and *star molecules* from five-way and six-way connectors. For example, from the double-connector fragments of tetrahedrane, as shown in Scheme 4.14b, we can make a chain of any number of tetrahedrane units and cap the dangling connectivities at the two ends with the single-connector fragment. We can also make a necklace of tetrahedranes from the same double-connector fragments. We can mix the LEGO fragments and generate mixed molecular worlds, and so on and so forth.

You can try to imagine how many molecules we can make just from the modular fragments discussed in this lecture. This process can be repeated for any molecule. We are talking about an infinity of infinite molecular worlds, made just from fragments that involve C and H atoms. As Bruno Schulz wrote:

"All matter flows out from endless possibilities … anyone may knead and shape it [matter]; it submits to all. All arrangements of matter are impermanent and loose, liable to retardation and dissolution."

The chemical matter is limitless, and it submits only to creativity.

4.7 SUMMARY

In this lecture, we built a variety of molecular worlds, chains, rings, cages, and molecular planes. All these were made from the LEGO fragments we generated from methane (Scheme 4.1), H_3C, H_2C, HC, and C. We witnessed the power of the concept, based on the following already familiar statements:

> ☏ Every molecule forms a basis for new modular fragments that are used for creating new chemical worlds.
>
> ☏ Chemistry is the poetry of mankind in matter!

4.8 REFERENCES

[1] M. Eliade, *The Forge and the Crucible*, University of Chicago Press, 1978.

[2] (a) J. Rigaudy and S. P. Klesney, Eds. *Nomenclature of Organic Chemistry*, IUPAC/Pergamon Press, 1979. (b) R. Panico, W. H. Powell, and J. C. Richer, *A Guide to IUPAC Nomenclature of Organic Compounds*, IUPAC/Blackwell Science, 1993.

[3] A. T. Balaban, J. W. Kennedy, and L. V. Quintas, *J. Chem. Ed.* **1988**, *65*, 304.

[4] G. Maier, S. Pfriem, U. Schäfer, and R. Matusch, *Angew. Chem. Int. Ed. Engl.* **1978**, *17*, 520.

[5] E. Heilbronner and F. A. Miller, *A Philatelic Ramble through Chemistry*, Wiley-VCH, Basel, 1998, pp. 3–4.

[6] L. A. Paquette, *Proc. Natl. Acad. Sci. USA* **1982**, *79*, 4495.

[7] Nobel Foundation announcement. Available at: http://nobelprize.org/nobel_prizes/ physics/ laureates/2010/

[8] http://www.nobelprize.org/nobel_prizes/chemistry/laureates/1996/

[9] Diamondoid molecules have been generated by: P. R. Schreiner, L. V. Chernish, P. A. Gunchenko, E. Y. Tikhonchuk, H. Hausmann, M. Serafin, S. Schlecht, J. E. P. Dahl, R. M. K. Carlson, and A. A. Fokin, *Nature*, **2011**, *477*, 308.

[10] See the blog by P. Ball: The name's (quadruple) bond? Available at: Chemistry World, June 2013, Chemistry World page 41. http://www.rsc.org/chemistryworld/2013/ 05/quadruple-bond-carbon-debate-shaik-hoffmann-frenking

4.R RETOUCHES

4.R.1 Atomic Weight, Isotopes, Atomic Mass Unit, and Molecular Weights

If you inspect the periodic table, you will find that in addition to the atomic number, near each element (usually on the right side) there is an additional number which is not an integer and which tells us the *atomic weight* of this atom. For example, Scheme 4.R.1 shows the atom C. It has an atomic number of 6 and an atomic weight of 12.011.

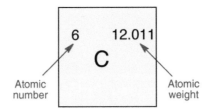

SCHEME 4.R.1 Carbon in its periodic table box with its atomic number (upper left) and atomic weight (upper right).

Let us refresh our memories. The atomic number tells us the number of protons in the nucleus of the atom. In the case of C, there are six protons (and hence also six electrons). However, the nucleus also contains electrically neutral particles called neutrons. The most abundant C atom contains six neutrons, which, together with the six protons, sum to 12 massive particles, which are accommodated inside the nucleus. This carbon atom, which is labeled as ^{12}C, constitutes 99% of all carbon atoms in the universe. However, there are also carbon atoms, almost 1% of all carbon atoms, with nuclei that contain seven neutrons, and which are labeled as ^{13}C, and there are very few (0.000000001%) carbon atoms with eight neutrons in the nuclei. These are labeled as ^{14}C. These three different carbon atoms are called *isotopes*. The term *isotope* is composed of two Greek words, *iso* (equal) and *topos* (place), and it tells us that the box for C contains different atoms which are located in the same place in the periodic table. Since the number of its electrons determines the chemistry of C, the three isotopes are "chemically identical" (almost, almost so) and are not distinguished in the periodic table. All atoms have isotopes.[1R] For carbon, both ^{13}C and ^{14}C are radioactive, and both decay and re-form at characteristic rates. The rate of decay of ^{14}C forms the basis of carbon dating of once-living creatures. (^{14}C is formed at a constant rate, so by measuring the radioactive emissions from once-living matter and comparing their activity with the level in living things, a measurement of the age of the dead object can be made.)

The relative weight of atoms, that is, the atomic weight, can be expressed by defining an *atomic mass unit* (*amu*), as the weight of a proton or neutron (which are nearly equal). Since the most abundant isotope of C has six protons and six neutrons, it has a *mass number* of 12 amu and is labeled as ^{12}C. Similarly, we have ^{13}C and ^{14}C, which weigh more than the most abundant isotope. The atomic weight is the average of these atomic masses (weighted by their percent abundance), and as might be expected, it reaches 12.011 amu, close to the mass of ^{12}C. Since all atoms have isotopes, their atomic weights in the periodic table are non-integer numbers; for example, for H, we find the number 1.0079, which arises from the weighted average of ^{1}H, ^{2}H (called deuterium), and ^{3}H (called tritium).

4.R.2 Avogadro's Number

Suppose we want to express the relative weights of atoms in gram (g) or kilogram (kg) units, which we use in our daily life. The actual atomic weight in these units is tiny; for example, one proton weighs 1.67×10^{-24} g, which means $0.00000000000000000000000167$ g. Certainly, it is very inconvenient to use these numbers!

A more convenient way is to express the atomic weights as relative weights of atoms in gram units. Thus, H would be 1.0079 g and C 12.011 g, etc. Since these numbers conserve the weight ratio of the atoms, then they contain the same number of atoms; for example, if we take 1.0079 g of H and 12.011 g of C, *we are taking the same number of atoms*. This number is a constant called *Avogadro's number* $(6.0221413 \times 10^{23})$, after the name of Amedeo Avogadro (shown in Figure 4.R.1), whose studies led to this idea and have put chemistry on a correct quantitative and qualitative track of understanding matter in terms of atoms and molecules.[2R]

Avogadro and the number carrying his name are depicted in Figure 4.R.1. Avogadro's number is huge, more than billions times larger than the population of Earth.

4.R.3   The Mole Concept

Since the quantity of matter that is expressed in Avogadro units conserves the ratio of the atomic weights, it is also *a fundamental quantity of matter*. This quantity is called a *mole*, and the weights are expressed as gram per mole and written as g/mol. Accordingly, a mole of H atoms weighs 1.0079 g/mol, and a mole of C atoms weighs 12.011 g/mol. As we are dealing with molecules, we should also express the molecular weight of molecules in the same unit. Thus, for example, the molecular weight of H_2 is simply twice the atomic weight, that is, 2.0158 g/mol. The molecular weight of CH_4 is the sum of the atomic weight of C and four times the weight of H, namely 16.0426 g/mol. By writing these weights in "g/mol," we mean to say that an Avogadro's number of H atoms weighs 1.0079 g, or an Avogadro's number of CH_4 molecules weighs 16.0426 g, and so on.

By the way, the major contribution of Avogadro to chemistry was his hypothesis that atmospheric gases are made of molecules, which is the name he used for the

FIGURE 4.R.1   Amedeo Avogadro's caricature (drawn by the historian W. B. Jensen and reproduced here with his permission) and Avogadro's number.

1 mole = 22.4 L

FIGURE 4.R.2 Stanislao Cannizzaro's caricature with little Avogadro alongside him (drawn by the historian W. B. Jensen and reproduced here with his permission) and the volume of a mole of a gas in litres.

combination of identical atoms.[2R] Thus, he hypothesized that atmospheric hydrogen existed as H_2, and oxygen as O_2, etc. He (and others) also stated that equal volumes of gases contain the same number of molecules. This hypothesis was furiously criticized in the early nineteenth century, since according to the theory of chemical matter used then, identical atoms had no affinity towards one another and could not bind. The greatest objections were expressed by Dalton, who is the father of the modern atomic theory, and by Berzelius, the inventor of the chemical language and one of the most influential chemists in the nineteenth century.

This objection held chemistry back for over 50 years, until Stanislao Cannizzaro (Figure 4.R.2), another Italian chemist, presented in a chemistry conference in Karlsruhe in 1860 and explained how the use of Avogadro's hypothesis resolves all the inconsistencies in the atomic theory of matter. Cannizzaro invented the term *mole* and also determined the volume of 1 mole of gas as 22.4 L. Thus, 1 mole of H_2 weighs 2.01158 g/mol and occupies at room temperature a volume of 22.4 L. The volumes of 1 mole of O_2, N_2, CO_2, etc., are all identical, with a value of 22.4 L. If you try to calculate the average space occupied by each molecule in the gas, you will find that large distances, much larger than their own size, separate the molecules.

4.R.4   Calculation of CO_2 Emission by a Car

It is interesting to calculate the volume of CO_2 emitted by an average car in a year. Suppose the car is driven 10,000 km each year, and while doing so, it consumes 1 L of fuel per 10 km; hence, it requires 1,000 L a year (a good number of trips to the gas station!).

If the gasoline is octane (and its C_8H_{18} isomers), then each liter of octane (1,000 cm^3) weighs 703 g (the density of octane is 0.703 g/cm^3). If we multiply by 1,000 L, we get 703,000 g of octane, which are consumed by this car every year. To calculate how much CO_2 is emitted, we have to write the chemical equation for the burning of alkanes (Scheme 4.6b) and using *the principle of atom conservation in a chemical reaction*, we get the following equation:

$$C_8H_{18} + (25/2)O_2 \rightarrow 8CO_2 + 9H_2O \qquad (4R.1)$$

So, one molecule of octane gives rise to eight molecules of CO_2. Translated to mole language, we would say that 1 mole of octane generates 8 moles of CO_2. Each mole of octane weighs 114 g (we rounded off the actual number by using 12 g and 1 g as the molar weights of C and H, respectively; all subsequent numbers are rounded off). Since the car burns 703,000 g of octane each year, this translates to 6,167 moles of octane every year. If we multiply by 8 to get the CO_2 emission, we get 49,336 moles of CO_2. Since every mole of CO_2 has a volume of 22.4 L, the total CO_2 volume emitted into the atmosphere for a year by a single car amounts to 1,105,126 L, namely more than one million liters! The weight of this emitted CO_2 can be obtained by multiplying the number of moles by the molecular weight of CO_2, 44 g/mol, giving 2,170,784 g, which is 2,171 kg or 2.171 tons of CO_2 every year! Since the numbers of cars in the world currently exceeds 1 billion, this means that only the cars in the world spew more than 2.171 billion tons of CO_2 into the atmosphere every year! This is a scary number.

4.R.5   The Molecule Benzene, Kekulé's Dream, and Resonance Theory

During construction of the molecule benzene in Scheme 4.12a, we overlooked the fact that we could click the bonds in two different ways, (a) and (b), which are now shown in Scheme 4.R.2. The two forms are indeed different but otherwise entirely equivalent, both containing alternating C–C/C=C bonds.

Are these isomers? The answer is no, and this answer has been known from the second half of the nineteenth century, when chemists discovered benzene and were wondering about its structure. The German chemist August Friedrich Kekulé, Figure 4.R.3a, who came up with this idea, tells in his memoirs that, one day, as he was sitting in front of the fireplace at home and thinking about this molecule, he dozed off and started dreaming. In his dream, he saw structures twining and twisting in a snake-like motion. Then suddenly, "one of the snakes had seized hold of its own tail, and the form whirled mockingly before my eyes ..." (see Figure 4.R.3a).

What precisely Kekulé was seeing in his dream is not entirely clear today, as electrons had not been discovered yet and chemists had only fuzzy ideas about the notion of structure. But what was clear to Kekulé from his dream (or, perhaps, he pretended to have a dream in order to add a flourish to his idea) was that benzene does not have isomers, even if the connectivities are distributed in two alternative manners. Later (in the twentieth century), after the electron was discovered by J. J. Thomson, the American chemist Linus Pauling (Figure 4.R.3b), who received the Nobel Prize in Chemistry for his work on the nature of the chemical bond, realized that the whirling entities are the three electron pairs of benzene in the C=C double bonds. These double bond components were not located in specific bonds, but rather were *delocalized* over all six C–C bonds, as shown in Scheme 4.R.3. To describe this structure with its six whirling electrons, Pauling placed the two structures of

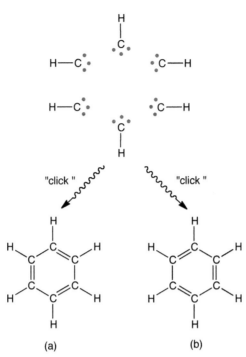

SCHEME 4.R.2   The two equivalent structures, (a) and (b), of benzene obtained by the alternative modes of clicking the connectivities of six HC fragments into a hexagonal molecule.

FIGURE 4.R.3   (a) Kekulé's caricature alongside the whirling snake in his alleged dream. (b) Linus Pauling, the originator of resonance theory. Kekulé's and Pauling's caricatures were drawn by the historian W. B. Jensen and are reproduced here with his permission. The whirling snake with benzene image is used from http://commons.wikimedia.org/wiki/File:Benzene_Ouroboros.png.

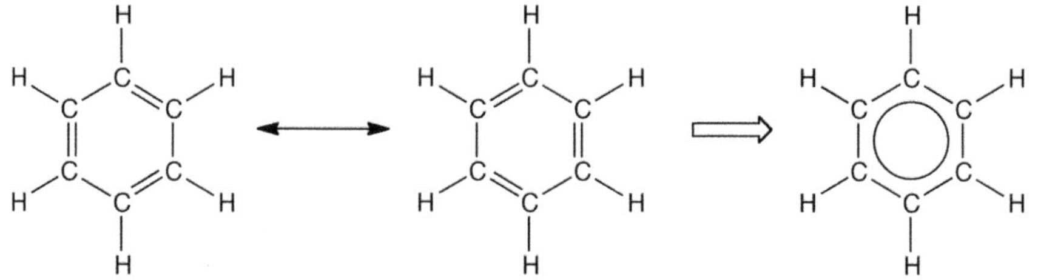

SCHEME 4.R.3   The structure of benzene displaying the resonance between the two local-ized electronic structures, indicated by a double-headed arrow. Combining the two forms together describes a single structure with six whirling electrons (far right), in addition to the six single bonds holding the molecule together.

benzene with localized bonds between them, and he added a double-headed arrow between them, calling this phenomenon resonance. His calculations using quantum mechanics (QM) showed that this electron delocalization or resonance further lowers the energy of the bonds in benzene.

Subsequent and more advanced QM calculations have supported the idea that delocalizing electrons (as in Scheme 4.R.3 in the far right) have an energy-lowering effect. However, at the same time, these advanced calculations revealed that when the electrons whirl in ring structures, the energy lowering depends on the number of electrons. It is important only when this number obeys the formula $4p + 2$, where $p = 0, 1, 2, \ldots$. Benzene follows this formula with $p = 1$ (i.e., six whirling electrons), and hence the electron delocalization is beneficial in this case. This makes for another great chapter of chemistry, which you will meet in more advanced courses.

4.R.6   Resonance Theory and Collective Bonding

The presentation of the whirling electrons of benzene in Scheme 4.R.3 in terms of resonance theory describes in fact *collective bonding*, namely all the whirling six electrons bind together the six carbon atoms. This bonding motif is very common in chemistry, and the C_6 fragment appears in an infinite number of molecules, starting from benzene and going all the way to graphene.

Collective (delocalized) bonding is most common in metals. All metals (except for mercury, which is a liquid) are solids, wherein the metal atoms are arranged in lattice structures that exhibit a repetitive order (called by chemists as periodic). An example is Scheme 4.R.4a, which displays a very common arrangement of metal atoms in a cubic form. Lithium (Li) adopts a cubic lattice. Inspecting the structure, it is apparent that we cannot see here pairs of Li atoms that are bonded, as would be expected from the fact that Li is a single connector. Because of the uniform distances between the metal atoms, we can click the connectivities in several alternative and equivalent ways. Thus, taking six Li atoms arranged in a triangular form, as in Scheme 4.R.4b, we can click the connectivities in two different manners (and, in fact, in more than just two), and the result is that the six electrons are delocalized over the six Li atoms and collectively bind these atoms.

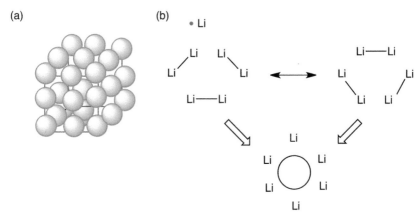

SCHEME 4.R.4   (a) A cubic lattice of metal atoms (shown as greenish spheres), as found, for instance, in lithium (Li). Due to the uniform distances between the atoms, the bonding in metals is collective. (b) An example of the collective bonding of Li. Li is a single-connector fragment, and, hence, Li_6 is collectively bonded due to the "resonance" between the two localized electronic structures, as the circle represents.

Many chemists refer to this bonding mode as metallic bonding because it is common in metals. But as we saw above for benzene and as might be deduced also in graphene, this collective bonding is by and large common in chemistry. What is special about metallic atoms is that they possess a small number of valence electrons. For example, Li has only one valence electron, and it can form the Li_2 molecule. Therefore, *the discrete molecules that metals can make will always be electron-deficient.* One way to ameliorate this deficiency is to arrange the atoms in lattices wherein each metal atom has a few neighbors (as in Scheme 4.R.4a), and they can enjoy thereby the added stabilization due to the delocalization of the electrons. This delocalized nature of the electrons is the root cause of the conductivity exhibited by metals and by analogous collectively bonded atoms as in graphene.

4.R.7   References for Retouches

[1R] S. S. Zumdahl and S. A. Zumdahl, *Chemistry*, 9th Ed., Brooks College, Cengage Learning, 2010, p. 54. Available at: www.cengage.com/global

[2R] R. Siegfried, *From Elements to Atoms: A History of Chemical Composition.* American Philosophical Society, Philadelphia, PA, 2002.

4.P   PROBLEM SET

4.1 Construct all the isomers of (a) C_4H_{10} and (b) C_6H_{14}.

4.2 Construct the single ring and cage isomers of $(CH)_4$ and $(CH)_8$. Note that the way they are written, these molecules are made only from CH fragments.

4.3 Construct from CH and C fragments the molecules $C_{10}H_8$ and $C_{14}H_{10}$, which contain only fused hexagons. Consult Scheme 4.11 in the text.

4.4 Use Scheme 4.10b and 4.10c to repeat the construction of the ring and the cage, and make from them new molecules by saturating the dangling connectivities with H fragments.

4.5 Make molecules containing two identical fragments using the three fragments shown in Scheme 4.13 on the left, as well as the fragment with an overall connectivity of three and one connection site on each carbon atom, which is implied in Scheme 4.13.

4.6 Make molecules containing two identical fragments using the first three fragments in Scheme 4.14.

4.7 Use the principle of conservation of the atom in a chemical reaction to balance the reaction of burning of C_nH_{2n} rings with oxygen.

4.8 Construct C_{60} from plastic models (this is a recommended bonus exercise).

CONSTRUCTING MOLECULAR WORLDS OF LIFE FROM LARGE LEGO FRAGMENTS

5.1 CONVERSATION ON CONTENTS OF LECTURE 5

Racheli: You barely touched most of the modular fragments. Quo vadis, Sason?

Sason: I see that our conversations made you fond of Latin …

Where am I leading? I am going to employ all those barely used modular fragments and construct molecules of life. This should connect to the message of Lecture 1.

Racheli: What message? You had so many …

Sason: The one about the unity of the chemical matter.

Racheli: I recall this presentation from the course I tutored in 2011. I was struck then by the story of the synthesis of an entire genome of a microorganism (*Mycoplasma genitalium*) by the Craig Venter team, who had deciphered previously the human and other genomes.[1] This synthesis is an assembling of a very large molecule, made from more than half a million DNA building blocks, much like the construction process you are trying to teach.

Sason: Mind you, this synthetic DNA produced self-replicating cells. This is a great achievement. It shows that the genetic machinery of a living organism can be made from "dead" pieces of matter. The synthetic DNA "software" builds its own *Mycoplasma genitalium* "hardware." This is exactly the message I just alluded to; we are made of chemical matter, the same matter as any other chemical matter. This

Chemistry as a Game of Molecular Construction: The Bond-Click Way, First Edition. Sason Shaik.
© 2016 John Wiley & Sons, Inc. Published 2016 by John Wiley & Sons, Inc.

has far-reaching implications for Mankind, and for the possibility of generating hybrid robotic systems.

Racheli: It is amazing, Sason! As we are conversing now, some scientists are busy making a whole cell from scratch.[2] But, are you sure all this is chemistry?

Sason: Of course Racheli! It is all about making more complex molecules from simpler ones.

Just think how far-reaching is chemistry. It started in the nineteenth century from the abstract concepts of atoms that no one could see. And, lo and behold, today, we can image atoms (see Figure 5.1), as well as the molecules we can construct from atoms, and the bonds which hold these molecules together.[3] Understanding bond making/breaking takes us as far as the synthesis of DNA "software" that generates its organism, and even the syntheses of whole cells. The future evolution is being shaped in front of our eyes without natural selection!

Racheli: This is scary!

Sason: But also full of opportunities … This and the next few lectures will show that Mankind is made from chemical matter, and He is Matter shaping new forms of matter.

Racheli: This is great preaching, Sason. But let's refocus: what *are you* going to teach precisely?

Individual surface atoms identified

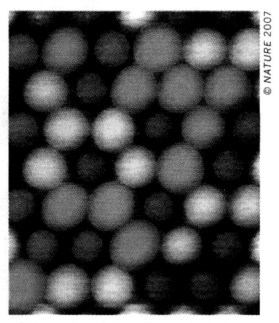

Surface atoms of alloy: silicon (red), tin (blue)
and lead (pale green)

FIGURE 5.1   Atoms on a surface imaged by atomic force microscopy. Adapted with permission from *Nature*, **2007**, *446*, 64.

Sason: I will start with families like alcohols, ethers, amines, acids, and esters, which can be made from the simpler modular fragments. Then, I intend to construct neurotransmitters, which control our behavior. Subsequently, I will mention fatty acids, cholesterol, cortisone, prostaglandins, triglycerides, etc.

Racheli: I am finding it hard to figure out how you would generate all the molecules of life using the *click* method and your small kit of modular fragments. What about proteins or peptides, which are made from amino acids?

Sason: Constructing these molecules will require a new element in our *click* procedure, which I call *clack-click*. *Clack* will allow me to break pieces from the amino acids, while with *click*, I will combine the so-formed pieces into peptides or proteins.

Racheli: Understanding these giant proteins is another great achievement of chemistry …

Sason: Which was recently acknowledged by the awarding of the 2013 Nobel Prize in Chemistry to Martin Karplus, Michael Levitt, and Arieh Warshel for developing computer-based methods that allow chemists to calculate the properties, reactions, and dynamics of proteins.

Racheli: There are zillions of protein molecules! Which ones will you teach?

Sason: Simple! Once the principle of construction is clear, I plan to show the students a few important proteins like oxytocin, hemoglobin, and the proteins that constitute the serotonin receptor, etc.

Racheli: But what is the point of teaching about molecules no one can draw, not even you?

Sason: Oh, Racheli, there is a point. Even if we cannot draw a specific protein on the board, because we do not know the sequence of amino acids, understanding its structural principles is essential. Here, the construction brings us understanding of the structural principles of proteins. The next revolution in science will be in proteomics.

Racheli: You did not define the term *proteome*. Let me do it for you. The proteome is the entire set of proteins that a given genome, for example, yours or mine, produces. These proteins constitute the fingerprints of our well-being and traits.

Sason: Proteomics already led to the development of immunotherapy for cancer. It turns out that the immune system does not recognize cancer cells as a threat, because these cells create proteins that mask them from the immune system. There are now synthetically made proteins that "unveil" the mask and cause thereby the immune system to destroy cancer cells.

So understanding the structural principles is hugely important.

Racheli: OK. You got me convinced. Then what about the "software" of life, the DNA and RNA molecules? Why not teach them in this lecture?

Sason: Not yet. I am going to need one more element in our game of construction, the hydrogen bond, which I teach in Lecture 7.

So, let's proceed with our lecture …

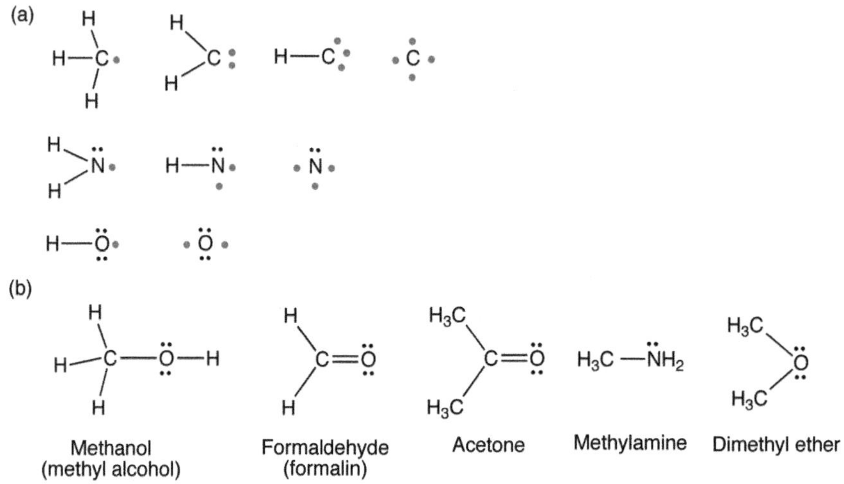

SCHEME 5.1 (a) Modular fragments made from CH_4, NH_3, and H_2O, along with their connectivities marked in red. (b) Simple molecules made from these fragments.

5.2 ALCOHOLS, ALDEHYDES, KETONES, ETHERS, AND AMINES

Scheme 5.1 assembles in part (a) the modular fragments which we generated before from CH_4, NH_3, and H_2O, along with their connectivities. In part (b), we show some molecules that we already made from these fragments in Lecture 3 (refer to Schemes 3.9 and 3.10), along with their names; methanol (or methyl alcohol), formaldehyde, acetone, and methylamine. A fifth molecule is made by clicking two single-connector CH_3 fragments with the double-connector O fragment to form the molecule called dimethyl ether. These molecules are representatives of the following molecular families: alcohols, aldehydes, ketones, amines, and ethers, which we shall now briefly describe. Many of these molecular families contain essential compounds produced by plants as a means of protection against infection damage, predators, DNA damage, inflammation, etc.

5.2.1 Alcohols

In Lecture 3, we constructed wood alcohol, methanol (Scheme 5.1b), which is the smallest member of and a representative of the family called alcohols. As shown in Scheme 5.2a, all the family members are characterized by an OH group bonded to a carbon, which in turn can be bonded to a variety of groups. The next simple alcohol in Scheme 5.2b is ethanol. Based on the fragments kit in Scheme 5.1a, ethanol is constructed by clicking together the single-connector $H_3C\bullet$ with the double-connector $H_2C:$ and the single-connector $HO\bullet$. Unlike methanol, which causes liver problems, blindness, and other ailments when drunk, ethanol, which is the alcohol used in wine and in spirits, is not at all harmful when drunk in moderate quantities. Of course, you cannot drink pure ethanol, but in the concentration it appears in wines (12–15%),

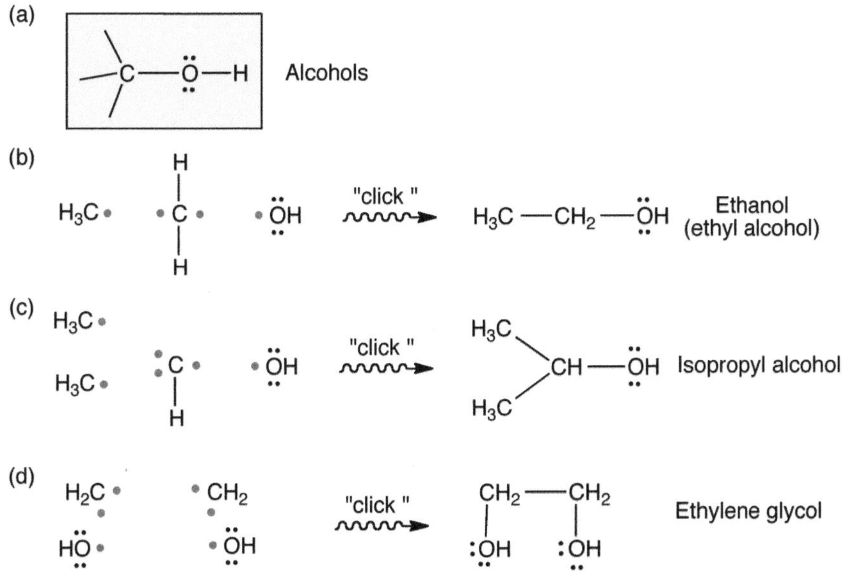

(a)

Alcohols

(b)

$H_3C \bullet$ $\bullet \overset{\displaystyle H}{\underset{\displaystyle H}{C}} \bullet$ $\bullet \ddot{O}H$ $\xrightarrow{\text{"click "}}$ $H_3C - CH_2 - \ddot{O}H$ Ethanol (ethyl alcohol)

(c)

$H_3C \bullet$ $\overset{\displaystyle }{\underset{\displaystyle H}{\ddot{C}}} \bullet$ $\bullet \ddot{O}H$ $\xrightarrow{\text{"click "}}$ $\overset{\displaystyle H_3C}{\underset{\displaystyle H_3C}{>}} CH - \ddot{O}H$ Isopropyl alcohol

$H_3C \bullet$

(d)

$H_2C \bullet$ $\bullet CH_2$ $\xrightarrow{\text{"click "}}$ $\overset{\displaystyle CH_2 - CH_2}{\underset{\displaystyle :\ddot{O}H \quad :\ddot{O}H}{|\qquad|}}$ Ethylene glycol

$H\ddot{O} \bullet$ $\bullet \ddot{O}H$

SCHEME 5.2   (a) The common structural motif of the family of alcohols. (b–d) Construction of a few simple alcohols using various fragments from Scheme 5.1a.

a few glasses cause most humans much pleasure. Drinking wine is also considered to be very healthy since some of the wine's ingredients have antioxidant properties. A difference of just one carbon and what a great difference does it make to human health! The human chemistry is most of the time hugely specific.

Schemes 5.2c and 5.2d show the construction of two other simple alcohols. One is isopropyl alcohol, which is used in cleaning liquids for television, computer, and cellphone screens, and also for massages. The second alcohol is ethylene glycol, which is used as an antifreeze agent. Many other alcohols, such as menthol, geraniol, and methyl salicylate, are produced by plants and cause muscle relaxation, anti-inflammatory effects, and so on. Please do not get intimidated by all the names. We note them so you could refer to them in case you need, but not every name must be memorized.

Scheme 5.3 shows two famous alcohol molecules. In part (a), we construct glycerol (also called glycerine) by clicking together two CH_2 fragments, a CH fragment, and three OH fragments. The Swedish chemist Scheele, whom we mentioned in previous lectures, discovered glycerol. It is oily and quite sweet, and it is therefore an ingredient of candies. Later, we shall discuss the role it plays in the human body.

The molecule in part (b) is cholesterol, which is drawn in two different manners. In the upper drawing, we show all the C's and H's, and one can easily identify the different modular fragments that constitute this molecule. In the lower drawing, we remove all atom labels, except for the OH group. This is the concise representation that organic chemists use to depict their molecules compactly without the "forest" of C's and H's. By consent, every chemist understands that what is drawn in Scheme 5.3b is the carbon skeleton of the molecule, wherein a terminal carbon represents CH_3, a

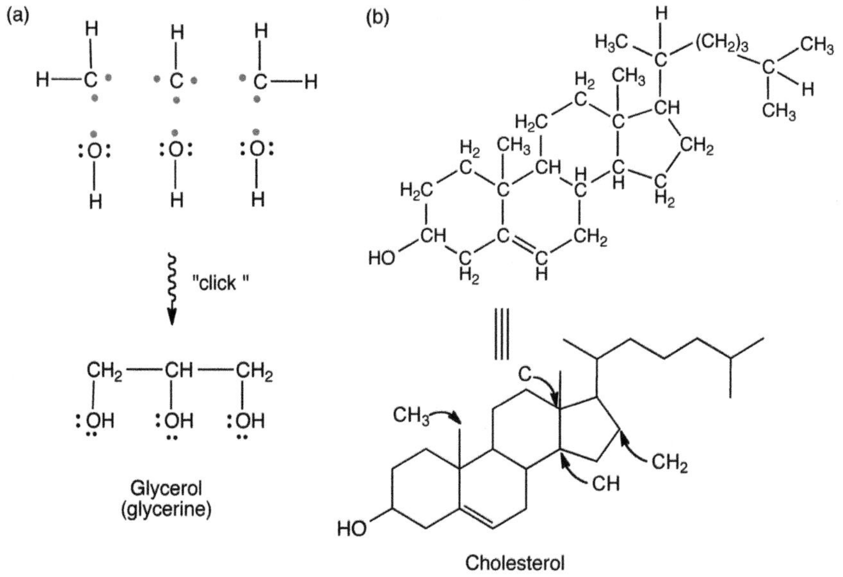

SCHEME 5.3   (a) Construction of glycerol (glycerine) from two CH_2 fragments, one CH fragment, and three OH fragments. (b) Cholesterol shown with an explicit formula (upper drawing) and shown with a schematic *implicit representation* (lower drawing).

carbon in the chain represents CH_2, and a carbon at the branching point is CH, and so on. It may be a good exercise to draw this skeleton and complete all the missing C and H groups. There are a few problems in the problem set (at the end of the chapter) that will encourage you to further practice this "translation" of the implicit drawing to the explicit molecular formula. Cholesterol is considered to be the "father of all evils," but later we will see that, in a manifestation of the Janus effect mentioned already, cholesterol is also good.

Of course, based on the above examples, in Schemes 5.2 and 5.3, it is easily seen that the number of possible alcohols is infinite. It is another molecular world, full of wonderful and less wonderful molecules.

5.2.2   Ethers

The structure of an ether is depicted in Scheme 5.4a. It contains an oxygen atom flanked by two carbon atoms that can be connected to a variety of groups. It looks like an alcohol in which the H in the OH group was replaced by a carbon fragment. This little difference makes a big difference. For example, alcohols are not so volatile, and they dissolve in water (later on we shall see the reason for that). By contrast, ethers are not water-soluble, and the smaller members are highly volatile compounds that have dizzying and anesthetic effects on our body. Ether anesthesia was used for the first time in 1841 when the American physician Crawford W. Long used it during

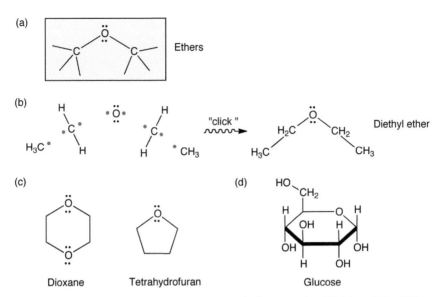

SCHEME 5.4   (a) The common structural motif of ethers. (b) Clicking CH_3, CH_2, and O fragments to form diethyl ether (once used for anesthesia). (c) Ethers that serve as solvents. (d) Glucose and its etheric ring. In parts (c) and (d), the molecules are described using the concise representation.

removal of tumors from the neck of one of his patients, who did not feel anything during the operation.

Scheme 5.4b shows the diethyl ether molecule, which can be constructed by clicking together two $H_3C\bullet$ fragments, two H_2C: fragments, and one $\bullet O\bullet$ fragment. The molecule has served for quite some time as an anesthetic during operations. But because of its highly volatile and inflammable nature, it had to be replaced eventually by halogenated ethers, wherein some of the hydrogen atoms in Scheme 5.4b are replaced by chlorine or other halogens to form less volatile and less inflammable anesthetics.

Less volatile ethers are used as solvents, and two of the widely used solvents are shown in Scheme 5.4c, where they are drawn in the implicit manner without explicit marking of C and H (see if you can complete the C's and H's). Larger ethers are used in perfumes and in cosmetics because they can dissolve the ingredients of cosmetics and additionally have pleasant smells.

As shown in Scheme 5.4d, the C–O–C structure of ethers also manifests itself in all the rings of sugars. For example, the six-membered ring glucose is an ether, but the OH substituents on the ring make it also an alcohol. Our cells use glucose as a source of energy, and our brain lives on glucose and gets addicted to it. As we shall show in one of the next lectures, The C–O–C unit also appears in the sugary components of DNA and RNA. Some plants like anise, basil, and tarragon generate ether molecules that are important essential oils.

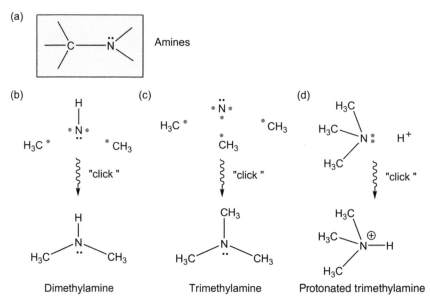

SCHEME 5.5 (a) The typical structural motif of amines. (b) Clicking two CH$_3$ fragments and one HN fragment to form dimethylamine. (c) Clicking three CH$_3$ fragments and one N fragment to generate trimethylamine. (d) The amine's lone pair of electrons on the nitrogen atom can form a bond with a proton (H$^+$): this ability to form a bond with a proton is called basicity.

5.2.3 Amines

Amines are kin to ammonia, and they are constituted from fragments of ammonia bonded to fragments of CH$_4$, as shown for the simplest member methylamine (Scheme 5.1b), which we formed by clicking the single-connector H$_3$C• and H$_2$N• fragments. The general structure of amines is depicted in Scheme 5.5a, containing a nitrogen and at least one carbon atom. Scheme 5.5b and 5.5c show the generation of two simple amines, which contain two and three H$_3$C• fragments that click with the double and triple connectors, HN: and N, respectively.

Like their parent, ammonia, amines are also smelly. Trimethylamine, as well as many other amines, has a fishy smell, not so pleasant! All amines are bases, and they are called so because the lone pair on the nitrogen enables these molecules to bind a proton, as shown for trimethylamine in Scheme 5.5c (showing this lone pair). Thus, since the proton has no electrons, it is electron-deficient, and it can utilize the lone pair of the amine to form a new N—H bond (Scheme 5.5d). Like *yin* and *yang*, the opposite of a base is an acid, and we shall show some acids in Lecture 8.

5.2.4 Biogenic Amines: Our Neurotransmitters

In Lecture 1, we mentioned neurotransmitters and their relationship to our psychological balance, addiction, love, and so on. All of them involve the typical C—N bond of amines. In this sense, neurotransmitters constitute a hugely important subfamily

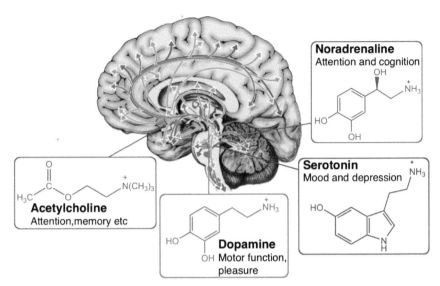

FIGURE 5.2   Some neurotransmitters that are generated in different areas of the brain; they serve as means of communication with the body. The depiction of the pathways in the brain is color coded and is adapted from Figure 4.4 in Neil V. Watson and S. Marc Breedlove, The Mind's Machine: Foundations of Brain and Behavior, Sinauer Associates, Inc., Sunderland, MA, 2012. Used with permission of Sinauer Associates to Wiley.

of amines, and they are called biogenic amines. Figure 5.2 shows part of the web of chemical communication between the brain and the body. These are neurotransmitters, which are generated in different areas of the brain and are transmitted to the body, causing a variety of sensations and exerting different kinds of control.

To describe these amines and construct them, we need to add some of the modular fragments we generated in Lecture 4 from the molecule benzene. This is repeated here in Scheme 5.6, which shows all the possible fragments. Since we are handling relatively large molecules, we have to use the implicit way of writing molecules. Thus, as shown in Scheme 5.6, we write benzene without its C and H atoms as simply a hexagon. Furthermore, in Scheme 5.6a we represent the whirling six electrons in the ring (see Retouches section 4.R.5 in Lecture 4) simply as a circle collectively binding the six carbons.

In the next step in Scheme 5.6b, we unclick C–H bonds, and we represent these sites with the electrons participating in connectivity in red. It is seen that there is one single connector C_6H_5 fragment, three different double connectors C_6H_4, and a few triple connectors, quadruple connectors, and so on. We can go on up to the sextuple-connector C_6, which we encountered in Lecture 4.

While each of these building blocks can generate a molecular world by combining with the various fragments we have in Schemes 5.1 and 5.6 (try generating some of these interesting molecules), we proceed here without much ado to the neurotransmitters. In Scheme 5.6c, we show how the molecule PEA, which is implicated in falling in love, can be constructed by clicking the single connector •C_6H_5 with two

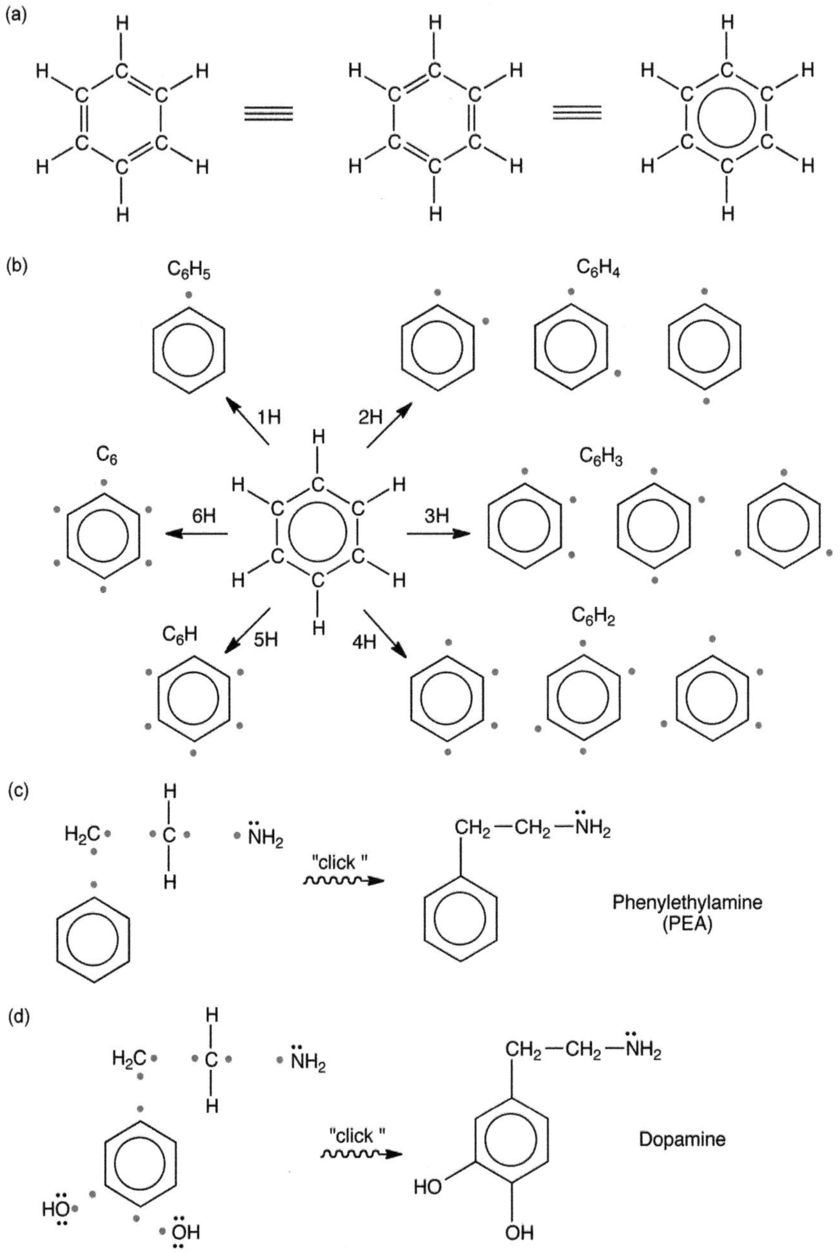

SCHEME 5.6 (a) Benzene and its concise mode of representation (far right). (b) Fragments of benzene obtained by the unclicking of different numbers of C—H bonds. The red circles indicate the connectivity of each fragment. (c) Construction of PEA. (d) Construction of dopamine.

double-connector $H_2C:$ fragments and a single-connector $\cdot NH_2$. Note that the tail $CH_2CH_2NH_2$ defines PEA as an amine.

In Scheme 5.6d, we show the construction of dopamine, which is implicated in addiction to pleasures, and as such it influences our ability to repeat actions that cause us pleasure: to smoke, drink coffee, do what we do best, and learn. Dopamine is also in charge of our fine motor movements. As shown in the scheme, dopamine requires a specific triple-connector C_6H_3 fragment, which clicks with the aminic chain components and with two single-connector $\cdot OH$ fragments.

We can, of course, continue to generate other neurotransmitters in a similar manner. But since by now the principle is clear, we simply show a collection of some of these biogenic amines, along with analogous molecules and drugs, in Figure 5.3. You can use these molecules to practice your command of the construction process.

The first molecule in the figure is serotonin, which was shown also in Lecture 1. Without actually constructing it, you probably see by now that it can be constructed from a triple connector benzene fragment linked to an OH fragment on one side and,

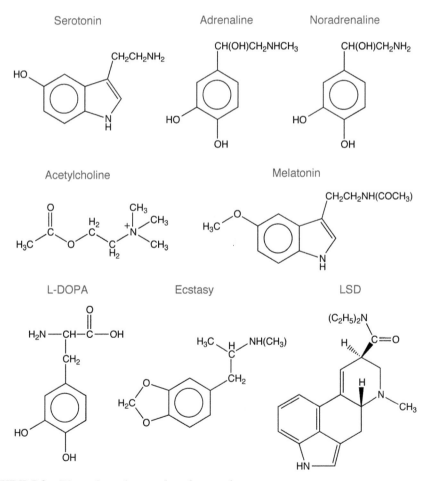

FIGURE 5.3 Biogenic amines and analogous drugs.

on the other, to a double-connecting fragment that makes a ring having an amine group in addition to the tail $CH_2CH_2NH_2$. As recounted in Lecture 1, and seen above in Figure 5.2, serotonin is formed in the brain and serves to control our psychological balance.

To the right of serotonin, we find the pair adrenaline and noradrenaline, which were mentioned in Lecture 1. These molecules can be reconstructed from a triple-connector benzene fragment, two OH fragments, and a single amino chain, which is slightly different in the two molecules. Adrenaline and noradrenaline (also called epinephrine and norepinephrine) form a pair that regulates the "fight or flight" instinct. Adrenaline is not a neurotransmitter. It is secreted by the adrenal gland, and it induces the synthesis of sugar, which is a source of energy needed in trying times and in stress. Noradrenaline focuses our attention/alertness and cognition in similar times and in times when we need to be efficient and competitive.

Acetylcholine is another neurotransmitter which is also involved in attention, memory, sexual arousal, etc. The structure of this neurotransmitter has only one part in common with the other biogenic amines depicted in this figure, and this is the amino component, $CH_2CH_2N(CH_3)_3{}^+$. Note that the terminal N sacrifices its lone pair and binds the electron-deficient fragment $CH_3{}^+$.

Further in Figure 5.3, we find melatonin, which has a structure similar to serotonin. In fact, serotonin is converted to melatonin, and the two amines control our sleep–wake cycle (thus forming an inner clock, called a circadian clock). The jet lag that most of us experience after long flights is a result of disruption of this cycle.

Finally, we show three drug molecules. The first is called L-DOPA, which has great structural similarity to dopamine. Indeed, L-DOPA is a medication given to Parkinson patients whose dopamine production is impaired. L-DOPA can reach the brain cells and be converted there to dopamine. The other two drugs are LSD and Ecstasy, which are psychedelic and hallucinogenic drugs that change our perception of reality.

We are clearly chemical creatures, motivated and controlled by molecules.

5.2.5   Aldehydes, Ketones, Acids, and Esters

These molecular families contain a C=O fragment linked to a variety of other fragments. As we have already constructed formaldehyde and acetone, we can use either one of them to generate all these families. Thus, as shown in Scheme 5.7a using formaldehyde, we can generate two new modular fragments by unclicking the C—H bonds. In Scheme 5.7b, we use these fragments to generate molecular families.

From the single connector HC=O fragment, we can generate the family called aldehydes by clicking it to a single-connector carbon-based fragment of any size, shape, or length. All aldehydes share the HC=O fragment.

Using the double connector C=O fragment, we can click it to two carbon-based single-connector fragments to form the family of ketones. Acetone is a ketone, and all ketones share the $(C)_2C=O$ unit.

If we use either one of the HC=O or C=O fragments in Scheme 5.7a and click to it one OH fragment, and in the case of the C=O fragment an additional

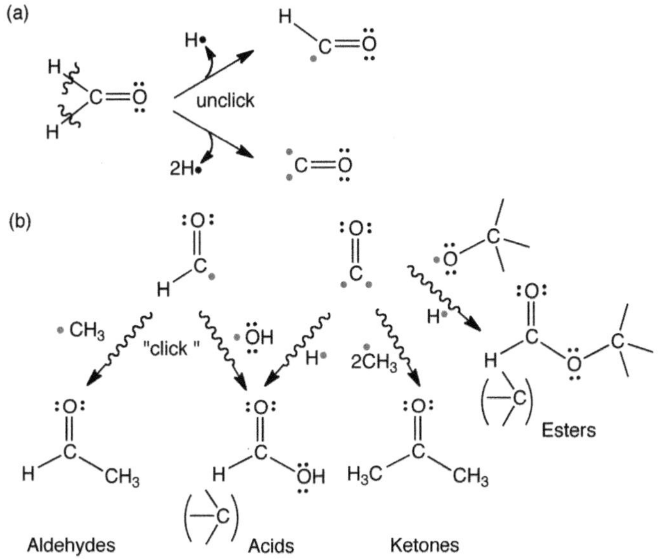

SCHEME 5.7   (a) Generating HC=O and C=O fragments from formaldehyde. (b) Usage of these fragments to construct the families of aldehydes, ketones, acids, and esters. Note that the CH$_3$ groups in the aldehyde and ketone and H's on the C=O in the acid and ester can be replaced with different carbon-containing groups, and the resulting species will also be in the corresponding family of molecules.

carbon-based single connector, we are going to obtain the family of *organic acids*, called carboxylic acids to indicate the COOH group. Finally, if instead of OH, we utilize an alcohol-based fragment, we are going to obtain the family of esters.

Figure 5.4 shows a few of the well-known members of these families. Since it is very easy to figure out the construction of these molecules from modular fragments, I skip these descriptions here, but you can do this exercise should you want to practice your construction skills.

Cinnamon and vanillin, in Figure 5.4a, are famous aldehydes that we use in desserts and cooking. Aldehydes appear also as essential oils in plants like melissa, lemongrass, and citronella, and they have antifungal, anti-inflammatory, and disinfecting effects. The raspberry ketone in Figure 5.4a is found in raspberries, as well as in other berries, where it gives the red color. It is used as a taste additive in foods. Ketones are generally toxic, but those present in hyssop, eucalyptus, and rosemary have therapeutic effects.

Figure 5.4b shows a few known organic acids. The simplest acid is called formic acid, or "ant's acid," since it is an ant's venom. The second molecule is acetic acid, or vinegar which finds daily use in our salads. Lactic acid is the acid that is formed in our muscles whenever we overexert them. It causes pain, and thereby it signals that the muscles need rest. Finally, butyric acid is a compound with the awful stench that forms when butter goes bad.

Acids are typified by their ability to release protons in water solutions. As we explained before in the section on amines, the released protons can be taken up by

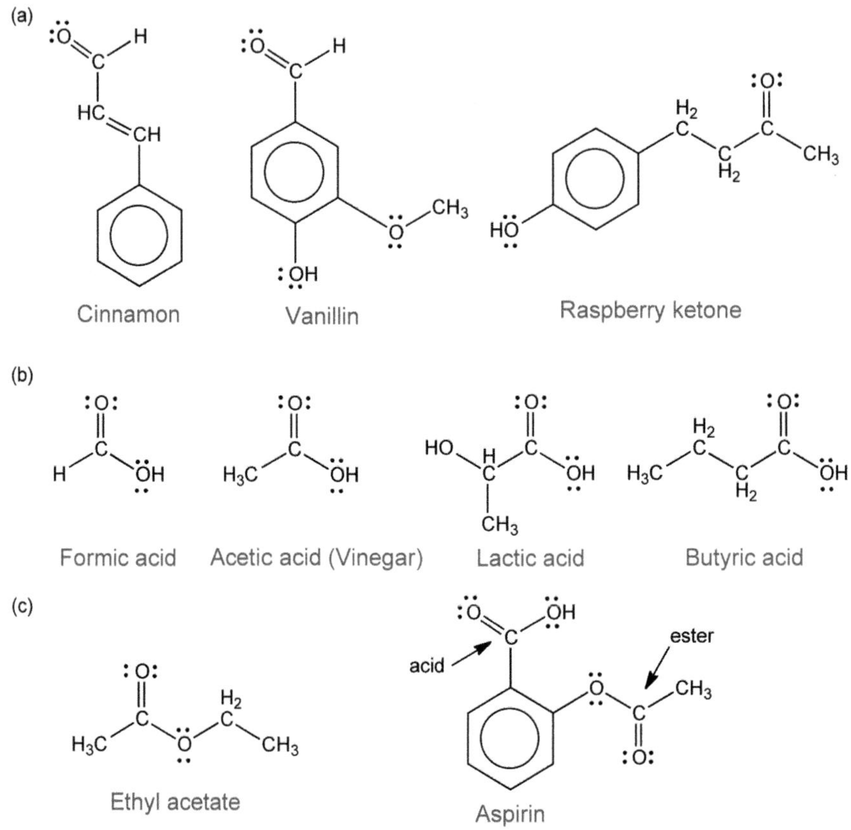

FIGURE 5.4 Some well-known molecules which are (a) aldehydes and ketones, (b) acids, and (c) esters.

bases, such as amines. Therefore, acids and bases are connected by proton transfer reactions (which you will see later). Chemists call these reactions neutralizations.

Figure 5.4c shows some esters, which are molecules that involve one part coming from an acid and another part coming from an alcohol. Since you will find this question in the problem set at the end of the chapter, you may want to think how an acid and an alcohol react to produce an ester. Esters are fragrance molecules that are found in fruits and leaves. They serve as solvents and lacquers. Ethyl acetate, in Figure 5.4c, is such a nicely smelling solvent. Near it, we see a very famous molecule, Aspirin. It can be seen that part of it is an ester, and part of it is an acid.

5.2.6 Fats (Lipids): Fatty Acids, Prostaglandins, Triglycerides, Cholesterol, Cortisone, etc.

Modern humans are concerned with fat consumption. To reduce this consumption, people would go a long way to lose/burn a little fat, would be very careful with their triglycerides and cholesterol, and would take "good" ω-fatty acids in their diets. On the other hand, fat is essential. Fat molecules are structural components of cell

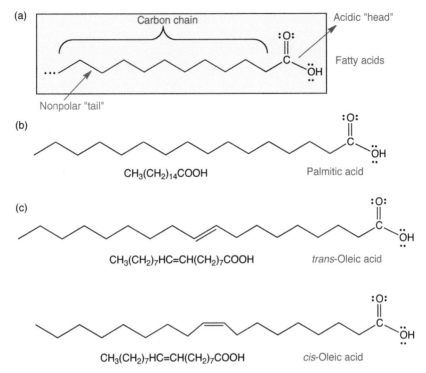

SCHEME 5.8   Fatty acids: (a) the general structural motif. (b) Saturated fatty acids, represented by palmitic acid. (c) *Trans* and *cis* unsaturated fatty acids, represented by oleic acid.

membranes. They store energy and participate in signaling in the biosystem (e.g., our bodies). Thus, each cell in the human body contains fat, and the human chemical laboratory produces some of the necessary fat and mines the rest from foodstuffs. The general name for the different fats is lipids, a word of Greek origin (*lipos* = fat). Lipids are generally oily or waxy. Let us see some of these molecules of fat/lipid.

Scheme 5.8a shows the general structure of fatty acids. The molecule has the acidic "head" COOH and a rather long "tail" made from a chain of carbon atoms with hydrogen atoms attached to them. The head dissolves in water, but the tail is oily, and it determines the properties of the acid.

Scheme 5.8b depicts a saturated fatty acid, called palmitic acid, which is represented in the implicit cartoon with a skeleton without explicit notation of the C and H atoms. The term "saturated" means that there are no double bonds in the chain, and as can be seen beneath the structure, the chain is constructed from a click of one single-connector $H_3C\cdot$ to 14 double-connector $H_2C:$ fragments. The dangling connectivity of the last CH_2 is in turn clicked to a single-connector COOH fragment. Palmitic acid is the most common fatty acid found in mammals, plants, and microorganisms. Excess carbohydrates in the body are converted to palmitic acid, and as a result, it is a major body component. As we drew it, the chain of palmitic acid is extended in shape, and therefore the many molecules of the acid can pack nicely together and form solids, which can clog our arteries and veins. This is why we are repeatedly being warned of saturated fat.

Scheme 5.8c shows two different forms of oleic acid. They have the same atomic constitution, but they differ in the arrangements of the $(CH_2)_7$ chains about the double bond. We shall learn later about the geometry and three-dimensional (3D) shape of molecules, but in the meantime, let us accept that the two acids are two isomers that differ in their geometry about the double bond, and they are called *cis-* and *trans*-oleic acids. The molecule in the upper component of Scheme 5.8c is called *trans*-oleic acid, and because of the *trans* geometry about the double bond, the entire $CH_3(CH_2)_7HC=CH(CH_2)_7$ tail is extended like in the saturated acid. As such, molecules of *trans*-oleic acid pack very well and easily form solid oil, which will precipitate and clog our veins and arteries. This is why we are constantly being warned not to consume foods that include *trans*-fatty acids. Margarine, for example, contains *trans*-fatty acids, and it is now considered to be hazardous to our health (when I was growing up, it was considered to be healthier than butter, but not anymore). On the other hand, *cis*-oleic acid has a kink in its tail, and it will not pack well with other molecules, and hence it will remain in a liquid state. Therefore, it will not precipitate in our veins. *Cis*-oleic acid is the natural ingredient of olive oil. This is why we are being told that olive oil is good for us.

Figure 5.5 shows a few saturated, *trans*, and *cis* fatty acid using molecular models. Note the extended saturated versus kinked *cis* chains. This is the major factor in the solidification of the former acids and the resistance of the latter to solidify. And this is the major difference in their impact on our health.

There are two *cis* fatty acids that are known by the name essential fatty acids (EFA), as they are essential to humans. Since humans cannot synthesize these EFAs from

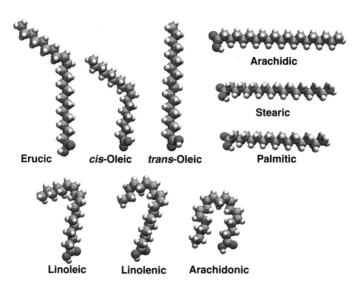

FIGURE 5.5   Some *cis*, *trans*, and saturated fatty acids, along with their names. Note the kinked *cis* molecules vis-à-vis the extended, almost straight saturated molecules (*trans* fatty acids also have straight chains).

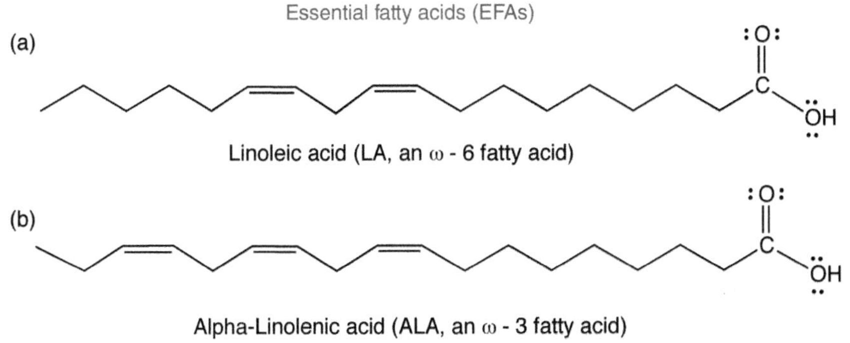

Essential fatty acids (EFAs)

(a)

Linoleic acid (LA, an ω - 6 fatty acid)

(b)

Alpha-Linolenic acid (ALA, an ω - 3 fatty acid)

SCHEME 5.9   Essential fatty acids (EFAs): (a) Linoleic acid (LA), and (b) alpha-linolenic acid (ALA). The number following the letter omega (ω) indicates the position of the double bond from the tail end of the molecule, where ω is the carbon atom at the tail.

other molecules, our diet must include these acids. They are shown in Scheme 5.9. These acids are also members of groups known, respectively, as ω-6 and ω-3 fatty acids, and we are constantly reminded that they are good for us.

These acids are called EFAs because they are essential for good health and for key processes in the body (see later). EFAs are essential for the lifetime of cardiac cells (and deficiency is responsible for death of cardiac cells), for prevention of illnesses such as dermatitis, for treatment of depression, etc.

Fatty acids are not found in Nature in a free state. Commonly, they exist as esters in combination with the alcohol glycerol (recall Scheme 5.3a). These esters are known by the name "triglycerides." An example of a triglyceride is shown in Scheme 5.10, where three different fatty acids form ester chains with the three alcohol groups of glycerol. The structure gives you another opportunity to practice the writing of the complete molecular formulae by completing the missing C and H atoms. Another

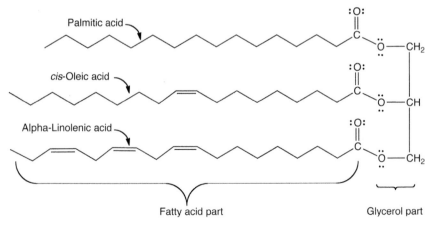

SCHEME 5.10   An example of a triglyceride made from glycerol and three fatty acids: palmitic, oleic, and alpha-linolenic acids.

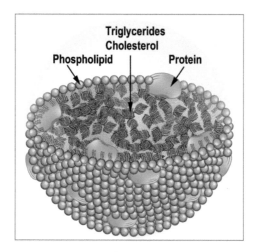

FIGURE 5.6 A schematic representation of a lipoprotein sphere (sliced in the middle) and its outer covering and contents of proteins and lipids (triglycerides, cholesterol, and phospholipids). Lipid (fat) is a generic name for materials that are oily and/or waxy. Phospholipids constitute triglycerides containing two fatty acid parts and a third component, which comes from an acid called phosphoric acid—to be taught later. The drawing was made by the artist Z. Cohen and copyrighted by the author.

interesting issue to think about is how to conceptualize, in terms of the *click* method, the reaction by which an alcohol and an acid form an ester.

Triglycerides are fatty acid transporters to sites where the fatty acid is needed as a source of energy and where enzymes break the ester bond and release the fatty acid, which in turn is carried into cells by protein transporters that enable the transportation of fat within the water around the cells and in the bloodstream. High levels of triglycerides generally cause oily hair, dandruff, eczema, hair loss, etc. In the intestines, triglycerides get mixed with cholesterol (see Scheme 5.3b) and proteins, forming spheres of different sizes called lipoproteins (*lipo* = fat), which are shown in Figure 5.6.

Subsequently, these lipoproteins are transported to the large blood vessels near the heart before mixing into the bloodstream. The ratio of protein to fat determines the quality of these lipoproteins. Lipoproteins with a high ratio of proteins to lipids are called high-density lipoproteins (HDL), while those with a low ratio of proteins to lipids are called low-density lipoproteins (LDL). HDL particles are small spheres that carry cholesterol from the cells to the liver (where cholesterol is broken down and excreted as waste), and they act also as scavengers (collectors) of cholesterol, and hence, they are not involved in the clogging of our vessels. Hence, HDL is called "good cholesterol." On the other hand, LDL carries cholesterol from the liver to the cells and is called "bad cholesterol." The LDL spheres, which are larger than the HDL spheres, can fall apart more easily and release the fat, which is subject to oxidation damage by O_2. This process causes the immune systems to send white plasma cells to destroy the damaged molecules, and the entire mixture (oxidized lipoprotein plus white blood cells) precipitates in the blood vessels and clogs them, in a situation which is called by the professional term *atherosclerosis.*

This may be the right place to say something good about cholesterol. Cholesterol is called a lipid because it is waxy/oily, but as you can recall (Scheme 5.3b), it is also an alcohol in terms of its molecular family. The body makes copious amounts of cholesterol, because it builds and maintains cell membranes, and it determines the cells permeability, namely which molecules would/would not be able to enter the cell. It is also a precursor of all the sex hormones, such as progesterone and testosterone, as well as of other hormones produced by the adrenal glands (like cortisol). Cholesterol insulates nerve fibers, and it protects the brain (against Alzheimer's disease). It is the raw material for making vitamin D, which is essential to the heart function. Cholesterol is converted to bile acids, which are necessary for neutralizing poisons and digesting fats in the bile. Thus, it is very important, as much as it may cause damage (atherosclerosis). It is a perfect Janusian molecule. The property that makes it useful in cell membranes (insolubility in water) makes it lethal elsewhere (e.g., in the bloodstream).

In the space left for this subsection, I would like to say a few words about another group of fatty acids, prostaglandins. Prostaglandins are generated from the combination of two molecules of fatty acids. For example, prostaglandin E1, which is shown in Scheme 5.11 (practice adding all the missing C and H atoms), is made from two molecules of linoleic acid (see Scheme 5.9a) by the enzyme cyclooxygenase (COX), which specializes in catalyzing this reaction. Later, when we discuss proteins, we will also talk about enzymes.

Prostaglandins function like neurotransmitters and are generated locally near the target cells. They regulate the following functions: blood pressure, action of smooth muscles (e.g., the heart's muscle is a smooth muscle, which we cannot control, unlike, e.g., our arm's muscles), formation of blood vessels, clotting of red blood cells, excretion of stomach liquids, regulation of body temperature, anti-inflammatory responses, hunger, labor contractions, miscarriage, high fever, and feeling of pain.

Prostaglandin E1 is a key member of the family. When we have an excess of it, we feel pain and inflammation, and we may have a high fever. When we take aspirin (see Figure 5.4c), it prevents the feeling of pain by deactivating the enzyme COX, and thereby, it reduces the amount of prostaglandin E1. Since prostaglandin E1 regulates blood clotting, taking aspirin also causes blood thinning.

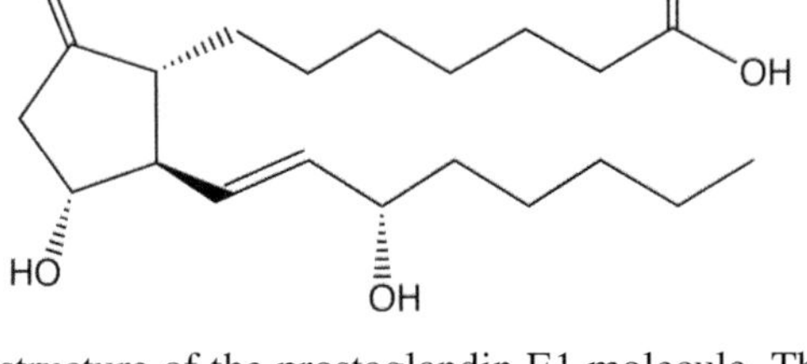

SCHEME 5.11   The structure of the prostaglandin E1 molecule. The bonds depicted with wedges and dashes carry 3D information, to be deciphered in Lecture 7.

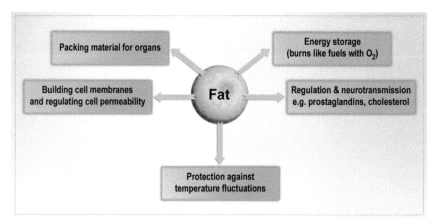

FIGURE 5.7   Roles and functions of lipids (fat) in biological systems. (Designed by Z. Cohen and copyrighted to the author)

Summary of the Roles of Lipids: Even though we did not show all the possible variations of lipids, we still discussed the topic of fat quite a bit. It is time now to provide an overview of the role of fat (lipids). Figure 5.7 summarizes the various roles of fats (lipids). Some fats build cell membranes and regulate the permeability of the cell. Generally, fat molecules serve as "packing material" for the inner organs of the body (imagine having to jump up and down without the packing of your heart, liver, etc.). Other general duties of fat are guarding against temperature fluctuations (our body cannot live long above 42°C and below 32–33°C) and functioning as a source of energy (our fats "burn" with O_2 sources, and our body uses enzymes to do that without catching fire). Some lipids like cholesterol and prostaglandins act as regulatory molecules and local neurotransmitters.

5.2.7   Amino Acids, Peptides, Proteins, and Enzymes

The human body contains quite a few organs that are packed by layers of fat. All organs are constructed from cells, which contain 70% water and a few organelles that constitute the chemical "laboratory" of the cell.

What gives shape and appearance to all these organs and to the body as a whole are molecules called proteins. Our skin, muscles, ligaments, nails, and hair are all made of proteins. This is the "hardware" of the human being. We all grow up by producing more proteins to supply to these organs. In a nutshell, these proteins are manufactured in the chemical "laboratory" of the cell, based on a master plan that is encoded in our genes, which are pieces of the molecule called DNA—the "software" for our body. The "heart" of the cell is the nucleus where this "software" resides and is kept safe. Copied pieces of this software are sent to an organelle called the ribosome, wherein these pieces are "read" and used to construct our proteins from molecules called amino acids, which we get by eating meat, legumes, eggs, and other kinds of foods that contain proteins.

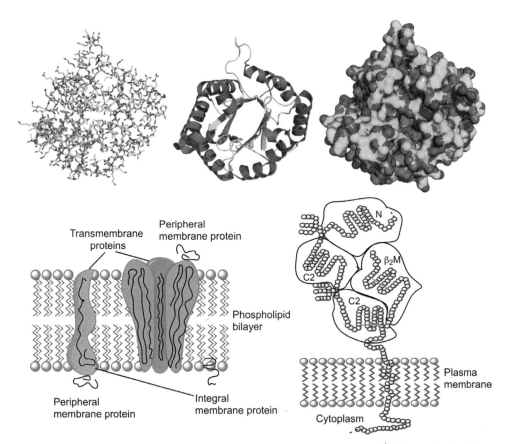

FIGURE 5.8 Some alternative ways of representing proteins and schematic drawings depicting different kinds of proteins in relation to membranes.

Proteins are large molecules, which may contain thousands of atoms (e.g., 4,000–5,000). It is therefore impossible anymore to write down all the atoms or even to represent the molecule by the implicit convention we use for C/H chains (see e.g., Scheme 5.11). We need more compact representations. Figure 5.8 shows some of these representations. Starting from the drawing in the top left, we see a mesh of spaghetti lines and ring structures. These lines represent the bonds between atoms, as we did for the C/H chains. Here, however, we have C, H, N, and O atoms, which are color coded; for example, all the oxygen atoms are in red. In the middle drawing, we see helices and sheets, which represent the general architecture of the respective pieces of the protein on the far right, we see a drawing where atoms are represented as color-coded spheres (N in blue, O in red, and C in green). Finally, on the bottom left, we see schematic wiggly lines, and on the bottom right, the proteins are represented as beads on strings. These are common drawings you will meet when you search for proteins on the Web or in books. Looking at the complexity of these drawings, it is clear that we are not going to request you to memorize such objects.

The amino acids are the building blocks of proteins and their smaller kin, the peptides. Scheme 5.12a shows how an amino acid is constructed from recognizable

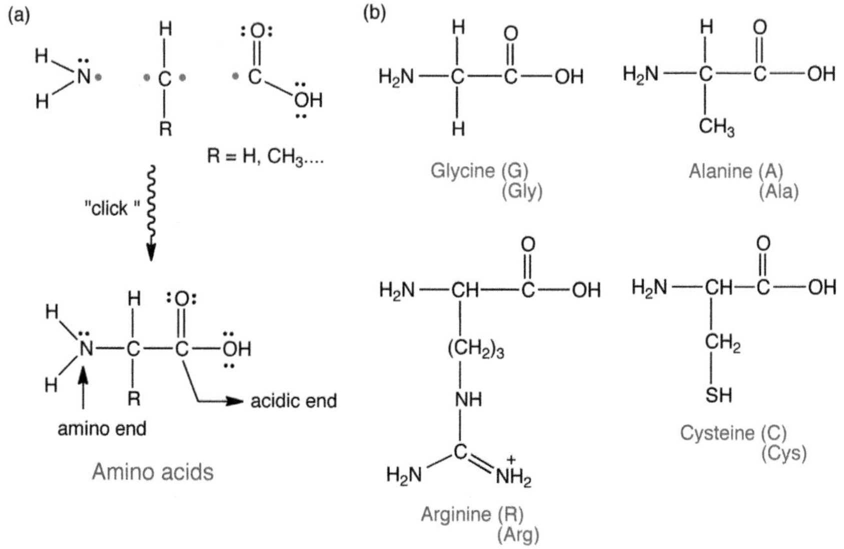

SCHEME 5.12 (a) Clicking three fragments to form an amino acid. R is a variable fragment. (b) Four out of the 20 natural amino acids. The full names of the amino acids are shown below the structures along with the shorthand notations in parentheses.

building blocks: the single-connector $H_2N\bullet$, the double-connector RHC: (where R is a variable single-connector group to be specified later), and the single-connector HO–C=O. Click, and the three fragments combine to form an amino acid molecule, which has an amino end (NH_2) and an acidic end (CO_2H) flanking the central CHR fragment.

The identity of the amino acid depends on the variable group R. In principle, R can be any single-connector group. However, as if in an act of magic, there are only 20 natural amino acids (NAAs), which appear in all forms of life and vegetation. All these NAAs are collected in Appendix 5.A.1, while Scheme 5.12b shows four of the set. It is seen that alongside the full names, there are shorthand notations. Thus, the simplest NAA is glycine, which can be written in shorthand notation either with three letters, as *Gly*, or with one, as *G*. The same convention is true for alanine, arginine, and cysteine, and for all other 16 NAAs, which are placed in the Appendix. As we shall see in a moment, these shorthand notations are used in expressing the molecular formula of proteins and peptides.

Proteins and/or peptides contain many NAA units. So, let us now construct such molecules from NAAs, using a new trick I promised to teach you, which I call the *clack-click* method in Figure 5.9. Here I lined up a few NAAs, with different R groups, all labeled as R, though they may be distinct. It is seen that we can combine these NAAs if we rip off a molecule of water from every successive pair of NAAs. Thus, we start by removing a single-connector OH from the acidic head of the first molecule and a single-connector H from the amino tail of the second molecule. We refer to this action as *clack*, that is, the sound of breaking a bond. After the *clack*, we are left with two single-connector NAA fragments, which *click* to form a new C–N

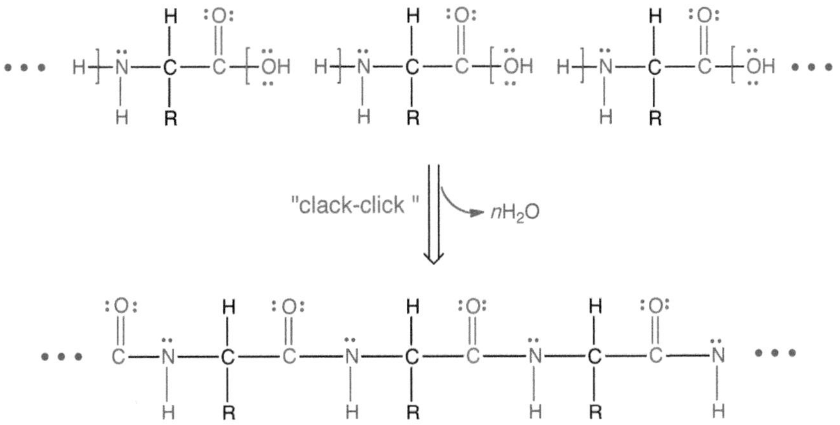

FIGURE 5.9   The *clack-click* method for constructing peptides and proteins. Note that for each *clack* action, we remove a molecule of water from a pair of successive NAAs. For n successive NAA pairs, we remove n H_2O molecules.

bond between these fragments. If we *clack-click* all along the train of molecules, we shall form a polymer made from NAA units (think about formation of esters with the *clack-click* method).

The difference between peptides and proteins is the number of NAAs in the polymer. When the number is small, we call the molecule a peptide, and when the number is large, it will be a protein. Figure 5.10 shows the peptide oxytocin written

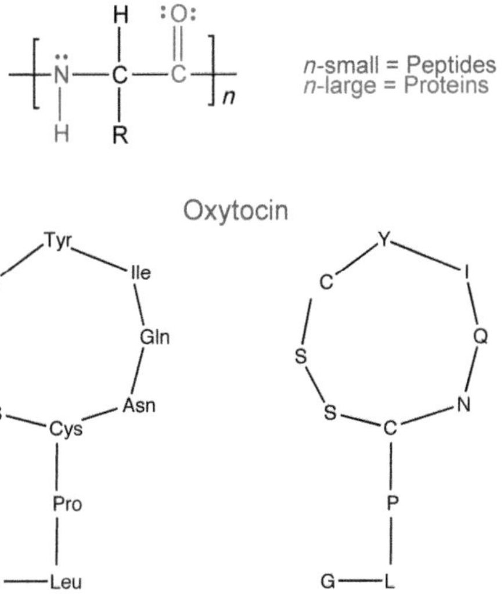

FIGURE 5.10   Peptides and proteins classified by the number (n) of NAAs in the chain or cycle of amino acids. Shown below is Oxytocin, written using two modes of shorthand notation for the NAAs, with three-letter names (on the left) and with one-letter names (on the right). The S—S in both drawings refers to a sulfur—sulfur bond between the tails of two cysteines.

in the two abbreviated formats. On the left-hand side, each NAA is represented by the three letters, and on the right-hand side, by single letters. Try to imagine how much writing would it take if you just had to write this relatively small peptide using all its atomic content!

You can recall that oxytocin is a neurotransmitter that is made in the brain in response to a hug, caress, baby nursing, and so on and so forth (go back to Lecture 1). It contains nine NAAs, and it is not a straight-chain peptide, but rather it forms a cyclic structure due to an S–S bond that is made between two cysteine (Cys/C) residues (see Scheme 5.12b). It is said that S–S bonds in our hair proteins cause curly hair.

Peptides and proteins are very important. Peptides serve as messenger molecules; for example, in the brain, they are the "soldiers" of our immune system, and unfortunately for us, also the "soldiers" sent by viruses. Proteins are the key hardware of our body, giving it shape, color, texture, and protection. They hold our skeleton erect, and they make our bones and teeth strong and flexible by interpenetrating into the material of the bones and teeth and they act as adhesives. Without the protein layering, our bone and teeth material will be brittle like pieces of chalk. An important protein in this respect is collagen, which involves three protein chains that are intertwined (by forces we shall discuss later, which are called hydrogen bonds) and form a flexible and mighty strong string. Collagen constitutes 30% of all bodily proteins. It is the major constituent of ligaments and cartilage, the inner linings of our blood vessels and our intestines, our muscles, etc. It is used in the food industry, wherein the protein is decomposed by boiling with water (opposite of *clack-click*) to give jelly. Some do not like even the sight of this quivering material, but actually it is rather tasty!

Because of their lengths, proteins tend to coil, and in some cases, the coiling forms ball-shaped objects which are called globular (from the word *globus*) proteins. Many of the globular proteins are enzymes that specialize in a certain reaction type and enhance the rate by which this reaction occurs in the cell (see catalysts and enzymes in the Retouches section of Lecture 2). Other globular proteins, called globins, hide within the protein a molecule that can carry oxygen and releases it into the blood.

Figure 5.11 shows a globular enzyme that belongs to a family called cytochrome P450. It is labeled as 3A4, a label that traces the gene which produces this enzyme, but which does not concern us. All P450s hide inside the protein an active molecule called *heme*, which is made from a ring with four nitrogen atoms pointing inside and binding a positive ion of iron, Fe^{3+}, as shown at the bottom of Figure 5.11. There is an additional arm, which is a sulfur (S) atom of the cysteine amino acid that holds Fe^{3+} (see bottom drawings of the figure). All the P450s have this same molecule inside their proteins, but their proteins are different, and hence they are different molecules!

The heme molecule of P450 takes up O_2 from the air, breaks the O=O bond, and creates a very reactive Fe=O bond, which can insert the oxygen atom into almost any imaginable molecule. In this manner, P450 3A4 performs metabolism of foods and drugs taken orally; for example, the green object above the heme in Figure 5.11 is a testosterone molecule (the male sex hormone) that will be metabolized in a minute by the active molecule. In the same manner, the enzyme protects the cells from poisons, since once the oxygen is inserted into a molecule of poison, this becomes an alcohol, which is soluble in water and is excreted in our urine. Other P450s

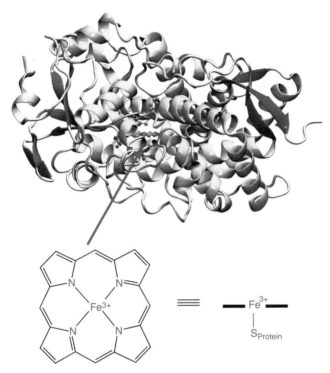

FIGURE 5.11   The human cytochrome P450 (3A4) enzyme, which is responsible for the metabolism of more than 50% of foods and drugs taken orally. An arrow pointing to the greenish object inside the enzyme indicates the heme molecule, which functions as the active molecule of P450. Below the arrow, we show an explicit molecular structure of the heme ring plane with Fe^{3+} in its center. On the right-hand side, we show a simplified representation of the heme, where two bold lines flanking Fe^{3+} represent the ring structure of the heme. Beneath the heme plane and bonded to Fe^{3+} is a sulfur atom of a cysteine amino acid, which is a part of the protein.

synthesize neurotransmitters in our brains, and still other P450s create molecules that have anticancer activity, and so on.[4]

Figures 5.12a shows another globular protein called myoglobin. We can look and marvel at the complexity of this molecular machine that bathes in the water molecules encircling the protein. The active molecule here is again a heme, and the only difference in the heme region with P450 is that the fifth arm that holds the iron ion (now Fe^{2+}) is not a sulfur atom of cysteine like in P450, but rather a nitrogen atom of the histidine amino acid. This difference and other differences in the respective proteins create together a completely different enzyme. When this enzyme takes up an O_2 molecule, it does not cleave its OO bond, which remains intact. Thus, myoglobin carries the O_2 molecule and releases it to the muscle cells, which need it for energy consumption.

Myoglobin is the simpler version of hemoglobin. Hemoglobin is the protein that takes up O_2, releases it to the blood, and thereby enables us to breathe and be alive. Hemoglobin, in Figure 5.13a, is seen to involve four entwined proteins that each

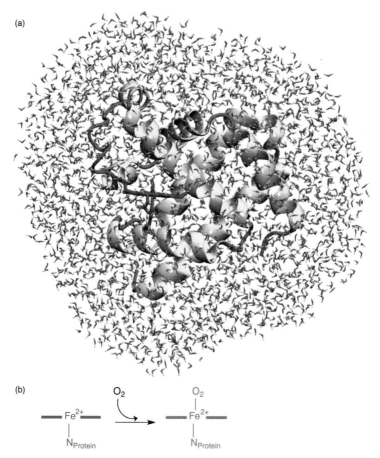

FIGURE 5.12 (a) Myoglobin bathing in water (the water molecules are the tiny specks surrounding the yellow protein coils). (b) The heme form (in the shorthand notation) with the five N arms that hold the Fe^{2+} ion is depicted in blue. When O_2 binds to Fe^{2+}, the color of the heme changes to red.

have one molecule of heme (in blue) inside the protein. The presence of four heme molecules maximizes the O_2 uptake. As shown in Figure 5.13b, the heme molecule which is not bound to O_2 has a blue color, and after the O_2 binds to the iron, the color changes to red. These are the true colors of the venous and arterial blood, respectively. Blue blood used to be a mark of nobility, but it is now known that due to interfamilial marriages, the nobles had poor hemoglobin that underwent mutation in the protein, such that the Fe^{2+} ion could not take O_2 as efficiently as the unmutated protein, and hence the blood looked blue. The blood disease sickle cell anemia is also caused by a mutated hemoglobin protein.

To some extent, hemoglobin was known for centuries. In 1668, the English physician John Mayow in Figure 5.14 described for the first time that the venous blood, which is blue, is converted to red arterial blood upon adding air or saltpeter (the compound KNO_3, which releases O_2 upon slight heating). Mayow did not

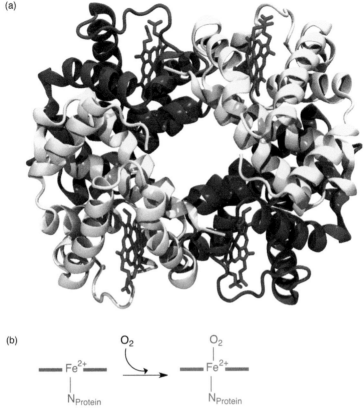

FIGURE 5.13   (a) Hemoglobin. The blue molecular structures inside the protein are four heme molecules. (b) The colors of heme before O_2 uptake (blue) and after O_2 uptake (red).

describe the situation in the manner I just did, because he was not aware of O_2, which was discovered only 100 years or so later by Priestley, Scheele, and Lavoisier (see Retouches section 5.R.2).[5]

In essence, Mayow unknowingly discovered the hemoglobin and oxy-hemoglobin forms drawn in Figure 5.13b for this protein. The blue and red colors of these complexes indicate how these molecules interact with visible light, which is the range wherein our eyes are also sensitive (Lecture 1). We shall talk about the interaction of molecules with light in Lecture 9.

Another protein type is one that makes receptors. We mentioned the word "receptor" in Lecture 1, when we discussed the action of serotonin on our psychological balance. The receptors are made from bundles of a few protein helices that are stuck in the membrane of the cell and sense molecules coming from the outside while activating inside signaling pathways that are translated, for example, to seeing, emotions, etc. In the case of serotonin, a stream of ions inside the neurons is the signal that brings about psychological balance. Figure 5.15 shows the receptor of serotonin in part (a) along with the serotonin binding (entrance) site, and in (b), the figure shows the result of serotonin entrance. It is seen that there occurs an opening of a

FIGURE 5.14 An engraving of John Mayow. This image is in the public domain, and it was obtained from http://commons.wikimedia.org/wiki/File:John_Mayow.jpg.

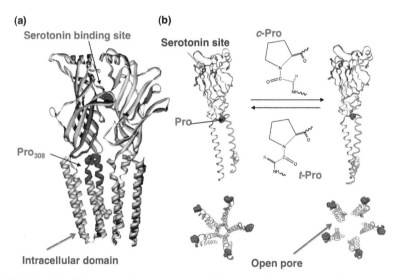

FIGURE 5.15 (a) The serotonin receptor and the site where serotonin enters and binds. (b) The opening of the pore involves a change in proline conformation and accompanies serotonin's entrance. Constructed and modified from Figures 1 and 5 of *Nature*, **2005**, *438*, 248.

pore between the protein helices, in the intracellular domain, such that ions can flow through and carry the message: "Grant us psychological balance!"

Among the mechanisms that were proposed for the effect of serotonin, one is particularly clear and was suggested by Dennis Dougherty, a chemist from Caltech.[6] It is depicted in Figure 5.15b above and below the arrows between the open and the closed forms of the receptor. The gist of this proposal is that the entrance of serotonin causes a proline (Pro; see Appendix 5.A.1) residue to undergo rotation around its N−C=O bond, and this motion brings about the opening of the pore. Thus, the combination of molecular interaction (serotonin enters the receptor) and the changes in the molecular architecture (the rotation in Pro) are responsible for our psychological wellness. That our psychological balance depends on a tiny molecular motion must be as mind-boggling to all of you as it is to me. Please go back to Lecture 1 to see that this is only one of many similar stories about the molecules of life.

5.3 SUMMARY

We can, of course, tell more fascinating stories about molecules of life, but we are not going to do so because the message is beginning to be clear by now, and it is the following:

☎ Mankind is made from chemical matter, which is identical to any other matter.

☎ Chemistry mediates the cognition that "living matter" has, namely the knowledge that "it" is made of matter.

☎ Hence, Chemistry is the window that Mankind uses to probe his own material being and come to terms with its limitations.

☎ Chemical matter involves infinite chemical worlds, since each and every molecule forms a basis for new modular fragments that create new chemical worlds. Chemistry is poetry in matter.

5.4 REFERENCES

[1] (a) For a highlight, see: U. T. Bornscheuer, *Angew. Chem. Int. Ed.* **2010**, *49*, 5228; (b) The synthesis is described in: D. G. Gibson, G. A. Benders, C. Andrews-Pfannkoch, E. A. Denisova, H. Baden-Tillson, J. Zaveri, T. B. Stockwell, A. Brownley, D. W. Thomas, M. A. Algire, C. Merryman, L. Young, V. N. Noskov, J. I. Glass, J. C. Venter, C. A. Hutchinson III, and H. O. Smith, *Science*, **2008**, *319*, 1215; **2010**, *329*, 52.

[2] (a) R. F. Service, *Science*, **2013**, *342*, 1032; (b) K. Amada and J. W. Szostak, *Science*, **2013**, *342*, 1098.

[3] J. Zhang, P. Chen, B. Yuan, W. Ji, Z. Cheng, X. Qiu, *Science*, **2013**, *342*, 611.

[4] http://en.wikipedia.org/wiki/Cytochrome_P450

[5] See the beautiful play, which retells the story of the discovery of O_2 by Priestley, Lavoisier, and Scheele: C. Djeracy and R. Hoffmann, "OXYGEN. A play in two acts," Wiley-VCH, Weinheim, Gemany, 2001.

[6] See the award announcement for this work in, *Chemical & Engineering News,* 2008, January 7, p. 27. See also, *Chemical & Engineering News,* 2006, October 2, p. 44.

5.A   APPENDIX

If time can be allocated, it is advisable to tutor this lecture and some material from previous lectures to tie up loose ends. The problem sets may be useful to these ends.

Next, we show the natural amino acids (NAAs).

5.A.1   The Natural Amino Acids (NAAs)

Figure 5.A.1 collects all the NAAs, their names, and the abbreviations of these names.

5.R   RETOUCHES

5.R.1   P450 Enzymes and Grapefruit Juice

The enzyme cytochrome P450 3A4, which we showed in Figure 5.11, is extremely important. It is present in many organs, but especially in the small intestine and the liver, as shown here in Figure 5.R.1. Whenever we take drugs orally, they will meet CYP 3A4 in the small intestine and the liver and undergo metabolism (e.g. insertion of an O atom into a C−H bond). It was found that residents of some nursing homes did not react well to medication. Research showed that the culprit was the grapefruit juice they were taking each morning with breakfast. It turns out certain molecules (so-called furanocoumarins) present in grapefruit juice inhibit the enzyme by occupying the space where the drug molecule would have to bind in order to be metabolized (see the place of testosterone above the heme in Figure 5.11). This inhibition increased the bioavailability of the drug, and this overdosing of the drug had some adverse effect. In other cases, it was found that drinking grapefruit juice had a beneficial effect because one could use lower dosages of drugs.

5.R.2   The Discovery of O_2

Until the middle of the eighteenth century, scientists did not consider gas to be a material, but rather "a celestial influence that trickled into earth from the heavens." Scientists like the Dutchman Drebbel and the Englishmen Mayow, Hooke, and Boyle, who produced oxygen and even saw its effects on burning candles and on venous blood, still failed to consider gas as a materialistic entity. What prevented the change of mind was the fact that there were no means of trapping gases and proving their materiality (e.g., by weighing them).

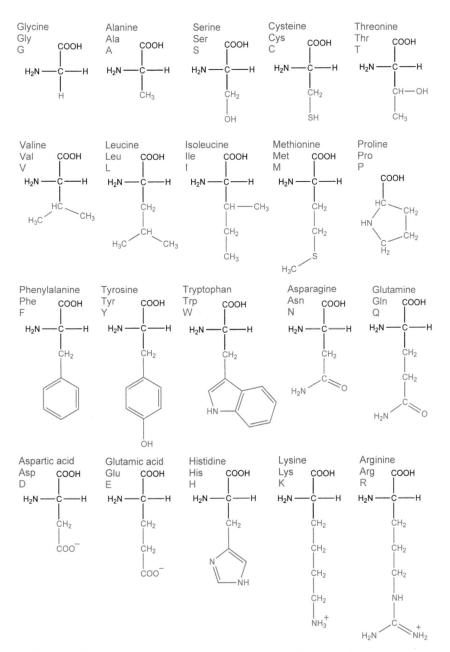

FIGURE 5.A.1 The twenty NAAs and the various ways of writing their names.

In 1727, an Englishman, Stephen Hales, invented the gas trough (shown in Figure 5.R.2), which could trap gas inside containers of liquids (e.g., water) and enabled determining the volume of the gas, its weight, and its chemical properties. Stephen Hales was intensively engaged in weighing and quantifying the gases he extracted from various objects (e.g., even from bile stones!). He asked "how much" instead of

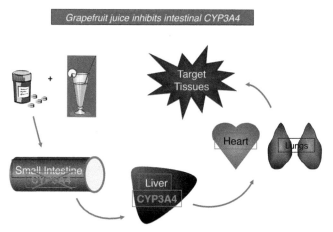

FIGURE 5.R.1   The arrows show the 'journey' of drug pills in the body, from the small intestine to the liver. In both places, they undergo metabolism by cytochrome P450 (CYP) 3A4 and then proceed to the heart, lungs, and target tissues. Drinking grapefruit juice incapacitates the CYP 3A4 in the small intestine. Reproduced with permission by Informa Healthcare, *Expert Opinion on Drug Metabolism & Toxicology*, **2007**, *3*, 67.

"what," and therefore he missed the discovery of the gases that make up our atmosphere. The great chemists Joseph Black, Henry Cavendish, Carl Wilhelm Scheele, Joseph Priestley, and Antoine-Laurent Lavoisier later discovered these gases because they asked the right question, "what?" The last three discovered O_2 and characterized its properties in one of the most exciting chapters of chemistry, which is characterized today as a scientific revolution.[1R] These three have interesting life stories, which I teach in a course on the History of Chemistry, and which can be found in many sources (e.g., Wikipedia, books on the history of chemistry, the play "Oxygen," etc.[5]). Let me tell their stories with extreme brevity.

The first discoverer of O_2 was the Swedish Scheele, but he did not publish his paper until 1777, and for many years, he has not been recognized as the first to discover the gas, even though he had told Lavoisier about his discovery before the latter discovered O_2 by himself. The second person was the Englishman Joseph Priestley, who discovered oxygen in 1774 and published his paper in the same year, writing on oxygen: "Hitherto, only two mice and myself have had the privilege of breathing it." But neither of the two recognized this gas as oxygen, because they were believers in a theory called the phlogiston theory, in which phlogiston was an entity that presumably resided in inflammable materials (like wood, sulfur, etc.), rather than a material body.

The last to discover oxygen was Lavoisier, who is shown in Figure 5.R.2 together with his wife-assistant Marie Anne Pierrette Paulze, who was a remarkable young woman, many years younger than Lavoisier, and who functioned as his scientific assistant. The play "Oxygen" even implies that she shared in his discovery. Lavoisier used to work with closed vessels and weigh the materials before and after the reaction. In this manner, he was able to substantiate the law of the conservation of mass.

Joseph Priestley

Stephen Hales   Gas trough

Carl Wilhelm Scheele   Antoine-Laurent Lavoisier with his wife

FIGURE 5.R.2 From left to right: Portraits of Stephen Hales along with his gas trough, Joseph Priestley, and Carl Wilhelm Scheele (in the middle). On the far right is Antoine-Laurent Lavoisier with his wife Marie Anne Pierrette Paulze in David's famous painting. Scheele, Priestley, and Lavoisier codiscovered O_2. (Stephen Hales' portrait is credited to William Ramsay's book *The Gases of the Atmosphere*. The gas trough is used with credit to Stephen Hales in *Vegetable Staticks*. Priestley's and Scheele's drawings are used courtesy of *Popular Science Monthly* Volumes 5 and 31. The painting of Lavoisier and his wife is used with assumed courtesy of Jacques-Louis David. All images are in the public domain.)

He called O_2 the "breathable" component of air. He then showed that "inflammable materials" (e.g., mercury) when burnt with the "breathable" gas gained weight (mass), and when subjected to heating gave off this gas and lost weight. He immediately understood that the gas that escaped these materials could not be phlogiston, because a material object cannot have a negative mass!

After toppling the phlogiston theory, Lavoisier realized that many of the acids that had been known then in chemistry contained this "breathable gas," which could be extracted from the acids. The chemistry of acids and bases had been the main area of established chemistry at that time (this chemistry eventually led to the development of the modern atomic theory via the weight relationships determined from combinations of acids and bases). Lavoisier, in a stroke of genius, decided to call this gas oxygen, namely "the generator of acids." As such, not only did he transform chemistry from a metaphysical to a material science, but also, in the same stroke, he linked his discovery to a large body of established chemistry. As such, his theory was immediately accepted and revolutionized chemistry. He himself recognized this and wrote to his disciple Chaptal, "The revolution is complete." The acceptance of his theory occurred despite the fact that, some years later, the great English chemist Humphry Davy showed that there are acids that do not contain oxygen. But it was too late: the revolution already occurred, and the name oxygen became part of the chemical vocabulary.

Scheele was a devout lover of chemistry and science and a discoverer of a great many compounds and elements. But in his life, he was a *schlimazel* (the Yiddish word for an unlucky person), who lost credit for a few discoveries (e.g., he discovered chlorine, but he described it in phlogistic terms, and when Davy rediscovered it, the credit went to Davy). In his thirtieth year, he bought a pharmacy in Köping, Sweden, from a widow of a pharmacist. He loved his pharmacy because he could do experiments all day long. At age 43, he decided to marry the widow of the pharmacist. He does so, and he dies within 48 hours! Do not get shocked! His death was not associated with his marriage. Scheele, like many chemists, used to taste/smell his discoveries. One of his discoveries was prussic acid (HCN), which is a lethal gas (with a smell of almonds). It is thought that this and his poor health were behind his death.

Priestley was a man of high culture, a man of God, and a great liberal who supported women's equal rights, liberty, and the French and American revolutions, etc. In 1770, when he intended to join Captain Cook's voyage, he invented soda water, which he thought would be a medication for scurvy (the ailment of sailors due to lack of vitamin C). In the end, he did not join the voyage, and hence, he abandoned his invention, which was taken later by Jacob Schweppe, who patented it in 1780. Avant-garde ideas, his non-Anglican religion, and his support of the French and American revolutions: all these caused a mob to storm his estate in Birmingham and burn it down in 1790. He subsequently fled to the United States and lived there in Pennsylvania. The highest award of the American Chemical Society is called the Priestley Medal.

The story of Lavoisier is tragic. He belonged to a family of low nobility. He made his money as a tax collector for the King of France, and in this capacity, he invented the city tax, which anyone entering Paris had to pay. He became very rich, and he served in high positions in France. As a member of the French Academy, and in charge of testing inventions, he rejected those handed to the Academy by Jean-Paul Marat. During the revolution, Marat became one of the leaders of the "popular committees" who were in charge of executing anyone who was suspected to have some counter-revolutionary sentiments. Being a tax collector, a rich person, and hated by Marat, Lavoisier got beheaded in 1794. The person who revolutionized chemistry died in a popular revolution …

5.R.3   References for Retouches

[1R] T. S. Kuhn, *The Structure of Scientific Revolutions*, University of Chicago Press, 2012.

5.P   PROBLEM SET

5.1 Phenol is an alcohol in which OH is bonded to a benzene ring. Use fragments from Schemes 5.1 and 5.6 to generate all the possible monophenols and diphenols. Is there anything interesting you can say about phenols (try searching on the Web)?

5.2 Using fragments in Schemes 5.1 and 5.6, construct adrenaline and noradrenaline (known also as epinephrine and norepinephrine, respectively). What functions do these molecules have?

5.3 Add the missing carbon and hydrogen atoms and the lone pairs in the drawings of the molecules in Figure 5.3.

5.4 Figure 5.4 shows many molecules. Construct from appropriate fragments the molecules Aspirin, cinnamon, vanillin and raspberry ketone. Identify the main molecular families (e.g., alcohol, ether, etc.).

5.5 Write the complete molecular formulae, showing all C and H atoms, for the fatty acids in Figure 5.5.

5.6 Add the missing H and C atoms in the drawings of the essential fatty acids (EFAs) in Scheme 5.9.

5.7 Construct ethyl acetate (Figure 5.4) from its constituents using the *clack-click* method.

5.8 Scheme 5.10 shows a triglyceride molecule. Draw its constituent molecular components.

5.9 Make a triglyceride with acetic acid as the only acid component.

5.10 Make an ester from proline and ethanol.

5.11 Construct all the dipeptides that can be made from the amino acids proline and alanine. Write the names of these peptides using the three-letter and one-letter labels of amino acids.

5.12 Practice drawing the ring of the heme molecule (bonus question).

LECTURE 6

ELECTRON RICHNESS, DNA AND RNA MOLECULES, AND SYNTHETIC POLYMERS

6.1 CONVERSATION ON CONTENTS OF LECTURE 6

Usha: I have been reading some of the lectures you produced. They are great fun! Would you let me in on this dialogue, Sir?

Sason: Usha, calling me Sir might be confusing for Racheli, who is not aware that I belong to any nobility. Before I answer your question, maybe you should explain this habit of calling me Sir to Racheli.

Usha: You see, Racheli, in India we respect a teacher by calling him/her sir/madam. Once a teacher, always a teacher ...

Racheli: We also respect our teachers. But the word "Sir" in Hebrew would be translated to *adoni* (אדוני). In modern-day Hebrew, we use the word *adoni* to set a limit with someone we are annoyed by. So, I would stick to using Sason ...

Sason: My answer to your first question, Usha, is, of course! You earned being a fellow trialoguer ...

Tell us what's on your mind.

Usha: I think, Sir, you have made quite a journey in our molecular universe using molecular building block fragments. Now you are reaching an ideal place to apply your construction principle.

Racheli: What exactly do you mean by "ideal," Usha? We are always in ideal places ...

Usha: Well, the DNA molecule, which Sir plans to tell about, is also made as a LEGO of molecules called *nucleotides*. You can combine them with the

Chemistry as a Game of Molecular Construction: The Bond-Click Way, First Edition. Sason Shaik.
© 2016 John Wiley & Sons, Inc. Published 2016 by John Wiley & Sons, Inc.

"clack-click" method to make the single-strand polymers of DNA. The DNA is "the software" that keeps generating us. Being so important, I thought …

Sason: Yes, Usha, DNA is the main topic of this lecture. But I have to introduce first a new feature of the electronic structure. Otherwise, I cannot explain the nature of the nucleotides.

Racheli: Sason, but RNA is as important as DNA. This is the molecule that is formed by copying information from the DNA.

Sason: How can I forget? DNA and RNA nucleotides differ very little. I will explain the differences and then proceed to use the "clack-click" method to generate the corresponding DNA and/or RNA polymers.

Usha: DNA chemistry has become so exciting! Now chemists make DNA origami by stapling single strands with small DNA pieces into all kinds of shapes.[1]

Racheli: DNA origami is a leading method in nanotechnology, and Sason could teach it easily with the "clack-click" trick had he wanted to.

Sason: Now, now, you are interviewing each other, Racheli and Usha! You are also digressing and adopting my bad habits …

As I was going to say before you interrupted me, I wish to start with a new concept, namely that some atoms of main elements in the periodic table can exist in different "states of Nirvana"; in one of these, the atom obeys the octet rule, and in the other, it adopts an electron-rich bonding environment.

Racheli: You probably mean the phosphate group in the nucleotide example in Scheme 6.1.

Sason: Yes, the nucleotide itself is made from three units, called sugar, base, and phosphate. Phosphate originates from phosphoric acid. I intend to construct this

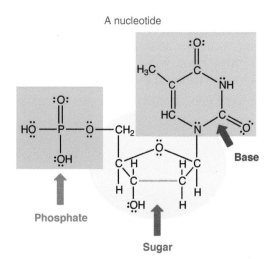

SCHEME 6.1 A nucleotide building block of the DNA molecule. The parts are indicated as sugar, base, and phosphate (under physiological conditions, phosphate is a negative ion, but we will use it in the form depicted here). Note that P is surrounded by 10 electrons.

acid using the electron richness idea. Recall, acids are sour. So, our genetic code is "sweet and sour"…

Racheli: It is just mind-boggling that a molecule will carry any genetic information and determine whether someone is blonde or prone to be a nervous wreck, and so on.

Sason: Racheli, previously you were digressing, and now you jumping ahead! This will be discussed in one of the subsequent lectures.

Usha: Isn't this going to be a very brief lecture, Sir?

Sason: Not really, Usha! The second part will be dedicated to man-made or synthetic polymers.

We can also construct many of these polymers by the "clack-click" method. But, here human wit has generated a practical method called *initiation*, whereby a very reactive species, for example, a free radical, generates in a molecule *a new connectivity center* that gets propagated over many molecules and zips them into a polymer.

Racheli: Unity of the chemical matter, from animate to inanimate, right?

Sason: Forget the unity for a moment, Racheli. Just imagine a world without polymers! You would not want to live in such a world …

Usha: If I might add, Sir, polymers have been known in the ancient world, in China, in Egypt, and in the Americas. Maybe even in India … I hope you will tell some stories about that.

Sason: OK ladies. You have given me so many instructions … How about letting me proceed with the lecture?

6.2 ELECTRON RICHNESS: A DIFFERENT STATE OF NIRVANA

The octet and duet rules that were introduced in Lecture 2 are very powerful. It turns out, however, that some specific atoms may occasionally exceed the octet and establish electron-rich molecules. Chemists still argue with passion whether this is the case or that "octet and duet are forever" (see Retouches section 6.R.1). Nevertheless, as electron richness is a very didactic concept for constructing some important molecules, one of which is phosphoric acid, which is a constituent of DNA and RNA, we shall not worry about the hot debate of the chemists, but instead use electron richness as an element in our molecular construction.

6.2.1 "Who Is Who in Electron Richness"

Scheme 6.2a shows the atoms which belong to the families 5A–7A in periods 2–4. As indicated in the scheme, the atoms N, O, and F in the second period are octet followers. If you wish, they possess a single state of Nirvana, in which the valence shell is saturated. However, the heavier elements in periods 3 and 4, while generally obeying the octet rule, can also occasionally exceed the octet. Thus, these atoms can

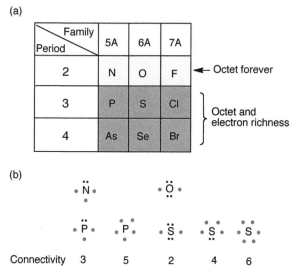

SCHEME 6.2   (a) Atoms belonging to families 5A–7A in periods 2–4 and the possibilities for rules governing their connectivity. (b) The connectivity states of S versus O and P versus N. Electrons participating in connectivity are marked in red.

have *several states of Nirvana* that saturate the binding propensity of these atoms, and as such, they will form side by side molecules that obey the octet rule with those that are electron-rich.

Scheme 6.2b exemplifies the states of Nirvana for N and O versus the heavier family members P and S, respectively. It is seen that O is always a double connector, whereas S has the freedom of using more of its valence electrons for the purpose of connectivity. S can therefore assume a connectivity of 2 like O, but it can also assume a four-way connectivity by utilizing two additional valence electrons or a six-way connectivity by utilizing all the valence electrons. Similarly, N is always a triple connector, while P can also be a five-way connector by utilizing all the five valence electrons. In other words, unlike N and O, which adhere to the octet rule and have fixed connectivities, P and S can have alternative states of connectivity by utilizing more of their valence electrons for creating electron-pair bonds.

6.2.2   Examples of Electron-Rich Molecules

Generally speaking, going down the periodic table, this tendency of the atoms to participate in bonding using different states of connectivity increases. Let us construct a few molecules to exemplify the new concept. We begin with Scheme 6.3 by constructing all the possible molecules that can be made from oxygen and as many fluorine atoms as needed. Then, we shall repeat the process by constructing all those molecules that can be made from sulfur and as many fluorine atoms as needed.

Thus, as seen in Scheme 6.3, since O is a double connector, it will require two single-connector fluorine atoms, and click, we obtain the molecule OF_2 that obeys

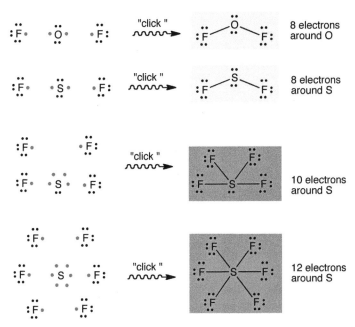

SCHEME 6.3   Molecules that contain either a single oxygen or a single sulfur atom and as many fluorine atoms as needed. The molecules following the octet rule are highlighted in yellow, and the electron-rich ones are highlighted in orange.

the octet rule. Sulfur can form the analogous molecule, SF_2, by following the octet rule. However, since sulfur is capable of attaining different states of connectivity, it can also form SF_4 by utilizing four of its valence electrons participate in connectivity and SF_6 by using all the valence electrons. As such, in the case of sulfur, we are going to find octet-obeying molecules existing side by side with electron-rich ones. The same is true for phosphorus and other atoms occupying the third period and below. By the way, like all Janus-type molecules, SF_6 is very useful, for example, as an insulating gas, but it is at the same time a greenhouse gas more potent than CO_2.

6.2.3   Phosphoric and Sulfuric Acids

Let us construct now two quite important molecules: phosphoric acid, H_3PO_4, and sulfuric acid, H_2SO_4. For both molecules, we qualify that the H atoms are not connected to P or S and that the two molecules do not possess O—O bonds. Scheme 6.4 shows the construction processes.

Based on the above qualifications, we must bind the H atoms to the O atoms, and since each O has a double connectivity, then in the case of H_3PO_4 in Scheme 6.4a, there remain collectively five dangling connectivities of the O atoms. Therefore, the phosphorus atom must use its five-connector state and click to form phosphoric acid. Similarly, in the H_2SO_4 molecule in Scheme 6.4b, after connecting the two H atoms to the O atoms, we are left with six dangling connectivities collectively on the oxygen atoms, and this makes sulfur use its six-connector state and click to form sulfuric acid.

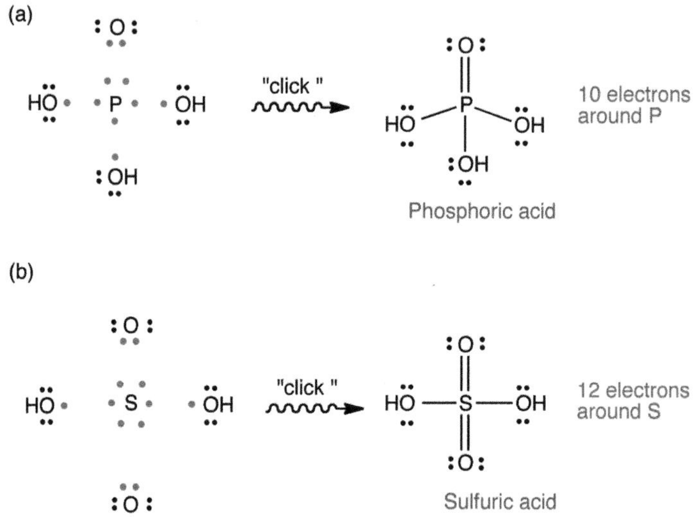

SCHEME 6.4 Construction of (a) H_3PO_4 and (b) H_2SO_4. Electrons participating in connectivity are marked in red. The specified numbers of electrons around P and S indicate the electron richness of these atoms.

Sulfuric acid is the most highly produced chemical in the world: for example, world production already in 2004 reached 180 million tons (40 of which are produced in the United States). It is amply used in various industries, such as detergent production, petroleum cracking catalysts, dyestuffs, etc., and it has a multitude of applications. It had been already known to the Sumerians in antiquity, in the fifth to third millennia BC. It was mentioned and made by all the great alchemists from Hellenistic times, like Zosimus, though to the Arab alchemists, like Jabir ibn Hayyan, and all the way to the European alchemists, who used to call it "vitriolic acid" or "oil of vitriol," since it was produced by cooking sulfate minerals, which were called vitriols, as they glisten like glass. The great German alchemist Johann Glauber produced it from the vitriol mineral Na_2SO_4 (still known today as Glauber's salt), in which Na replaces the hydrogen atoms of H_2SO_4.

Terrestrial sulfuric acid is generated naturally due to oxidation of materials containing sulfur, and due to acid rain, which is formed when SO_2 is oxidized in the atmosphere to SO_3, which together with water forms the acid. There is also extraterrestrial sulfuric acid, due to the oxidation of sulfur, for example, on Venus. Some marine algae use concentrated H_2SO_4 as a means of protection.

H_2SO_4 is a very strong acid, and when it is concentrated, it also decomposes living tissues quite easily (e.g., breaking down the proteins to amino acids and further to smaller fragments, which are also oxidized). It has to be handled with extreme care! Like many chemicals, H_2SO_4 is also Janusian, useful and dangerous at the same time. There are no "good" or "bad" molecules; this depends on how they are used …

By the way, H_2SO_4 is a molecule that "starred" in a movie. This is Luc Besson's movie, *La Femme Nikita*, which is about a convicted woman who, instead of serving

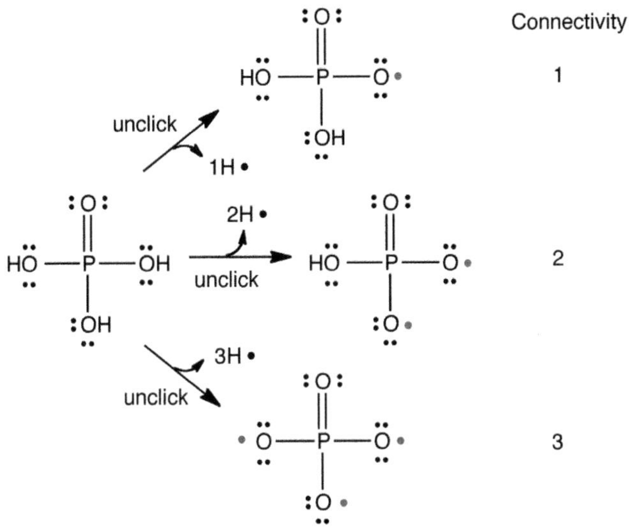

Connectivity

SCHEME 6.5 Modular fragments made from H_3PO_4 by unclicking O—H bonds. The electrons available to participate in connectivity after unclicking are highlighted in red.

her term in jail, becomes an assassin/spy for the state. When she does one killing too many, there is a need to get rid of the body, and the wonderful actor Jean Reno, who plays Victor Nettoyeur ("Victor the Cleaner"), comes with two tanks of sulfuric acid on his shoulder to get rid of the body ...

Phosphoric acid is also an amply produced chemical worldwide. It serves as a food additive, and it is present in all soft drinks, such as colas, giving them this sour tangy taste. It is widely used in cosmetics, fertilizers, the leather industry, and explosives. Importantly, its modular fragments appear in living organisms, as constituents of tooth enamel and the skeleton, in signaling molecules, and also in DNA and RNA. Since we are interested in these molecules, we present in Scheme 6.5 the modular fragments that can be generated from phosphoric acid by unclicking O—H bonds. These fragments can be used to generate new molecules and molecular worlds.

6.3 DNA AND RNA STRANDS

DNA is a hugely long molecule that contains hundreds of millions of nucleotides in a single strand (there are 3 billion or so nucleotides in total in the genome overall) linked together in a manner we will immediately discuss. A DNA strand can reach a length of meters if completely stretched. However, inside the nucleus of the cell, the DNA is compacted by a wrapping around proteins called histones to a size of 1 millionth of a meter (0.000001 m), a length that is called one micron. Since the DNA resides in the nucleus and it contains fragments of phosphoric acid, it is also called "nucleic acid" (NA). RNA is a related molecule, which copies genetic information

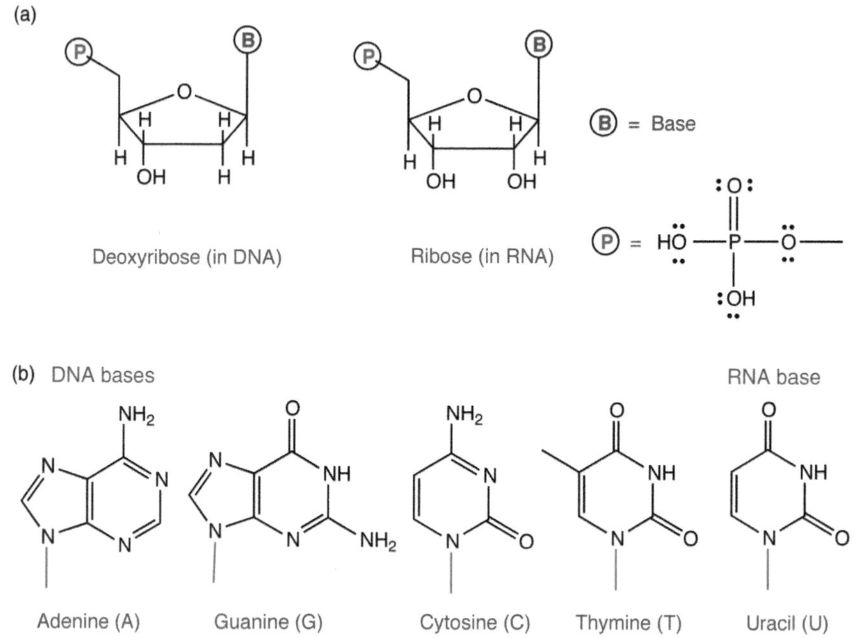

(a)

Deoxyribose (in DNA)   Ribose (in RNA)

(B) = Base

(P) = (phosphate group structure)

(b) DNA bases   RNA base

Adenine (A)   Guanine (G)   Cytosine (C)   Thymine (T)   Uracil (U)

SCHEME 6.6 (a) The nucleotides of DNA and RNA drawn in the implicit manner. The circled group **P** is a phosphate, drawn explicitly on the right-hand side. The circled group **B** is a base. (b) The four DNA bases are A, C, G, and T, while, in RNA, the T is replaced by U. The red lines indicate the sites of connection of these bases to the sugar fragment.

from the DNA and goes out of the nucleus to enable protein synthesis based on the copied information (see later in Lecture 7).

Scheme 6.6a shows the fundamental nucleotides of DNA, where we now use the compact molecular drawing convention (not showing C and H). Both nucleotides contain a five-membered ring sugar molecule. In RNA, the sugar ring contains two OH groups, and it is called ribose. The DNA sugar has one less OH, and hence it is called deoxyribose (the initial "de" means "without"). As such, both RNA and DNA are nucleic acids (NA), and the initial letter R or D specifies if the sugar is ribose (R) or deoxyribose (D).

In addition, both sugar rings carry a CH_2–**P** tail, where the **P** symbolizes phosphate, which is shown on the right-hand side of Scheme 6.6a, and you can recognize this constituent fragment of the electron-rich phosphoric acid. The additional group on the sugar is **B**, which is the shorthand notation for a *base*. DNA contains only four bases, which are symbolized by the first letters in their chemical names as A, C, G, and T. In RNA, the base U replaces T. Scheme 6.6b shows these different bases. There are no other bases in all DNA and RNA molecules in all life forms in nature, and the possible reason for that is that these are the only bases that survived the radiation damage during evolution (there are good reasons for this survival: quantum mechanics (QM) theory shows that when these molecules are hit by radiation, they do not undergo bond breakage).

6.3.1   Formation of DNA and RNA Strands

To learn how to construct strands of DNA nucleotides, let us first recall how we constructed proteins from amino acids. This was done by excising an H_2O molecule from each two successive amino acids and connecting the dangling connectivities, hence "clack-click." To illustrate this, we placed in Scheme 6.7a two nucleotides, where the base moieties are indicated as generic **B** and **B'**, which can be identical or different. It is seen that the OH of the phosphate in the right nucleotide encounters an O–H of the sugar in the left nucleotide. Since OH and H are single connectors, by removing them as a water molecule ("clack"), and by simultaneously "clicking" the remaining dangling connectivities, we create a dinucleotide with a *sugar-phosphate* C–O–P *link*. In the cell, the "clack-click" we are doing here on paper is performed by an enzyme (called polymerase), which facilitates the extraction of water and thereby stitches the DNA strands and/or the RNA strands.

To appreciate the information content in a strand, consider now four nucleotides, from which we excised an H_2O molecule from each successive pair and created thereby a tetranucleotide, as in Scheme 6.7b. For convenience, we labeled the bases as B_1–B_4, to mark the DNA bases in the tetranucleotide. However, the bases can be identical, or be present in any other combination and ordering of the four available bases, so the number of the molecules possible in this small DNA piece is not small. It reaches 256 different molecules!

Generally speaking, if the number of nucleotides which participate in this process is n, then we create a DNA strand that contains n nucleotides connected via C–O–P links. As we just said, in the cell, the number n can reach 200 million or so. How many different polymers can we make from hundreds of millions of DNA nucleotides? Well, since there are four different bases, then for a chain with a hundred million nucleotides, *we can make 4 to the power of a hundred million* different DNA strands, which possess different base sequences. This is an unimaginably large number, which you can try to calculate and compare with the Earth's population.

As we shall learn later, the genetic code is determined by the identity and sequence of the bases, and since we have so many different DNA molecules that we can make, and which differ in the sequence of the bases, there are enough possibilities to create many different forms of life and many different individuals of each life form. Since RNA strands are copies of DNA pieces, their possible number is also hugely large. Clearly, the sequence of bases on a DNA strand is information, and the large size of the strand generates a huge amount of information, which is sufficient to create complex organisms. In the next lecture, we shall learn that the mechanism by which DNA stores the information in a strand uses weak bonds, which also enable the transfer and translation of the information to generate the "hardware" (the proteins).

6.3.2   DNA and RNA Nucleotide-Based Drugs

A virus is a piece of DNA or RNA wrapped in a protein. When it invades the host cell (e.g. a human cell), the virus tricks the cell during its division into producing the

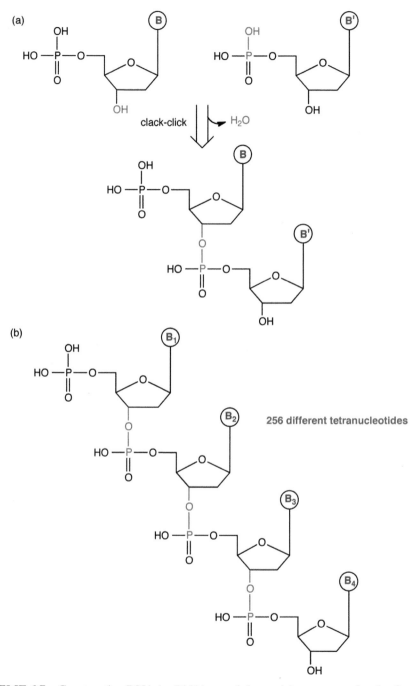

SCHEME 6.7   Constructing DNA (or RNA) strands by excising water molecules from pairs of nucleotides: (a) a case of two nucleotides with bases **B** and **B′**, (b) a case of four nucleotides with bases **B₁**–**B₄**. The C—O—P linkages are highlighted in red. The process can recur to form a DNA strand with as many as 200 million nucleotides (a total of 3 billion nucleotides are in the genome). In the cell, this is performed by enzymes, which stitch the nucleotides together into a strand.

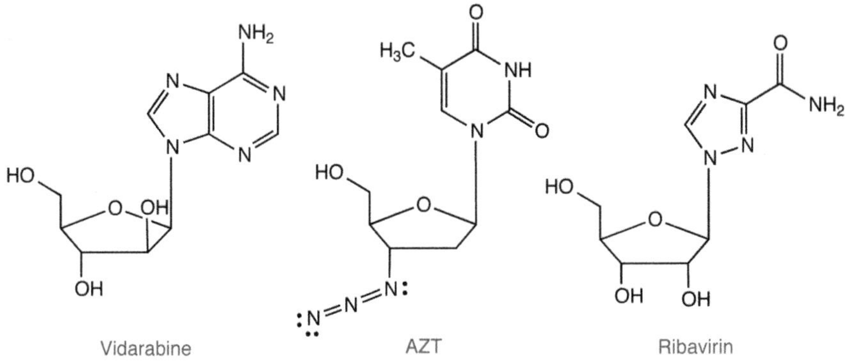

SCHEME 6.8 Nucleotide-like drugs (missing the phosphate).

viral DNA/RNA and starting to make the protein that is encoded in the DNA/RNA of the virus. It is sensible therefore to think that nucleotide-like molecules can serve as antiviral drugs by interfering with the synthesis of the viral DNA and RNA. This therapeutic approach uses the strategy "matter-anti-matter,"[2] in which the molecular cause of the disease is combated with molecules looking like the perpetrator.

Scheme 6.8 shows three well-known nucleotide-based drugs. The first antiviral nucleotide-like drug in Scheme 6.8 is Vidarabine, which was synthesized in the laboratory in the 1960s, and it has been used since the 1970s in the treatment of many viral infections. The second molecule, which is dubbed AZT, is the first drug that was administered successfully to AIDS patients. It was made for the first time in 1964, and only in 1984, when the human immunodeficiency virus (HIV) had been definitely identified as the cause of acquired immunodeficiency syndrome (AIDS), was the drug incorporated in AIDS treatment. Because of the N_3 group, which tends to break apart and spew the stable molecule N_2, AZT can bind to the enzyme that HIV uses to make its DNA and inhibit its ability to catalyze the DNA formation reaction. Because the HIV develops resistance to AZT, the drug is now administered in a cocktail of drugs, and it is included in the List of Essential Medicines of the World Health Organization. New and better drugs continue to be developed.[3] The third molecule in Scheme 6.8 is the RNA-like nucleotide Ribavirin, which is given as a treatment for the hepatitis C virus, which is an RNA virus.

Concluding Remarks about DNA and RNA Strands: At this point, we have made a long and impressive excursion into molecular space, starting from the small H_2 molecule and all the way to giant molecules, such as proteins and DNA. During this journey, you must have witnessed the marvelous unity of matter in the universe. We are going to meet DNA and RNA again when we discuss the mechanism of preservation and duplication of the information stored in these molecules, and the copying, reading, and translation of the information to make proteins of living systems. Right now, we want to shift to another topic, which, in terms of construction principles, is very much related to the giant molecules we just discussed.

6.4 SYNTHETIC POLYMERS

Look at the world around you, and you will not cease seeing synthetic polymers. These are man-made giant molecules, which are light, strong, and amazingly colorful (recall the demo of coloring glass by usage of "water glass" in Lecture 3). Almost everything you touch, wear, sit on, or eat from is made of synthetic polymers. If you use pencils for writing, you will be using rubber polymers as erasers. If you drive a car, you should know that its tires and many of its parts are made of polymers. Stents in open-heart surgery are made of polymers. If you know someone who happened to undergo some cosmetic operation, they would end up with polymer (called silicone) implants. If you are using a laptop, its parts will be made by and large from polymers. Fiber optic cables are made from silicone. Your baby's pacifier is made of silicone polymers. If you go shopping, you will use polymer-based plastic bags to carry the stuff you bought. Polymers are used in buildings, construction, plumbing, sealants, water repellents, adhesives, lubricants, molding material, three-dimensional (3D) printing materials, and whatnot! Synthetic polymers are one of the greatest achievements of chemists.[4]

Mankind has been using for millennia natural polymers, such as wood (cellulose), cotton, leather, silk, lacquer, etc. The first polymer that was discovered and produced by the Chinese as early as 1300 BC was the lacquer that was used for producing tableware, cooking utensils, furniture, and works of art.[5] Lacquer was obtained from the sap of the tree *Rhus vernicifera*. Around 200 BC, the Chinese learned how to control the degree of polymerization by dropping a crayfish into the sap. A characteristic component of the sap is a class of phenols collectively called urushiol, with formulae such as $C_{21}H_{36}O_2$ (we constructed some phenols in Lecture 5). It turns out that the protein of the crayfish inhibited the enzyme that was responsible for the polymerization of the phenol. The Egyptians of antiquity used varnish, a sap of pine trees, for protecting and decorating the coffins of their mummies. But the origins of the varnish appear to be in India. In Central and South America, the Olmecs (1440 BC) and then much later the Mayans, the Aztecs, and the Incas used the sap of rubber trees (*Castilla elastica* and *Hevea Brasiliensis*) to produce rubber by mixing the sap with alcohol and heating it. The sap was also used in rituals (drinking), while the rubber product was used for making balls for playing (e.g., in the Mayan culture). Another sap is the caoutchouc, which was brought to Europe by French explorers from South America in the eighteenth century.[6] Priestley (the co-discoverer of O_2) discovered caoutchouc's usage in removing pencil marks, and hence it became known as rubber. As their consistency, hardness, and elasticity could be controlled, caoutchouc products began to be manufactured at the beginning of the nineteenth century. The chemical formula of the monomer unit, which makes up the natural rubber polymer, was identified at the end of the nineteenth century.[6] However, the actual structure of the rubber as a long molecule, which we now call a polymer, was not really known until the German chemist Hermann Staudinger (Figure 6.1) published in 1922 the key scientific paper "On Polymerization," in which he proposed for the first time that these rubbery creatures consisted of long chains of atoms held together by covalent bonds. It took over a decade for his work

FIGURE 6.1 Hermann Staudinger, the father of modern polymer science, holding a model of a polymer. Used with permission of the Deutsches Museum to Wiley.

to be widely accepted, and after a boom in the field, he was awarded the Nobel Prize in 1953. The twentieth and twenty-first centuries have been marked by great discoveries in the chemistry of polymers and the production of tailor-made ones, which now dominate our material world.

6.4.1 Constructing Polymers Using the LEGO Principles

In this section I will limit myself to the LEGO principles for construction of polymers from their building blocks. To make matters clear, let me state at the outset that our principle is *a reverse construction* of a giant molecule, with an aim of conceptualizing it. We are not describing the actual chemical mechanisms whereby these molecules are made, although the resemblance is there.

The word "polymer" means many (*poly*) units (*meros*), and chemists call the individual unit a monomer (*mono* means one). Let me then teach you two simple principles of constructing polymers from their monomer units. These principles are outlined in Schemes 6.9 and 6.10.

The first construction principle is shown in Scheme 6.9 for the generation of polysiloxane (silicone). This archetypal principle was already demonstrated when

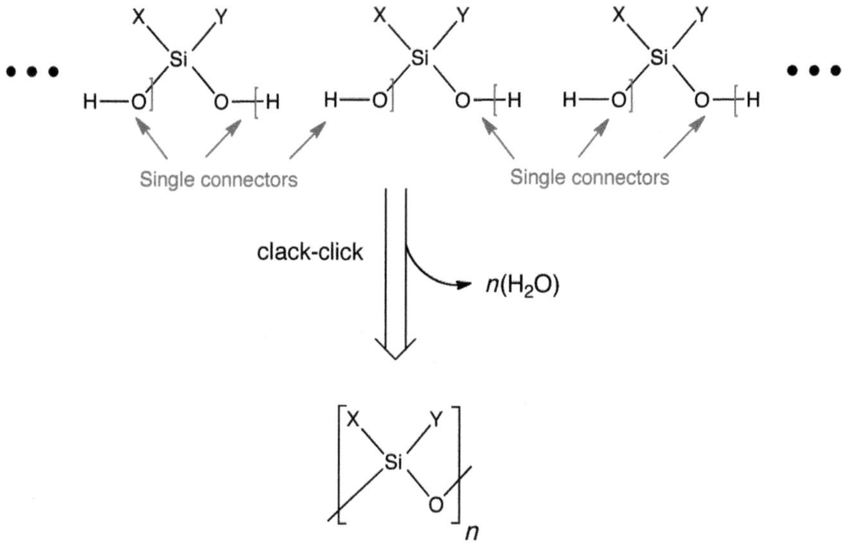

SCHEME 6.9 Archetypal principles for constructing polymers: Illustration of the condensation method (here clack-click) using the formation of polysiloxane (or silicone). X and Y can be any single connector chemical groups.

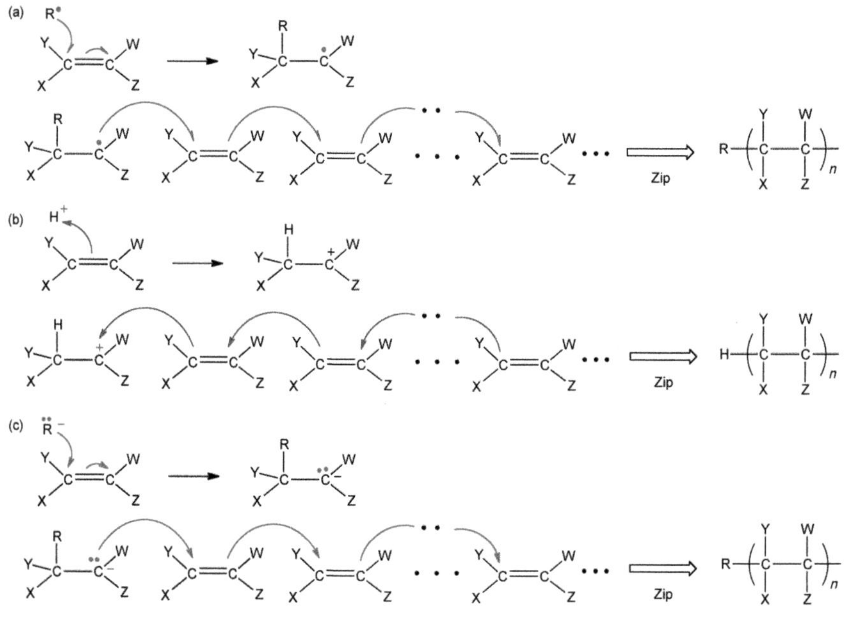

SCHEME 6.10 Illustration of the initiation method using the construction of a polyalkene with: (a) a free radical initiator, R•, (b) a proton initiator, H+, or (c) an anion initiator, R:−. The arrows indicate the directions in which the electrons migrate in order to propagate the dangling connectivity (indicated in red or blue to distinguish the two different directions of flow). Single-headed arrows indicate single electron migrations, while double-headed arrows indicate migrations of electron pairs. (Note that, in all of these, the focus is on the initiation step and migration of electrons, and, thus, what happens to the last monomer to react is not marked explicitly.)

we generated proteins from amino acids and DNA strands (or RNA strands) from nucleotides. In polymer chemistry, this process is called *condensation*. We referred already to this construction principle by the name "clack-click" to emphasize the bond reorganization process. Thus, by excision ("clack") of n molecules of H_2O from n successive pairs of siloxane molecules, we create dangling connectivities on the remaining molecular fragments, and these can "click" to generate here polysiloxane, also known as silicone, which is an example of the huge family of silicon-based polymers that adorn our world.

The second principle is called *initiation*, and it involves generally alkenes as monomers. It should be pointed out that many of the alkenes are byproducts in the petroleum industry. Thus, by cracking the black ugly oil we pump from the depths of the earth, we get these small molecules and make from them many useful polymers. Scheme 6.10 illustrates the initiation method, where a molecular fragment or a catalyst that has a dangling connectivity center, or an electron-deficient center, or still a lone-pair–rich center attacks the monomer, thus breaking one of the components from its double bond, and thereby *transferring the dangling connectivity to the monomer*. This activated monomer then attacks another, and the latter attacks a third molecule, and so on, thus *propagating the dangling connectivity* and zipping all the molecules (Note there are several different kinds of dangling connectivities, namely radical centers, positively charged sites, and lone pairs.)

A common initiator is a free radical $R^{\bullet}$ that forms a bond with one of the electrons of the electron-pair bond in the alkene. This leaves an unpaired electron on the adjacent carbon, as shown in Scheme 6.10a. The formed species then attacks another double bond and repeats the activation pathway. This process is propagated, and like a zipper, it zips all the molecules into a polyalkene. The end of the chain can then be capped by a single-connector atom or fragment, such as $H^{\bullet}$, which is abstracted from the reaction mixture (e.g., from the solvent).

As further seen in Scheme 6.10b, the initiator can also be an electron-deficient species like a proton that is supplied by an acid. By definition, an electron-deficient fragment is a connectivity center since it is seeking an electron pair to bind with. Thus, being electron-deficient, the proton appropriates the electron pair of the alkene to make an H–C bond, and this leaves a cationic (positive) center on the adjacent carbon, which in turn can further attack double bonds and propagates the electron-deficient center until all the monomers are zipped into a polyalkene (where the end group is capped by reacting with any negatively charged fragment from molecules in the reaction mixture). The same process can be initiated by an electron-deficient transition metal catalyst that can form a bond with the alkene and activate the double bond, which then zips into others.

Finally, the third archetypal initiator in Scheme 6.10c is an anionic fragment $R:^{-}$. The electron pair of the anion is a connectivity handle that can be transferred to make bonds with other centers. Thus, the anion attacks the double bond, makes with it an R–C bond, and shifts the electron pair to the adjacent carbon. The so-created electron pair further attacks another alkene and propagates the connectivity handle, thus zipping the molecules into a polyalkene.

FIGURE 6.2   Karl Ziegler (left) and Giulio Natta (right), the developers of a highly useful alkene polymerization catalyst. Ziegler's photograph is from *Nachr. Chem.* **2003**, *51*, 12, and Natta's photo is from http://www.natta.polimi.it/Natta/Galleria/gallery.htm.

A most famous metal-based initiator is the Ziegler–Natta catalyst, which is made from a titanium-based catalyst and an electron-deficient $Al(C_2H_5)_3$ cocatalyst. Since this used to be patented, naturally, there are now many Ziegler–Natta-type catalysts based on different metals. The Ziegler–Natta process is used today to produce polyalkene plastics and elastomers, which are produced on a scale of over 100 million tons a year. Karl Ziegler and Giulio Natta (Figure 6.2) were awarded the Nobel Prize in Chemistry in 1963 for their invention, which altered our world for better and for worse.

The properties of polymers depend, of course, on their molecular constitution, and for the case of polyalkenes, these properties are regulated by the groups X, Y, W, and Z on the two carbon atoms (Scheme 6.10). Figure 6.3 shows a few famous polymers along with the corresponding monomers. First of all, note that the polymer is called by the monomer's name with a prefix "poly," such as polyethylene. Also, the groups that cap the ends of the chain are not drawn anymore, because they are not important (negligible compared with the long chain), and they can anyway vary depending on the reaction conditions.

The first polymer, shown in Figure 6.3a, is polyethylene (PE). When PE contains ~500 monomers, it is used for common plastic bags. Around 10,000 monomers create a rigid plastic, which is used for other purposes. The world production of PE has reached 10 million tons a year and is getting higher.

Figure 6.3b shows polyvinyl chloride (PVC), which is used in floor tiles, raincoats, pipes, and phonograph records (yes, these are becoming popular again …). Its production is 2 million tons a year.

Figure 6.3c shows Teflon, which is used for making gaskets, bearings, pan coatings, and so on. Figure 6.3d shows polystyrene (PS), from which we make coolers for

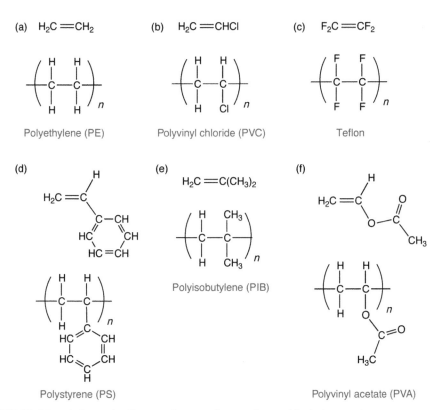

FIGURE 6.3 A few polyalkene polymers shown along with their constituent monomers.

water and food, and insulating material for buildings. Its production is 2 million tons a year.

Figure 6.3e and 6.3f show two elastomers, which have many uses. One of these uses is worthy of a story, since it is connected with chewing gum. Mankind was chewing gum already in the Neolithic period (5,000 years back). It is known that the Olmecs chewed gum, and so did the ancient Greeks. These gums were saps of various trees, such as the mastic tree used by the Greeks. Commercialization of chewing gum started in the nineteenth century when New England settlers picked up the habit from the Native Americans. Modern chewing gum was developed in the 1860s when natural rubber called chicle was used. Later, chewing gum started to be based on polyisobutylene (PIB) (Figure 6.3e), an oil byproduct that is cheap and highly available. Recently, chewing gum producers started adding some polyvinyl acetate (PVA) (Figure 6.3f), also an oil product.

I was drawn to inquire more about chewing gum after I read a story about the city of Barcelona and its gum problem. It turns out that the city workers of Barcelona had to remove from the streets every day 1,800 chewing gum pieces at a cost of 85,000 euros/year. Removing PIB-based chewing gum turns out to be problematic because PIB sticks strongly to surfaces. Therefore, so goes on the story, the Spanish government issued an ordinance to manufacture chewing gum containing more PVA,

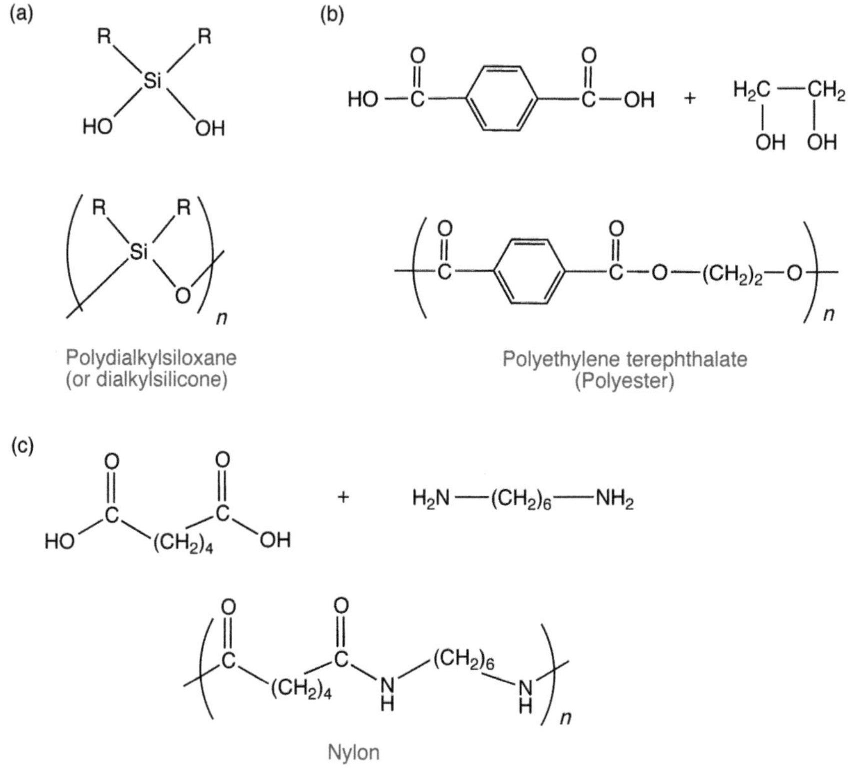

(a)

(b)

Polydialkylsiloxane
(or dialkylsilicone)

Polyethylene terephthalate
(Polyester)

(c)

Nylon

FIGURE 6.4   Three polymers formed by condensation (these can be constructed using the clack-click approach): (a) Dialkylsilicone. (b) Polyester. (c) Nylon. The latter two are copolymers including two different monomers.

which is less sticky. However, PVA is suspected to be cancerous. The same applies to all the above polymers; they serve good purposes, but, at the same time, when discarded, they pollute the Earth. As usual, we find that matter is Janus-like, good and bad at the same time ...

Figure 6.4 shows three polymers that are obtained by the condensation method (Scheme 6.9). The one in Figure 6.4a is already familiar to us. This is silicone (or polysiloxane), which is made here from dialkylsiloxane monomers by condensation, wherein an extraction of a water molecule from each two successive monomers clicks the remaining fragments to silicone.

The condensation method can be also applied to two different molecules, and the respective process is called *copolymerization*. For example, as shown in Figure 6.4b, usage of a mixture of a di-acid molecule and ethylene glycol (di-alcohol) leads to a certain version of polyester, which is used for making very durable clothing known by the names polyester, Dacron, and Terylene. The British chemists John R. Whinfield and James T. Dickson invented Terylene (a common type of polyester).

Finally, in Figure 6.4c, you can see the polymer nylon (as present in nylon stockings ...), which is made by condensation of a di-acid with a di-amine. The inventor of nylon was Wallace Carothers, who made the polymer while he worked in DuPont. He also invented a type of polyester, but DuPont chose to focus on nylon, because the latter was considered as a replacement for the natural polymer silk (a protein).

Perhaps the most versatile polymer is silicone, the jack-of-all-trades of polymers. The British chemist Frederic Kipping invented silicones. The performance and reliability of millions of modern products depend on silicones. For example, silicones give personal care products essential qualities, which we take for granted today. But they are also found in thousands of industrial applications. In aerospace applications, for instance, silicone products increase the life span of vital components, while in railway locomotives, they provide tough, long-lasting motor insulation and lubricants for bearings. Silicones are used as coatings to protect facades and historical monuments and are also used in window and bathroom seals. Silicones are the basis for coolants in transformers, protective encapsulating materials for semiconductors in computers, and foam-control agents in laundry detergents. The electronics and telecommunications industries need silicones to produce optical fibers and silicon wafers and chips. Some of their many other uses include promoting adhesion in glues, sealants, pigments, paints, textiles, and wire and cables, and serving as strengthening agents to reinforce rubber. Figure 6.5 shows silicones' uses in medicine. Even if one does not understand all the medical terms in the figure (I don't …), still the impression is overwhelming; almost any organ can be made from or be fixed by silicones.

6.4.2   Polymers and Additives

Raw polymers can be groomed and improved by additives. Some additives make the polymer more durable. For example, the tires of your car are made of rubber. But rubber as such is not durable. In the nineteenth century, the American Charles Goodyear was experimenting with rubber, trying to improve its qualities. In the 1840s, he finally succeeded in treating rubber with sulfur, which made the polymer more durable, more elastic, and less sticky. This invention is called vulcanization, and in due course, it was found that the sulfur links different polymer chains via −S−S− bonds, thereby improving the rubber for commercial uses. Later, it was found that adding carbon to rubber made it more durable to chafing and abrasion. Further durability of the rubber is achieved by adding antioxidants (had the rubber been human, the antioxidants would have been vitamins C and E!). All these and many other additives prolong the lifetime of the tires in your car.

Other additives to polymers make them colorful and beautiful. Just think about the stunningly colorful fabrics that are made from polymers, or the colorful kitchen utensils, house paints, and so on. A world without polymers? Hard to imagine!

Concluding Remarks on Polymers: We are in the Age of Polymers. There can be no question about that. Polymers are and will continue to be the materials of choice. Polymers will be used for conduction and storage of electricity, heat, and light, data storage and processing, medicine, housing, transportation, etc. In days to come, mankind will encode information in polymers, which will serve as software for making other materials or complex systems and for self-correction of chemical

Silicone medical devices and the body

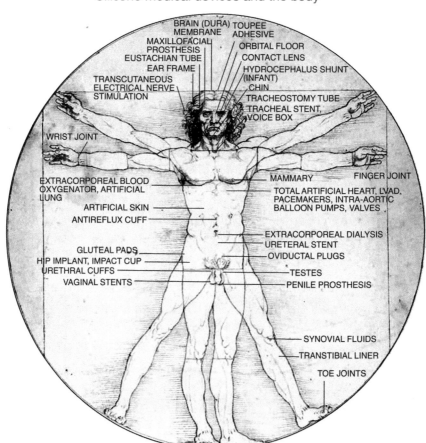

FIGURE 6.5   Medical uses of silicones, showing both silicone devices and body parts which can interact with silicone material implants. The initial picture of Leonardo da Vinci's artwork, used to create this image, is in the public domain and was obtained from http://commons .wikimedia.org/wiki/File:Leonardo_da_Vinci-_Vitruvian_Man.JPG. The details about the uses of silicone materials were obtained from an image provided by Professor Y. Apeloig.

damage. Just as Nature selected protein and DNA polymers as the materials of choice for propagating genetic information in living systems, so mankind manipulated molecules and created the Polymeric Age, which will eventually usher the revolutions to come.

6.5   SUMMARY

In this lecture we learned some new principles:

☞ Unlike atoms belonging to the second period (O, N, and F), which adhere always to the octet rule, those of the third period and below (S, P, Cl, etc.)

can maintain several states of Nirvana and generate molecules which obey the octet rule side by side with electron-rich ones, which exceed the octet (see also Retouches section 6.R.1). Chemists call electron-rich centers *hypervalent.*

☞ DNA and RNA polymers can be constructed from nucleotides, like proteins from amino acids. Thus, by "clack" (excision of a water molecule from each successive pair of nucleotides), we create dangling connectivities that "click" to form DNA and RNA strands, which are polynucleotides.

☞ Synthetic polymers can be constructed following two principles: (a) the condensation method, which is a "clack-click" construction, and (b) the initiation method, which uses a reactive species that creates a new connectivity center in a monomer. *This connectivity center in turn propagates to many monomers and zips them up to a polymer.*

The bulk of the lecture focused on DNA and RNA polymers and synthetic or man-made polymers. In so doing, we witnessed the marvelous unity of matter in the universe. All matter, be it inanimate or animate, is made from LEGO fragments that recur and give rise to infinite molecular worlds which constitute our universe. Another aspect of looking at polymers is the appreciation of the information content of complex molecules. The Polymer Age, which includes the nanotechnologies of DNA/RNA origami,[1] nanowiring,[7] information encoding, and also synthetic DNA/RNA forms,[8] is a revolution in the making …

6.6   REFERENCES AND NOTES

[1] See, for example: K. Sanderson, *Nature*, **2010**, *464*, 158.

[2] It turns out that incorporation of RNA nucleotides into DNA strands is risky, even for the same species. See, K.W. Caldecott, *Science*, **2014**, *343*, 260.

[3] See cover story on HIV treatments in: *Chemistry & Engineering News*, 2014, September 1, pp. 14–21.

[4] Read in the wonderful little book: P. Atkins, *What Is Chemistry?* Oxford University Press, England, 2013.

[5] Another wonderful book, that tells this story and others is: E. Heilbronner and F. A. Miller, *A Philatelic Ramble through Chemistry*, Wiley-VCH, Weinheim, Germany, 2004 (paperback edition).

[6] G. Patterson, *A Prehistory of Polymer Science*, Springer, Germany, 2012.

[7] (a) The Lund-based (Sweden) Sol Voltaics company has recently invented a new technique to generate nanowires, which dramatically improves the efficiency of solar energy cells. See *China Daily*, 2013, December 10, p. 18; (b) For the description of the technique and the produced nanowires, see: M. Heurlin, M. H. Magnusson, D. Lindgren, M. Ek, L. R. Wallenberg, K. Deppert, and L. Samuelson, *Nature*, **2012**, *492*, 90.

[8] Making artificial nucleotides is a rather simple task. Will they generate new forms of life? See: S. Everts, *Chemical & Engineering News*, 2012, April 23, p. 5.

6.A   APPENDIX

6.A.1   Proposed Demonstrations

Our proposed demos for this lecture are restricted to artificial polymers.

- **Polymer slime**

This experiment demonstrates the effect of a cross-linker agent on a solution of a polymer, turning it into slime, with interesting viscoelastic properties. The cross-linker holds the individual polymer chains together by weak forces, which cause the polymer to behave sometimes like a liquid (viscous) and other times like a solid (elastic), depending on the cross-linking one applies to it. In the demonstration,[1A] a solution of polyvinyl alcohol (related to polyvinyl chloride [PVC] by replacing Cl with OH) in water is mixed with a borax solution with a stick until it solidifies. Then, a volunteering/volunteered student is asked to put on gloves, lab coat, and goggles, and to explore the viscoelastic properties of the polymer. For example, what happens when one tries to pour the slime from one cup to another? It spills all too slowly, like a very viscous liquid. Then talcum powder is added to the slime, and the mixture is rolled into a ball and dropped on the table. Is it bouncy? Yes! Here it behaves like an elastic solid. The student is then asked to pull apart the slime fast and then slowly: what is the difference? When the polymer is slowly pulled, a thin film is formed. This is because the polymeric chains have sufficient time to flow, one on top of the other, and to rearrange to the stretched form. But, when the polymer is pulled abruptly, this rearrangement is not enabled, and hence the polymer tears apart.

- **"Vanishing" water**

This is a visually amazing demonstration in which water is poured into an opaque disposable cup, and, when the cup is turned upside down, the water "vanishes." What the students do not know is that inside the cup there is sodium polyacrylate powder, the same powder that is used in disposable baby diapers. Sodium polyacrylate is a water-loving (what chemists refer to as hygroscopic) polymer that absorbs water, and as a result, turns into a gel, which does not pour out when the cup is turned upside down.

To make the demonstration even more spectacular, it is advised to use three cups, where the polymeric powder is only present in one of them. The water is poured from one empty cup to another empty cup, and the places of all three cups on the table switch. The students are then asked to follow the water and tell, at each pouring and switching step, which cup has the water in it. Then, the demonstrator pours the water from one of the two cups into the third one, which contains the polymeric powder, switches the cups, and asks the students about the whereabouts of the water. When the students point to the third cup, the demonstrator turns over all three cups, and the water suddenly seems (to the students) to vanish! The experiment is repeated, but this time with transparent cups, and the swelling ability of the super-absorbent sodium polyacrylate is discussed.

- **"Vanishing" cup**

Following the previous experiment, the demonstrator causes a cup to mysteriously vanish. Here, a disposable cup made of polystyrene is placed within a beaker, and a box that is open from its top and bottom sides is placed over it. Then, a transparent liquid, which the students assume is water but is actually acetone, is poured into the beaker. Subsequently, the demonstrator asks the students: "Where is the water?" The box is removed, and it is seen that this question is irrelevant! The right question is, where is the cup? The experiment is repeated without the box, and the dissolution of the polystyrene in acetone is discussed. Another important lesson that the students learn is that *not every transparent liquid is water!*

The last two experiments were both originally designed by Dr. Y. Aouat.

6.A.2   References for Appendix 6.A

[1A] K. Hutchings, *Classic Chemistry Experiments*, The Royal Society of Chemistry, London, 2000, p. 195.

6.R   RETOUCHES

6.R.1   To Be or Not to Be in Octet? This Is the Question

As I mentioned in the text, chemists still fight over this question. Let me try to explain why and what I personally think about the fight. I will do so using H_3PO_4 as an example. The way we drew this molecule in the text is by pairing all the valence electrons of phosphorus (five electrons) with connectors of the oxygen atoms. Once we knew the formula of the molecule, and the fact that it does not have O—O bonds, the electron-rich formulation of bonding was self-suggesting. And once you were told and you accepted that atoms like P (from the third period on) can form electron-rich molecules side by side with ones obeying the octet rule, then construction of such molecules became a simple matter.

However, this is not the only way to pair up the electrons in the molecule. If you were a staunch adherent of the octet rule, you could have shifted one of the electron pairs of the P=O bond in the electron-rich formulation and placed these two electrons on oxygen. Since oxygen gained now an electron, and phosphorus lost one, the P=O bond will become a single bond, but its atoms will be charged, which we can represent as P^+—O^-. Now, all the atoms obey the octet rule (and H obeys the duet rule), and we all have peace …

So why are chemists fighting over this issue? This is because of quantum mechanics (QM), which we mentioned already in Lecture 1. Thus, in advanced chemistry courses, we learn that the electrons reside in pairs inside *orbitals*, which are little clouds that describe the probability of finding the electrons in the space surrounding the nuclei. These orbitals also have energy levels. The number of these orbitals is limited, and it depends on the period to which the atom belongs. The atoms that belong to the second period in the periodic table, like O, N, and F, have *only four orbitals* in their valence shell. Therefore, when they pair electrons by forming bonds with other atoms, they will obey the octet rule, since the four orbitals can accommodate only eight electrons.

As we move to the third period onward, the number of orbitals in the valence shell increases to nine. And here comes the fight! Those who adhere to QM and use it qualitatively (i.e., counting the number of available orbitals in the valence shell and multiplying by two to get the total number of electrons allowed in this valence shell), say, "Phosphorus (and other third-row atoms) will have a propensity to form electron-rich molecules." By contrast, those who use QM quantitatively (they use computers to calculate the electronic structure of molecules) look at the additional five orbitals of phosphorus and see hardly any electron population in them. They understand the reason: this is because these additional five orbitals have significantly higher energies than the first four orbitals, and hence populating high-energy orbitals with electrons is not beneficial for the molecule. Therefore, they conclude, "The octet rule is forever." Scientific disagreements are fine and, in fact, are desirable, because conceptual tensions lead to progress in science. So, do not feel that there is anything wrong here!

Where do I stand on this issue? I can give you a hint: I use QM quantitatively … Still, I am also a teacher. And teaching the electron-rich form in Scheme 6.R.1 is easy, compared to the one obeying the octet rule. My goal is really to teach students to construct the molecule and know how to make from it more and more molecules. Furthermore, at any level of study, when you have two clashing theories, you must ask yourself, "Do they have a different explanatory power? Does one predict something more than the other?" We already know that a double bond such as P=O should be stronger and shorter than a single bond. But the P^+-O^- bond is not really a simple single bond, because it has, in addition, also a strong attraction between the opposite charges, and therefore it is predicted to be shorter and stronger than a single bond. So, at our level of understanding, the two descriptions of H_3PO_4 have identical explanatory powers, and they predict similar features of this bond. Therefore, as a

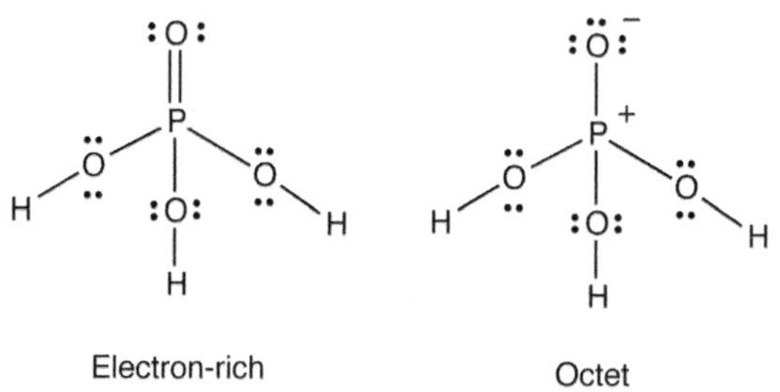

Electron-rich   Octet

SCHEME 6.R.1   Two electronic representations of H_3PO_4. One is electron-rich, and the other obeys the octet rule.

teacher, I will teach the easier way of describing this molecule. And since there is no additional information you can gain from the octet-following form in Scheme 6.R.1, you as students can use the electron-rich form with no concern.

Before moving on, I would like to point out that electron richness infiltrates also into the "lonely" existence of noble gases and makes some of them participate in bonding with other atoms. For example, xenon forms compounds like XeF_2, XeF_4, and so on. Nevertheless, such molecules are rare, and the chemistry of noble gases remains essentially atomic.

6.P PROBLEM SET

6.1 Construct the following molecules based on the LEGO principle we learned in this lecture. (a) H_2SO_3 as an acid, (b) SO_3, and (c) SO_2. None of these molecules contains O—O bonds.

6.2 $HClO_4$ is called perchloric acid. Given that it has no O—O bonds, construct it using the principle we learned in this chapter. Is there an analogous acid with the formula HFO_4?

6.3 How many possible dinucleotides exist for DNA nucleotides? List them with letters, such as AT, and show the molecular structure for one of your choices. Are dinucleotides like AC and CA identical? Support your answer with appropriate drawings.

6.4 Construct an artificial RNA nucleotide made from a fragment derived from the ribose sugar and modular fragments of sulfuric acid and an unnatural base. Restrict your choice to a base that contains nitrogen (see the natural bases in the text). Show your chosen fragments and the construction process.

6.5 Use the "clack-click" method we learned in this lecture to construct an ester made from phosphoric acid and an alcohol of your choice.

6.6 Acrylonitrile is made from ethylene, where one of the hydrogen atoms is replaced by a CN fragment (called nitrile). Construct acrylonitrile from appropriate fragments. Then construct the polymer that can be formed from acrylonitrile by initiation with a free radical. Acrylic fibers are made mostly of this polymer.

6.7 The molecule $(CH_3)_3C-N=N-C(CH_3)_3$ decomposes upon gentle heating and creates an initiator for polymerization and a molecule of gas that exists naturally in the atmosphere (the initiator and the gas are two separate molecules). Write down the chemical equation describing the decomposition process. If you can identify the initiator, then use it to polymerize styrene (consult Figure 6.3 to identify the molecule styrene).

6.8 Try to draw the chemical formulae of the copolymers that can in principle be formed from ethylene glycol (a di-alcohol; consult the text) and the following acids: H_2SO_4 and H_2CO_3 (carbonic acid). Figure 8.7 presents the structures of the two acids.

THE 3D STRUCTURE OF MOLECULES, ELECTRONEGATIVITY, HYDROGEN BONDS, AND MOLECULAR ARCHITECTURE

7.1 CONVERSATION ON CONTENTS OF LECTURE 7

Usha: I do not see Racheli with us.

Sason: Yes, Usha, Racheli is getting married soon. She is extremely busy. So, you are going to interview me on the contents of this lecture.

Usha: It would be my pleasure, Sir. I understand you are going to talk about the three-dimensional (3D) structure of molecules and the architecture of chemical matter. I thought you would never reach this central topic—I have always been fascinated with the 3D structures of molecules.

Sason: You are right, Usha. I kept delaying this, with statements like: "Later we shall learn how to predict the 3D structure of molecules." Later is now …

Usha: As far as I remember, Sason Sir, it was van 't Hoff who first came up with the idea that molecules were 3D, in 1874. Why did it take chemists so long to come up with this idea? After all, our world is 3D, and it seems logical to assume that material objects will have 3D shapes too, no matter how small.

Sason: This is a good question, Usha. But, before I answer it, let's do away with the "Sir." Also, I have a surprise for you. Roald Hoffmann is going to join us at some moment. Roald also loves molecules, and I have yet to see anyone who has a sharper eye than he does for the beauty of molecules.

Usha: I can't, Sason Sir. This is a great surprise! Roald Sir was your and Jemmis's teacher. Jemmis was my PhD adviser, and you are my postdoctoral adviser, and

Chemistry as a Game of Molecular Construction: The Bond-Click Way, First Edition. Sason Shaik.
© 2016 John Wiley & Sons, Inc. Published 2016 by John Wiley & Sons, Inc.

here we are talking about a textbook for students. So now we have a teacher, a student, and a student of a student, all eager to teach more students. This is really marvelous! I am glad and honored that I have a chance to interview the two of you.

Roald: Let me try to answer the question you asked, Usha, before Sason interrupted you to tell you that I am joining the dialogue. Why did it take so long to go into 3D? By the middle of the nineteenth century, chemists had an idea that in compounds, real matter, atoms were *somehow* connected to other atoms. Not that all people even believed in atoms because they were invisible. Chemists had the idea that certain groups of atoms were preserved in chemical reactions—say a CH_3, C_6H_5, OH.

Usha: So why didn't they just draw bonds connecting those fragments?

Roald: They did—look at this collage of mid-nineteenth-century chemical formulas in Figure 7.1.

You see lines, blocks, linkers, spheres, and overlapping rectangles …

Sason: I get this overwhelming sense of people desperately trying to show connections. Recall, the electron was not yet discovered. So there was no physical definition of a bond, and the connections were abstract fuzzy ways of trying to describe molecules.

Usha: But did they believe in them?

Roald: If pushed, they would say: "No, we don't know what's in there, down deep. The lines are just a symbol for 'this is connected to that.'" But you know, draw

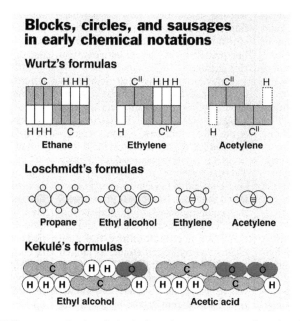

FIGURE 7.1 Different methods of depicting the structures of molecules in the 1860s. Adapted from "A History of Chemistry (Volume 4)" by J. R. Partington, Macmillan & Co. Ltd., London; St. Martin's Press, New York, 1964.

a line often enough, and it hardens. As my friend Emily Grosholz says, the lines became "the furniture of the mind."

Sason: So here's the answer to your question, Usha, of why it took so long to go 3D—people didn't know how atoms were arranged in space. They could not see atoms, even as they felt they were there. And they did not have the techniques to extract spatial information from their macroscopic observations of chemicals.

Usha: So what was the trigger, Sason and Roald Sir?

Roald: As usual, the trigger came from experimental facts and worrying about handed (or chiral) molecules.[1]

Usha: Sason Sir already talked about these in Lecture 1—these are molecules that are related to one another like our right and left hands or as an object and its mirror image.

Roald: But the experiment that set people off thinking about handed molecules had to do not with microscopic molecules but with the properties of solutions that contained an immense number of such molecules—actually molecules found in wine.

Sason: And, of course, it happened in France ... Roald, with you and I being together, we are going to have many digressions ...

The property of handedness was discovered by Louis Pasteur, who was investigating crystals of a salt which is related to an acid. This salt is found in grapes and hence also in wine (called tartaric acid). Though Pasteur worked on the salt form, the same conclusions also apply to the standard acid form. Under the microscope, he could see four different crystal forms, two of which were asymmetric (see Figure 7.2) and looked like an object and its mirror image, while the other two looked symmetric.

Boldly, Pasteur concluded that these crystals were built of isomers of the same molecule.

Usha: But those crystals are not molecules; they're assemblies of zillions of molecules arranged in an ordered framework.

Sason: This was Pasteur's incredible insight—from the fact that crystals were shaped as mirror images, and that solutions of the crystals turned the plane of polarized light to the left and to the right, he concluded that it was the small molecules that formed the crystals and that were swimming in the solution they studied—that these tiny creatures actually had a handedness.

Roald: The molecule tartaric acid is like ethane (H_3C-CH_3), in which two of the H's on the two carbons were replaced by OH and COOH fragments, that is, (HOOC)(HO)HC–CH(OH)(COOH). If the molecule were planar, it would have had a different number of isomers. There were similar problems of isomerism, which could not be accounted for by the usual "structure" models, as in Figure 7.1.

Usha: So, how did van 't Hoff solve this puzzle?

Sason: van 't Hoff was a young man when he published his first paper in Dutch in 1874, "... *extension of the formulae in use now in chemistry into space* ..." In this paper, he showed that if the valencies (directions of attachments) of carbon

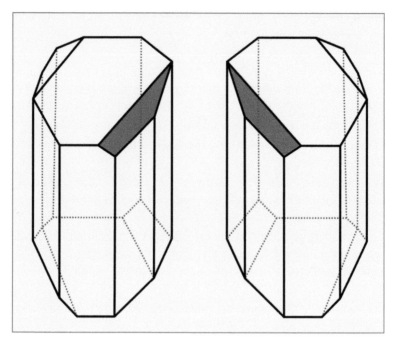

FIGURE 7.2 Schematic representations of the two crystal forms of tartaric acid salt, which look like an object and its mirror image. This image, which is in the public domain, was obtained from http://commons.wikimedia.org/wiki/File:Pcrystals.svg.

were tetrahedrally distributed, one could solve the puzzle raised by Pasteur's observations. Young men are daring …

Figure 7.3 shows the tetrahedral models he designed in order to test his idea of the handed molecules and the non-handed isomer. Going 3D seemed to be the answer to the puzzle. This is a remarkable idea given the fact that electrons were not yet known, and *valencies* were abstract undefined entities. Usha, if you recall,

FIGURE 7.3 van 't Hoff and his tetrahedral models. The van 't Hoff caricature is used courtesy of W.B. Jensen. The image of the tetrahedral models comes from the Museum Boerhaave in Leiden and is used with permission of the museum to Wiley.

in Lecture 4, I used spatially directed connectivities. So, I was letting the concept of 3D into the construction story before I explained why could I do so.

Usha: I assume this idea was accepted with flying colors.

Sason: Wait, wait, you are jumping to conclusions too fast! In 1875, van 't Hoff published a second paper in French, entitled "La Chimie dans l'Éspace" (chemistry in space). In this hefty paper, he explained the Pasteur problem showing that, if one adopts a 3D structure of tartaric acid, then there can only be three isomers, not four as Pasteur thought.

Roald: But there were four compounds Pasteur had.

Sason: Wait. Two of these were made up of molecules of the two handed forms (Figure 7.2); the third crystal was of a form wherein the handedness of the two carbon atoms "cancelled" one another, and the fourth was not a real isomer but simply a 1:1 mixture of the two handed isomers, which together assembled into a crystal structure. Three molecules, four compounds.

Furthermore, when tartaric acid lost the two hydroxyl groups, it did not give one isomer but two, called maleic and fumaric acids, which had the same molecular formula, $(HOOC)(H)C=C(H)(COOH)$, but nevertheless, had different properties. van 't Hoff argued that the double bond (double valency in the nineteenth-century language) between the two carbon fragments could be represented by two tetrahedra connected via an edge, which prevents rotation around the $C=C$ bond, giving rise thereby to two isomers (as we shall see, these are called *cis* and *trans*). He also described what we view today as a triple bond like in acetylene as face-sharing tetrahedra.[2]

Usha: This is an amazing intuition! But you still didn't answer my question—were people thrilled to see this explanation?

Roald: Not everyone. Many important chemists were not that enamored of the idea of unseen objects (molecules) endowed with 3D extent and that this feature, in turn, could lead to tangible physical consequences. One of these was Hermann Kolbe, a highly influential chemist, who expressed his objection in a satirical essay in 1877:[1a]

"A Dr J.H. van 't Hoff of the veterinary school at Utrecht, finds, as it seems, no taste for exact chemical investigation. He has thought it more convenient to mount Pegasus [a mythological winged horse], obviously loaned at the veterinary school, and to proclaim in his *La chimie dans l'éspace* how during his bold flight to the top of chemical Parnassus [the name of the mountain sacred to Apollo and the home of the Muses], the atoms appeared to him to have grouped themselves throughout universal space."

Sason: Kolbe, fortunately, by this time was an outlier, as influential as he once had been. van 't Hoff's theory explained so much, so simply. People accepted it.

Usha: Tell me, Roald Sir, why then he did not get the Nobel Prize for the work? And what is the story of Joseph Achille Le Bel, who has supposedly published the same idea?

Roald: Well, actually van 't Hoff was the first Nobel Prize winner (in 1901). He did many other things in addition to his youthful extension of chemistry into three

FIGURE 7.4 A photograph of Joseph Achille Le Bel. This image, which is in the public domain, was obtained from http://commons.wikimedia.org/wiki/File:Le_Bel.jpg.

dimensions, and his award was given to him for his "laws of chemical dynamics and osmotic pressure in solution." It's OK, the Swedish Academy of Sciences, the body that gives out the Nobel Prizes, doesn't always get it right. For instance, they forgot to give a Nobel Prize to Mendeleyev.

Le Bel (Figure 7.4) was an assistant of the renowned French chemist Wurtz. As was van 't Hoff! Just a month after van 't Hoff, Le Bel published his own explanation of what Pasteur had seen. Le Bel's paper was formalistic, mathematical. Most importantly, it lacked drawings of 3D molecules, even though he was speaking about 3D objects. Well, the mind is geometrical. Chemists like drawings. van 't Hoff's models caught on much more quickly; they captured the geometrical imagination of chemists.

Usha: Sason Sir, what else do you plan to teach in this lecture? We are both very curious.

Sason: I want to teach about the architecture of matter. Chemical bonds create 3D molecules with strong forces holding the atoms together. I also want to teach about the hydrogen bond ...

Usha: But I know that these are weak. Why bother with them?

Sason: Let me complete my sentences, Usha. The answer to your question is because much of the architecture of aggregates of molecules or of macroscopic, human-sized matter as we see it is determined by weaker forces.

Roald: And these add up.

Sason: Yes. I will also have to teach a new concept before talking about hydrogen bonds. This concept is called *electronegativity*. It measures the ability of a bonded atom to attract the electron pair of the bond in its direction and hence acquire some excess negative charge. If I will manage to drive home this concept, then I can teach about hydrogen bonding (H-bonding) as an element of architecture.

Roald: No justice in the molecular world; the strong wins out.

Sason: Not entirely. There's a balance. But not all atoms are equal.

Usha: And what comes of electronegativity and the H-bond?

Roald: He'll speak about the structure of liquid water, of ice, and ...

Sason: Hey, whose book is this? I will speak also about the architecture of proteins as shaped by H-bonds. We are all made of proteins, and our skin is a protein wrapping. The way we appear then is again a function of H-bonding. Of course ...

Usha: What about DNA, Sir?

Sason: Usha, you are learning our bad habits of interrupting one another!

Of course, this will be the pinnacle of this lecture. I will show how the DNA bases pair up by H-bonds and how they recognize one another *selectively* through H-bonds. This is a neat way of storing the information in the sequence of the bases.

Roald: But is it the only way? This will be the greatest fun of meeting up with the bug-eyed monsters on an exoplanet that can support evolution of complex matter. How will they transfer genetic information?

Sason: See the previous lecture. This complex matter would be made of synthetic polymers that contain transmittable information!

Usha: Gentlemen, let us not digress. Isn't H-bonding also the mechanism whereby the DNA "gives orders" to the cell to make proteins?

Sason: The DNA does not give any orders. It is a pretty passive giant molecule that contains and stores all the information. RNA contains DNA information, and then RNA gives the orders ...

Roald: Don't forget to talk about the puzzle of why all living systems use handed (chiral) molecules of a given handedness (chirality), left or right.

Sason: Sure, Roald. Proteins in living organisms and plants are made of left-handed amino acids, never of right-handed ones. So marvelous is this choice of nature that even in these days, there are heated arguments about how we ended up with this uniform chirality (called homochirality).

But the mechanism of recognition by handedness is not so difficult to demonstrate. In Lecture 1, I use the handshake trick with students and show them how two right hands snuggle nicely in the handshake, while using a right hand and a left hand is awkward. I learned that from you many years ago. It is so much fun to shake hands with students in class. They wake up ...

Usha: Perhaps you will mention chiral drugs. Right? Drugs of the wrong chirality can kill ...

Sason: I will mention thalidomide ...

Usha: Thank you, Roald Sir, for participating in this interview. Sason Sir is looking at me with one of his severe looks. I guess this is a hint he wants to start the lecture.

Sason: Yes, Usha. Let's stop here since we have ahead a very long lecture. I preferred to group the 3D and architectural topics in a single lecture because of the affinity of the topics. But, I think the teachers who will use this book should consider splitting it into two lectures and starting the second one with intermolecular interactions. This second lecture may be accompanied by some demos described at the end of the lecture.

7.2   3D STRUCTURES OF MOLECULES

Natural objects prefer to be in their lowest possible energies. You can experience this by yourself if you hold, say, a heavy stone high above the floor and then let go of it. It will fall to the floor, and you will sense the impact of the energy release. You will be able to feel the amount of energy that was released in the process by letting the stone fall on your foot. (I'm just joking. Please don't do that.) Like all natural objects, molecules too prefer to be in their lowest possible energy state. Electron pairing makes bonds, which lower the energy of the molecule, and hence atoms tend to make as many bonds as they can to satisfy the Law of Nirvana.

A bonded atom is generally surrounded by electron pairs (bonds and lone pairs). These electron pairs repel one another, because all the pairs have negative charges (see however, Retouches section 7.R.1). Therefore, the molecule has an additional mechanism to lower its energy. It will keep the electron pairs around each atom at a maximum distance from each other, *and hence the bonds of a given atom will be distributed in space to lower the molecular energy. This driving force for minimum energy is the root cause of why molecules acquire spatial shapes, and are often 3D objects.* The actual shape depends ultimately on the number of electron pairs surrounding a given atom.

Before proceeding, let me make a practical recommendation. As the number of electron pairs increases, the structures become less easy to visualize. I highly recommend using molecular models (which can be purchased), which can help visualization. Alternatively, websites (e.g., http://www.chemtube3d.com/:select VSEPR) can also be very useful here. Some teachers may choose to skip cases with five and six electron pairs, depending on the circumstances. I usually skipped these for humanities and social science students, but neither for chemists nor for advanced high school students.

7.2.1   Selection Rules of 3D Molecular Structures

The selection rules for predicting and understanding molecular shapes are the following:

☎ Bond pairs and lone pairs around the atom will be distributed in space so as to maximize their distances and minimize thereby the pair–pair repulsion.

☎ The preferred molecular shape will depend on the number of pairs that have to distance themselves around a given atom.

The rules are known in chemistry as the VSEPR rules, where VSEPR is the acronym of "valence-shell electron-pair repulsion," and they were devised in 1957 by Gillespie and Nyholm,[3] two young scientists—one English, Gillespie, and the other Australian, Nyholm—who were both students of Sir Christopher Ingold, one of the greatest British chemists in the post-Second World War era.

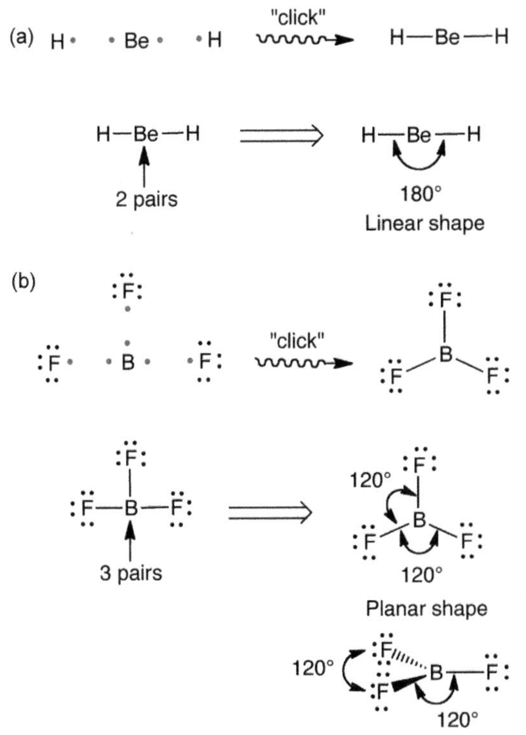

SCHEME 7.1   Predicting molecular shapes for cases with two and three electron pairs around a central atom: (a) BeH_2, with two pairs around Be, is linear, with an H—Be—H angle of 180°. (b) BF_3, with three bond pairs around B, is planar, with F—B—F angles of 120°. BF_3 is shown using two perspectives. In the first, the molecule lies in the plane of the page, and you are viewing it from above. In the second drawing, beneath the first one, the molecule is drawn with its plane being perpendicular to the page. The bonds coming out of and going behind the plane of the paper are drawn using, respectively, a wedge and a dash.

Let us now apply these rules and predict the shapes of some molecules. As examples, we are going to use simple molecules we have constructed already in Lecture 2.

The first such molecule is made from a beryllium atom (Be) and as many hydrogen atoms as needed. Since Be is a double-connector atom and each H is a single connector, then as shown in Scheme 7.1, the atoms "click" and generate the familiar molecule BeH_2. What shape will BeH_2 prefer? In order to predict the shape, we simply count the number of pairs around the atoms. Since Be is surrounded by two pairs, the maximum distance between the pairs will be obtained when the molecule adopts a linear shape with a bond angle of 180° as shown in Scheme 7.1a.

Consider now a molecule made from one B atom and as many F atoms as needed. As shown in Scheme 7.1b, since B is a triple connector and F a single connector, we need three F atoms to bind to B and "click," we generate the molecule BF_3. It is seen that we have several electron pairs around both B and F. However, since F is connected by one bond, the "geometry" around it is meaningless, and hence we disregard its lone pairs for our present purposes. But, as B has three bonds, it is meaningful to consider the geometry around it. To maximize the distance between the three pairs around boron, BF_3 adopts a planar shape, where all the FBF bond angles are 120°.

To challenge our artistic talents, we want to draw BF_3 in two different perspectives. The one shown first in Scheme 7.1b has the molecule lying flat on the plane of the paper. The one shown below it places the plane of the molecule perpendicular to the page, such that one of the B—F bonds comes out of the plane toward you, and another goes in and is behind the plane of the paper. To symbolize this perspective, we draw the B—F bond coming out of the page using a wedge, while for the other B—F that goes behind the page, we use dashes. The third B—F bond is in the plane of the page. Here again, we meet another element of the language of chemistry. Chemists communicate by these drawings, and they all recognize what the drawing means. Just imagine you had to write an essay with 100 different molecular shapes. If you had to explain the structure of each molecule in words as I just did, you would certainly drop the assignment. Languages must be learned because they provide efficient ways of communication.

After preaching to you, I can proceed to a more challenging case with four electron pairs around a central atom. We have already constructed methane many times, so let us skip the "click" step and focus on the molecular shape. As shown in Scheme 7.2a, methane has four pairs around the carbon atom. We might have thought that the molecule CH_4 could be planar and have HCH angles of 90°. But this is not the

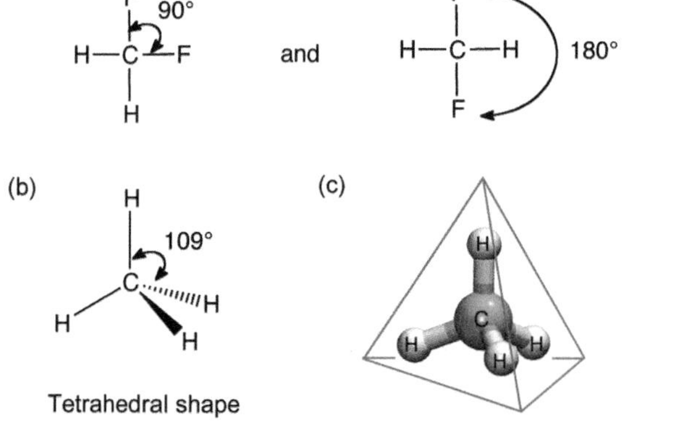

SCHEME 7.2 A case with four electron pairs around a central atom, using methane as an example, to understand the chosen geometries of such molecules by exploring different possible structures. (a) CH_4 is not planar! If it were planar, then the planar CH_2F_2 would be predicted to have two isomers, but CH_2F_2 actually has a single isomer. (b) CH_4 is tetrahedral, with H—C—H bond angles that are close to 109° (the precise angle is 109.47°). The 3D shape is drawn with wedged and dashed bonds. (c) A 3D structure of methane with balls (representing atoms) and sticks (representing bonds) placed inside a tetrahedron.

true structure. How do chemists know that? Very simple; if CH_4 were planar, the replacement of two H's by, for example, two F's would have resulted in two isomers (Scheme 7.2a), one in which the two C—F bonds form an angle of 90° and the other where they are opposite to one another with an angle of 180°. However, the molecule CH_2F_2 does not have isomers, so the structure of CH_4 and other molecules of its type with four pairs around the carbon *cannot be planar*. This is because the planar structure does not keep the C—H bonds at the maximum possible distances. There is a preferable structure in which the four pairs are further apart! It is called a tetrahedral shape, where a tetrahedron is a body that has four triangular faces (in Greek, *tetra* = four, *hedron* = face).

Scheme 7.2b shows that in order to maximize the distances among the bond pairs, the structure assumes a tetrahedral, 3D shape, in which the HCH bond angle is 109°, *significantly larger than the 90° in the planar structure*. Note that we drew this tetrahedral shape, with two C—H bonds in the plane of the page (regular lines), while one bond is coming out of the page and drawn as a wedge, and the other going behind the plane is drawn with dashes. Scheme 7.2c shows a computer-generated structure of CH_4, where the atoms are represented by spheres and the bonds by sticks (so-called "a ball-and-stick model"), and the structure is enclosed inside a tetrahedron, which is traced around the molecule using red lines. We shall use the drawing in (b), but you are advised to build CH_4 from molecular models, rotate it, and see how to draw it from other perspectives (or go to the above website, which allows you to rotate the molecular images).

Scheme 7.3 contains two electron-rich molecules, wherein the central atoms have five and six electron pairs. With five electron pairs around P in PCl_5, the maximum distances among the pairs will be obtained in the structure in Scheme 7.3a, wherein three of the P—Cl bonds are in the same plane with Cl—P—Cl angles of 120°, and the two others form a Cl—P—Cl axis perpendicular to this plane. This 3D structure is called a *trigonal bipyramidal structure*, since it is derived from the corresponding bipyramidal body, drawn on the right-hand side using red lines; note that the trigonal bipyramid is made of two trigonal pyramids sharing a face. Thus, if we place P in the middle of the shared face and the five Cl's in the corners of the bipyramid, while connecting all the Cl's to the P, we shall obtain the 3D molecular drawing on the left.

Scheme 7.3b illustrates the already familiar molecule SF_6. With six pairs around S, the maximum distance among the pairs will be found in a structure where all the F—S—F angles are 90°, and which is called octahedral. On the right-hand side, we show the body octahedron using red lines that connect the F atoms. The body (it has eight triangular faces) is made from two square pyramids (like those in Egypt) sharing their bases. If we place S in the middle of the square base and the F atoms in the six corners, and draw bonds (in black) from S to F, we shall generate the 3D structure on the left-hand side.

7.2.2 Lone Pairs Count in 3D Structure Determination

When both bonds and lone pairs surround a central atom, the lone pairs repel one another as well as repel the bond pairs. Thus, when we predict the 3D structure of

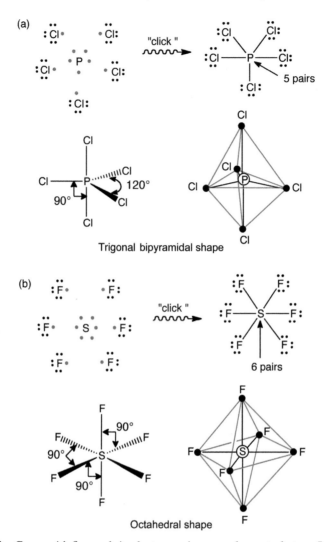

SCHEME 7.3   Cases with five and six electron pairs around a central atom. In both cases, the molecules are first constructed using the "click" method. Then, one counts the electron pairs around the central atoms, and one proceeds to predict the 3D structures: (a) With five pairs, PCl_5 has a trigonal bipyramidal shape. (b) With six pairs, SF_6 has an octahedral shape. The red lines in (a) and (b) are not bonds, but rather lines that serve to outline the corresponding polyhedral shapes.

such a molecule, we must count also the lone pairs. Take, for example, NH_3 and H_2O, which are shown in Schemes 7.4a and 7.4b. Since we already constructed these two molecules, we know that, in both cases, the central atoms are surrounded by four electron pairs. *Therefore, both molecules will assume shapes that are derived from a tetrahedral structure.* This is reflected in the bond angles, which in both cases are close to the tetrahedral value of 109° (very few degrees smaller).

To appreciate the role of the lone pairs, we could consider the situation if the lone pairs were not geometry-determining. Thus, discounting the lone pair for NH_3, we

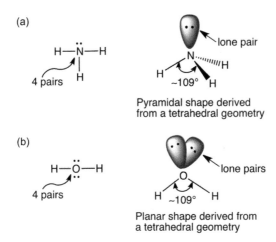

(a)

4 pairs

lone pair

~109°

Pyramidal shape derived
from a tetrahedral geometry

(b)

4 pairs

lone pairs

~109°

Planar shape derived from
a tetrahedral geometry

SCHEME 7.4   Cases including lone pairs that participate in determining the 3D shape of the molecules: (a) NH_3 and (b) H_2O. Note that, in both cases, the central atom is surrounded by four electron pairs (considering bonding pairs and lone pairs). Hence, the molecule's structure and bond angle are derived from those for a tetrahedral structure. For convenience, the lone pairs are placed in pinkish lobes that look like little ears occupying space (chemists call them "rabbit ears").

would have counted only three pairs around N and hence would have predicted a planar molecule with an HNH angle of 120°. Likewise, discounting the two lone pairs in H_2O, we would have counted only two pairs around O, and have predicted a linear structure with an angle of 180°. Obviously, neither of these latter two predictions is correct. The lone pairs must be counted for the prediction of geometry!

Note that in Scheme 7.4, we placed the lone pairs in lobes that look like little ears occupying space. For our purposes in the text, this may be simply considered as convenient cosmetics for appreciating the tetrahedral parenthood of the molecules. But actually these little lobes have some meaning in quantum mechanics (QM) theory. Whoever is interested in learning more about the nature of these lobes may consult Retouches section 7.R.2.

7.2.3   A Multiple Bond Counts as a Single Space Unit

In some molecules, the atoms share a multiple bond between them, such as a double or triple bond. How do we consider a double or a triple bond while predicting the 3D shape? Well, since the individual pairs of the multiple bonds between two atoms cannot be freely separated in space, they have to be counted as a single unit. Scheme 7.5 shows ethylene and acetylene, C_2H_4 and C_2H_2. It is seen that by counting the double bond in Scheme 7.5a as a single "pair," we would predict that each of the CH_2 units would be planar with an angle of 120°. The precise H–C–H angle is 116.6°, which is very close to the ideal 120°. Similarly, for acetylene in Scheme 7.5b, each C is surrounded by two "pairs," which lead to H–C–C bond angles of 180° and a linear molecule.

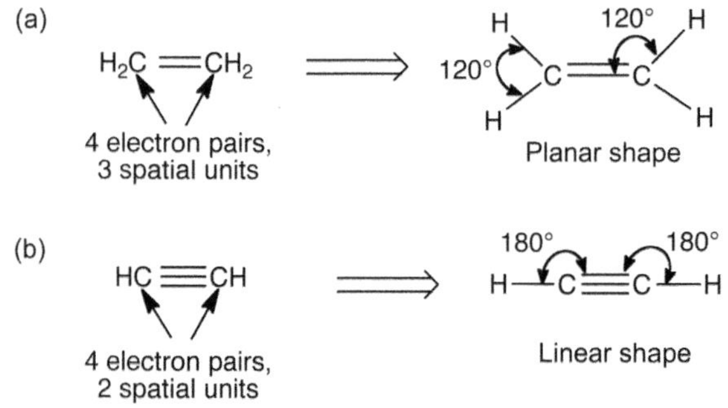

(a)

$$H_2C = CH_2$$

4 electron pairs,
3 spatial units

Planar shape

(b)

$$HC \equiv CH$$

4 electron pairs,
2 spatial units

Linear shape

SCHEME 7.5   Molecules with multiple bonds between two carbon atoms. As a rule, a multiple bond is counted as a single spatial unit: (a) The C_2H_4 molecule, with three "pairs" around each C: the geometry around each C is planar, and it is depicted with ideal H—C—C and H—C—H angles of 120°. (The precise angles are 121.7° for the H—C—C bond angles and 116.6° for the H—C—H bond angles.) (b) The C_2H_2 molecule, with two "pairs" around each C: the geometry around each C is linear, with H—C—C angles of 180°.

7.2.4   Isomerism in Double-Bonded Molecules

C_2H_4 is interesting. Its two CH_2 moieties are planar, but we have no way of predicting whether the entire molecule lies in the same plane, as was drawn in Scheme 7.5a, or with the two CH_2 units occupying two different planes. We know for a fact that the entire molecule is planar, and rotating the CH_2 moieties to occupy mutually perpendicular planes breaks one of the components of the double bond, which becomes a single bond. Here again QM theory accounts very well for this behavior (see Retouches section 7.R.3). Let us see at least a heuristic reason.

Since the bonds and lone pairs in a molecule are distributed in space, this necessarily means that *the atomic connectivity can also be regarded as being distributed in space*. As shown by the red patch at the top of Scheme 7.6, *the fourth connectivity of each carbon in the H_2C—CH_2 moiety lies in a plane perpendicular to the H_2C—CH_2 plane*. In this geometry, the two fourth connectivities of the two carbon atoms occupy the same plane and can click to form an additional bond between the two atoms. As indicated by the two red patches on the bottom left of the scheme, when the CH_2 groups twist by 90° relative to one another, the connectivity of each carbon remains perpendicular to the CH_2 plane, and therefore the double bond breaks and becomes a single bond. This bond breakage means that twisting the double bond requires the investment of a lot of energy (Retouches section 7.R.3).

These features of a double bond can then be summarized in the following statements:

☞   A double bond between two atoms creates a planar molecule.
☞   Rotation around the double bond by 90° breaks one of the bonds. Hence, the rotation around the double bond is not free.

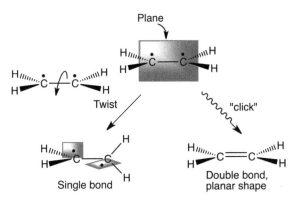

SCHEME 7.6   The two CH_2 groups of $H_2C{=}CH_2$ are shown with their fourth connectivities oriented in the red plane that bisects the two H—C—H angles. In this orientation, these connectivities can "click" to form an additional bond between the C atoms, which forms $H_2C{=}CH_2$, with a double bond overall between the C atoms. However, twisting the CH_2 groups to mutually perpendicular orientations breaks one of the components of the double bond and results in a single C—C bond (the two unpaired electrons on the C atoms in the two H—C—H units are no longer in the same plane, as shown by the two red patches).

The major outcome of these features is *a new kind of isomerism*, which is illustrated in Scheme 7.7. Thus, if we replace two of the H's of ethylene by two Cl fragments, we are going to get three different isomers. One isomer is rather simple; the two Cl's replace two H's on the same carbon; this isomer is called *syn* ("together," in Greek).

The interesting isomers are the ones that arise by replacement of H's on the different carbon atoms. One isomer is called *cis* ("on the same side," in Latin), where the two Cl's are on the same side of the double bond. The other is called *trans* ("beyond one another," in Latin), where the two Cl's are on opposite sides of the double bond. Since there is no free rotation around the double bond, *cis* and *trans* isomers coexist and are stable molecules having different properties (refer to Section 7.1 about van 't Hoff and the isomerism of maleic and fumaric acids).

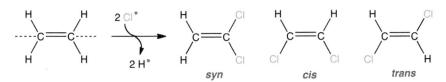

SCHEME 7.7   The replacement of two hydrogen atoms (single connectors) by two chlorine atoms (single connectors) in ethylene leads to three isomers, called *syn*, *cis*, and *trans*. There is no free rotation around the double bond, so the *cis* and *trans* isomers cannot be converted to one another, unless a significant dose of energy is invested to break one of the components of the double bond.

7.2.5 Nature's Usage of *Cis* and *Trans* Isomers

To convert *cis* to *trans* or vice versa requires quite a bit of energy, such as radiation energy from light. When light hits the *cis* or *trans* isomers of a double-bonded molecule, one of the components of the double bond breaks, and it can rotate to the 90° twisted form, which is singly bonded and has two unpaired electrons (Scheme 7.6). Continuing the rotation, the single connectivities are brought again to the same plane and can reclick and give the other isomer.

Nature has made beautiful use of *cis* and *trans* isomerism as a switch for sensing light to create the sense of "seeing." Thus, as we already mentioned in Lecture 1, in the retina of our eyes, there exist photoreceptors called rhodopsins (opsin is a protein), which process information and transmit it down the optic nerve to the brain, which in turn converts it to "seeing." Rhodopsin contains a molecule called retinal, which is shown in Figure 7.5 and is sensitive to light in the region called visible (we will learn later about light). In the "off state," this molecule has a *cis* configuration of the double bond marked in red. When light hits the eye, the retinal molecule absorbs the light and undergoes isomerization to the *trans* form marked in fuchsia-red. This molecular change triggers a sequence of chemical events, which are detected, transmitted, and converted to "seeing." After "seeing," the *trans* retinal slowly reverts back to the *cis* form in a few steps and the receptor restores the "off state," which is ready to receive another signal. If too much light falls on our eyes in a given time (e.g., a strong sunlight), then all or most of the retinal molecules are converted to the *trans* form, and suddenly we feel "blinded."

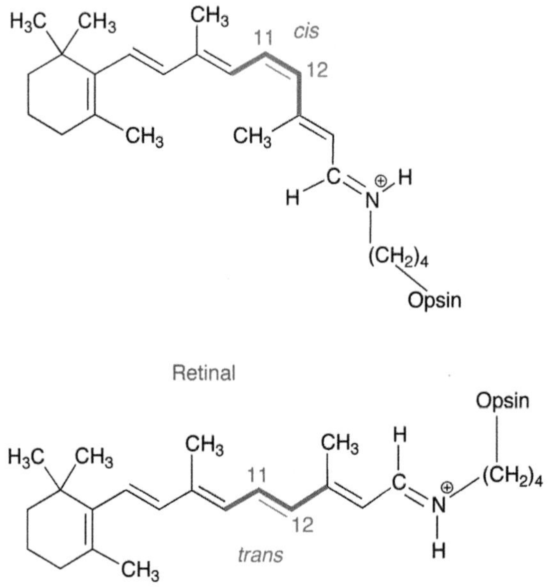

FIGURE 7.5 The structure of *cis* retinal (positions 11 and 12) and its *trans* isomer formed upon sensing of light. Retinal's attachment to the opsin protein is also indicated.

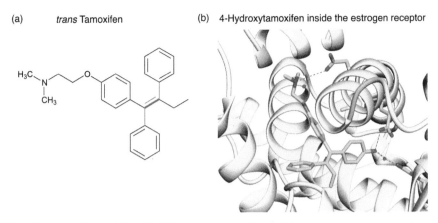

(a) *trans* Tamoxifen

(b) 4-Hydroxytamoxifen inside the estrogen receptor

FIGURE 7.6 (a) Tamoxifen. (b) 4-Hydroxytamoxifen inside the estrogen receptor (structure was drawn using 3ERT.pdb).

Since *cis* and *trans* isomers have different geometries about the double bond, other molecules can sense these differences by shape matching/mismatching. An example is the recent anticancer drug called tamoxifen, which offers rather efficient therapy for breast cancer. The active drug is the *trans* isomer (the *trans* fragments are highlighted) in Figure 7.6a. As shown in Figure 7.6b, the *trans* isomer has the shape that can snuggle into the estrogen receptor and bring about the therapeutic effect (in a rather complex process). The *cis* isomer is ineffective.

7.3 HANDEDNESS (CHIRALITY) AND ISOMERISM

As we discussed in Section 7.1, chirality or handedness is a special kind of isomerism, which arises whenever an object is not identical to its mirror image. Instead of saying "not identical," chemists use the adjective non-superimposable. If you place your *right* hand in front of a mirror, you will see a *left* hand in the mirror (please try). If you try to rotate, say, the left hand in any fashion imaginable to make it "a right hand," you will find that this is impossible. The two hands have an inherent difference, an internal direction (from the thumb to the little finger) either clockwise or anticlockwise, and therefore the right and left hands are not superimposable (unless your hand "crosses the mirror" like in *Alice in Wonderland*). So, our hands are not superimposable because they contain some asymmetry (clockwise vs. anticlockwise from the thumb to the pinkie). Therefore, when molecules have this property we call it handedness or chirality (from Greek origins).

Chirality (handedness) in molecules occurs, for example, when an atom is bonded to four different groups. Let me exemplify this with carbon in Figure 7.7. Thus, if we replace three of the H's of methane by F, Cl, and Br, we get first, say, the molecule we labeled as "object." If we place the "object" in front of a mirror, we will get the mirror image molecule. Now try to superimpose them. No matter what you try, you will fail to superimpose them (unless you will break two bonds and replace their

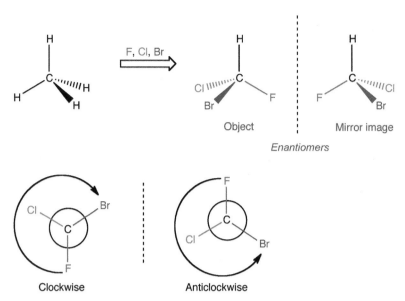

FIGURE 7.7   Replacement of three H's of CH_4 by three different halogens gives rise to two molecules that relate to each other as an object and its mirror image (the mirror is represented by the dashed line). The two molecules are not superimposable, and hence, they are isomers: specifically, they are called enantiomers or chiral (handed) isomers. Below the molecules, we draw the C(F, Cl, Br) components as we see them when looking from the bottom of the molecules on the top right, with the C—H bond being away from us, pointing into the page. The direction from F to Br via Cl is clockwise in one enantiomer and anticlockwise in the other.

positions. But then, these will not be anymore the object and its mirror image but simply two identical molecules, which you generated by switching two bonds). As such, the two molecules labeled in Figure 7.7 as an object and its mirror image are different molecules. They are isomers, which chemists call *enantiomers* ("having opposite parts," in Greek) or simply chiral (handed) isomers.

To see the inherent asymmetry in the two enantiomers, I drew below the molecules in Figure 7.7 the positions of the three halogens when looking at the molecules from the bottom viewpoint such that the C—H bond is away from me. Now, let me trace the direction in going from F to Cl and then to Br. As you can see, the direction in one of the molecules is clockwise, and in the other enantiomer it is anticlockwise. These clockwise/anticlockwise rotation arrows are not superimposable. This is the asymmetry that is the root cause of the handedness of the two molecules. Another way to identify chiral molecules is if there is no way to cut them into two identical halves …

Pasteur discovered handed isomers in the nineteenth century when he passed polarized light through a solution of chiral molecules. Polarized light is a form of light that is seen only in a certain plane, while all other directions are dark. When such a light hits a chiral molecule, the molecule rotates the plane of polarization and the outcoming light is polarized in a rotated plane. Enantiomers rotate the plane of

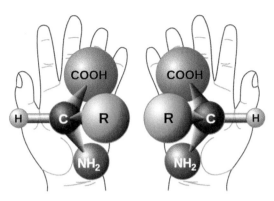

FIGURE 7.8   Amino acids are chiral and snuggle nicely with their respective left and right hands. Proteins are made only from left-handed amino acids. This image was created by NASA. It is in the public domain, and it was obtained from http://en.wikipedia.org/wiki/File:Chirality_with_hands.svg.

polarization in opposite directions. Chemists still use this technique today to identify chiral molecules.

7.3.1   Handedness (Chirality) in Nature

Nature makes ample use of chirality. Let me mention a few such usages.

Figure 7.8 shows the enantiomers of amino acids. It is clear that amino acids have a carbon atom which is bonded to four different groups: hydrogen, an acidic group COOH, an amino group NH_2, and a variable group R. Thus, amino acids are handed (chiral), and the figure shows them being snuggled in the right and left human hands, and being an object and its mirror image which are not superimposable (you can also verify the opposite senses of rotation from COOH to NH_2 via R [or from NH_2 to COOH via R] or the lack of two identical halves in the molecules).

As you recall, proteins are made from amino acids and we have constructed proteins using the "clack-click" method. It turns out that the proteins of living systems contain exclusively left-handed amino acids (chemists call this "homochirality"). This is quite exciting because it would seem as an act of magic that a molecule could recognize the chirality of another one, and exclude all the right-handed amino acids. However, molecular recognition in this case is quite a simple matter. Select one of your neighboring classmates and hand him/her your left hand (or a teacher hands a hand to a student). Your neighbor will use his/her left hand for the handshake and the two hands will snuggle nicely in the handshake. Now ask him/her to shake your right hand with his/her left hand, and lo and behold, the shake will be awkward and inconvenient. *You can recognize your friend's left from his/her right, just by a handshake*!

Handedness is the basis of molecular recognition during protein synthesis. As the amino acid molecules are stitched, by the cell's machinery (see more in Retouches section 7.R.6), to form a protein by excision of water molecules, adjacent pairs of amino acids experience mutual interactions that are similar to the handshake. Thus, a left-handed amino acid will always interact better with another left-handed molecule, so that in the end, the entire protein will be made from left-handed units.

The question of why have left-handed amino acids been chosen in the first place over the right-handed ones is a much more difficult question that continues to occupy the imagination of scientists. There are many speculations, starting from "symmetry breakage" by the "weak force" (which is a chiral force, and hence capable of creating a slight excess of the left-handed amino acids, which is amplified by the protein formation process), going through "chiral stardust" to ideas about some evolutionary accident.

Figure 7.9 shows in (a) and (b) chiral molecules. They are depicted using the implicit drawing representation, and while these do not show C and H atoms, you can

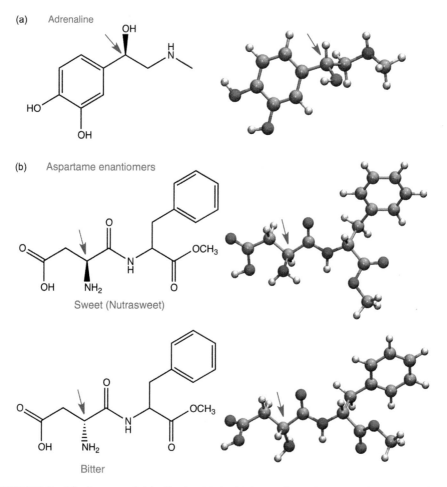

FIGURE 7.9   The impact of chirality in: (a) the biologically active adrenaline (epinephrine). (b) The two enantiomers of aspartame—only one of them is sweet, and it is sold under the name Nutrasweet. The other enantiomer of aspartame is bitter. All the molecules are shown also with 3D ball-and-stick representations. The arrows point to chiral centers of interest.

recognize the chiral carbon, which is marked by an arrow. The molecule in Figure 7.9a is the hormone adrenaline (epinephrine). Epinephrine binds to receptors on heart muscle cells (so-called beta-adrenergic receptors). This binding induces an increase of the contraction rate of the heart, and it ultimately leads to increased blood supply to the tissues in the body. We talked about the natural role of epinephrine in the fight-or-flight instinct (Lecture 1), but it has other usages. Epinephrine is injected to patients in CPR (cardiopulmonary resuscitation) treatment. Interestingly, only the specifically handed molecule drawn in Figure 7.9a is effective. Thus, our biosystems are inherently chiral!

Figure 7.9b shows aspartame. You can see here two isomers with a chiral carbon pointed out by the arrow. It turns out that one of these chiral isomers is sweet and is a component of the artificial sweetener Nutrasweet, and the other is bitter. As receptor proteins (which are chiral) mediate our sense of taste, our taste is inherently chiral. Thus, some, if not all, of *our senses are chiral!* For example, lemons and oranges may contain the two different enantiomers of a molecule (called limonene), which gives these fruits their distinctive but different smells. So our sense of smell is chiral, and likewise our sense of touch could also be chiral since it utilizes our chiral hands.

The molecule thalidomide is associated with a great human tragedy. The drug based on this molecule was administered during the 1950s and 1960s to pregnant women who suffered from insomnia and headaches. In the 1960s, Dr. Frances Oldham Kelsey was appointed to a post in the FDA (Food and Drug Administration) of the United States, and her first assignment was to review the drug thalidomide, because of some reports that it caused numbness of limbs to patients. By her sheer persistence to determine whether the medication was safe to take during pregnancy, it was discovered that women who have taken thalidomide gave birth to children with severe limbs malformation.[4]

Where is the punch line of our story? Well, if you look at Figure 7.10, you will see that thalidomide has a chiral carbon. It turns out that only one of the enantiomers causes the birth defects. Chirality of biological systems strikes again! However, in this case, the two enantiomers transform to one another quite rapidly, and hence administering the "right" thalidomide enantiomer will give rise to formation of the wrong one … However, as in every Shakespearean story, the thalidomide story is full of twists and turns, as recently, it was found to be efficient in cancer therapy.[5] If we needed another proof of the Janusian character of molecules, then here it is.

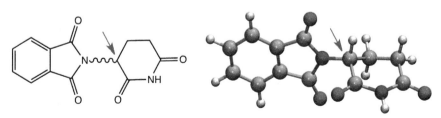

Thalidomide

FIGURE 7.10   Thalidomide with its chiral center indicated by an arrow. The ball-and-stick representation is shown on the right.

7.4   EXTENSION OF THE 3D RULES TO CONFORMATIONS

It is possible to extend the VSEPR rules to repulsions between bond pairs on adjacent atoms. The extension is straightforward; the best structures would be the ones that place these bonds as far apart as possible. As an example, consider the already familiar molecule ethane, H_3C–CH_3. Since each carbon is surrounded by four electron pairs, it will assume a geometry derived from a tetrahedron, with angles of 109°, as shown in Scheme 7.8a.

But what would be the relative orientations of the C–H bonds on the two CH_3 groups? Scheme 7.8a shows two structures with different C–H/C–H orientations. As we already pointed out in Lecture 4, the rotation of the groups around a single C–C bond does not require much energy investment. It occurs very fast, and it interconverts the structures more than 10 billion times per second. Chemists call these structures *conformers* to distinguish them from isomers, which coexist as stable molecules.

Thus, the molecule can attain two extreme conformers, one on the left of Scheme 7.8a, where each C–H on one carbon bisects the HCH angle on the other, and the

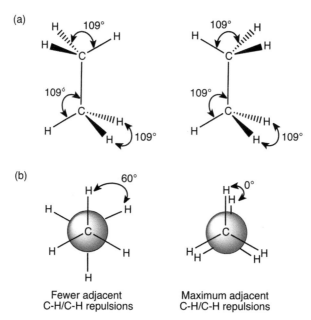

SCHEME 7.8   (a) The two conformers of H_3C–CH_3. The bond angles around each carbon are tetrahedral (109°), but the relative orientations of the C–H bonds on the two CH_3 groups are different in the two conformers. (b) These conformers can be understood by looking at the two conformers from the bottom carbon (in a) toward the upper carbon. We are drawing the C–H bonds on planes parallel to one passing in the middle of the C–C bond (the C–C bond is drawn going into the page here). This plane passing in the middle of the C–C bond is drawn as a circle, highlighted in blue, separating the front CH_3 group from the one in the back.

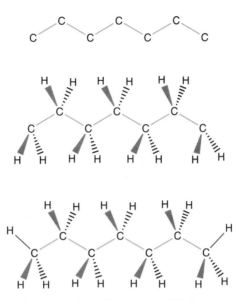

SCHEME 7.9   Drawing the extended conformation of C_7H_{16} in three steps.

other on the right side, where the C–H bonds of the two carbons are parallel. If you construct a model, you can see these relations very well.

Scheme 7.8b can assist you to express the relative C–H/C–H orientations. Here we draw the molecule by looking along the C–C bond and tracing the C–H bonds of the two groups on a plane that bisects the C–C bond and is drawn as a blue circle. We clearly see that one conformer maintains an angle of 60° between the adjacent C–H bonds, while in the other, they are parallel. It is very clear that the structure that maintains 60° angles between the adjacent C–H bonds is preferable over the one where the bonds are parallel (see Retouches section 7.R.4).

These considerations can be applied to longer alkane chains. Thus, the conformation that places all the adjacent carbon-carbon bonds as distant as possible from each other will be the preferred or one of the preferred conformations of the chain (this is true up to a certain length beyond 10 carbon atoms). To give an example, I present in Scheme 7.9 the drawing of this conformation for C_7H_{16} in three steps. Initially, draw the C_7 backbone as a zigzag of the C–C bonds that exist in the molecule (here six of them). Then draw with wedges and dashes the C–H bonds in the CH_2 groups and two of the three C–H bonds on each terminal carbon, which alternately point up and down along the chain. In the last step, complete the two in-plane C–H bonds on the two terminal carbons. In this manner, we created an extended chain where all the C–H and C–C bonds on adjacent carbons are as distant as possible. All other conformations involve chains that are coiled and not extended.

Extended-chain conformations are common and easily accessible. When many molecules are present together, these extended chains tend to pack nicely and hence they can form solids quite easily. This is one of the problems of saturated fatty acids, about which we talked in Lecture 5.

7.5 THE ARCHITECTURE OF MATTER AND ITS ORIGINS

Molecules do not roam around being lonely. They exist in populous "societies," wherein they form aggregates that are held by forces between molecules (called by chemists *intermolecular forces/interactions*), which endow these aggregates with 3D architecture, which in turn determines the architecture of the macroscopic matter (macroscopic objects are ones we can see with the naked eye). Figure 7.11 shows some of these architectures.

The upper left and middle drawings show the architecture of ice, which is seen to contain many molecules hugging each other via H•••O contacts. The architectural element involves an aggregate of six water molecules, $(H_2O)_6$, which has a shape of a reclining chair, shown on the left-hand side. This determines the global architecture of ice (in the middle drawing), as a composition of many reclining chairs in space. On the right-hand side, we show the structure of diamond, which we discussed in Lectures 1 and 4; you can see the same architectural motif of reclining chairs appearing in diamond and in ice.

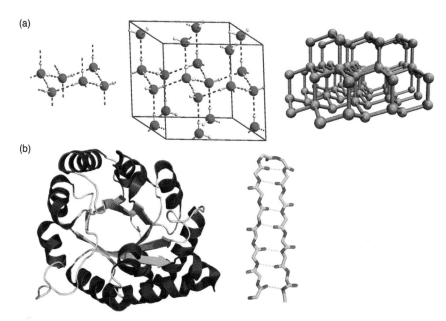

FIGURE 7.11 Architecture of chemical materials: (a) Ice ("solid water") and diamond. On the left, we show six water molecules held in a structure looking like a reclining chair, as occurs in ice. In the middle, we show the resulting airy architecture of ice on a larger scale, and on the right, we show the structure of diamond. This crystal structure of hexagonal ice is a public domain image obtained from http://commons.wikimedia.org/wiki/File:Hex_ice.GIF. (b) Some architectural motifs that appear in proteins. On the left, we see a protein molecule with coiled parts, which are called α-helices (alpha-helices, shown in purple) and sheet parts, which are called β-sheets (beta-sheets, shown in green). On the right, we see two protein parts in extended conformations that form NH•••O contacts with each other. The hydrogen atoms in the NH•••O contacts are not shown.

The drawings on the bottom show architectural elements that appear in proteins. The drawing on the left-hand side shows that protein chains can assume a helical shape, called an α-helix (as shown in purple), or a sheet shape, called a β-sheet (as shown in green). The drawing on the right shows two protein parts that maintain NH•••O contacts, which is the architecture that appears in silk.

What are these contacts which bring about intermolecular forces that hold molecules together? This is our next subtopic.

7.5.1   The Electronegativity of Atoms

To understand the nature of the intermolecular forces, we must first learn a new concept, called *electronegativity* (EN), which is defined as follows:

> ☎ The electronegativity (EN) of an atom is its ability to attract the bond-pair electrons from its bonding partner and acquire some negative charge at the expense of its bonding partner.

What is pulling the electrons is the positively charged nucleus of the atom, so EN gauges the capability of the atomic nucleus to attract the negatively charged electron pair toward it. Scheme 7.10 illustrates the consequences of EN on the bonding pair. Part (a) shows a bond between two identical atoms A—A. In the middle of the bond, I traced a dashed line and represented the bond-pair electrons by two heavy dots. It is seen that no matter what is the EN of atom A, since the bond partners are identical, they apply the same pull on the electron pair and it stays therefore "midway" between the atoms, such that neither of them acquires any excess charge at the expense of the other. By contrast, in Scheme 7.10b, we show a bond made from two different atoms A and B. Since different atoms have different EN properties, then the more electronegative

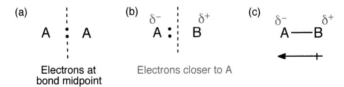

(a) A ⋮ A Electrons at bond midpoint

(b) δ^- A ⋮ B δ^+ Electrons closer to A

(c) δ^- A—B δ^+

SCHEME 7.10   Illustrations of the effect of electronegativity on the position of the bond-pair electrons between the bonded atoms; the bond midpoint, marked by the dashed line, serves as a reference for an equally shared bond pair. (a) In a bond made from the same fragments (A—A), the electron pair is equally shared and is located midway between the atoms, and (b) in a bond made from different fragments (A—B), the bond pair shifts a bit off the bond midpoint toward the more electronegative atom, A, which acquires a small negative charge (marked δ^-). The less electronegative atom, B, acquires in turn an identical (in magnitude) positive charge (δ^+). The bond becomes polarized, also called polar. (c) An arrow directed from the positive charge to the negative charge defines the direction of the bond dipole (which is a measurable molecular property).

atom, say A, pulls the electron pair, which shifts from the center closer to A. The result is that A gained some electron density and became slightly negative, while B lost the same amount of density and became slightly positive. Chemists symbolize a small fractional charge by the Greek letter δ and add the sign of the charge (+/−). As such, the A—B bond will be polarized, as shown in Scheme 7.10b, with atom A having a slight negative charge (δ^-) and B a slight positive charge (δ^+). Thus, the difference in EN of the atomic constituents of the bond creates a *polarized* bond.

The polarity is associated with a molecular property called a *dipole* (or *dipole moment*). For the case of the molecule A—B, the dipole is defined in Scheme 7.10c by an arrow that points in the direction from the positive charge toward the negative charge. Chemists can measure polarity of molecules using electrical and light refraction instruments. But, you can conduct your own "measurement" quite easily! When you next take a shower, fill the bathtub halfway with water. If there is a window in the bathroom, there is sufficient light. Look at your legs in the water, and they will seem to you not aligned with the parts of the legs that are outside of the water, as if they were broken. This happens whenever a ray of light passes from the air to a polar medium such as water; the polar water bends the ray of light, and you see your legs as if they are broken.

Most of the periodic tables include within the square of each atom also its EN value. These values can be used to gauge the relative power of the atom to attract electrons from its bonding partners. The values range from 4.0 down to about 0.7. For example, F has the largest EN value, and O has the second largest value, while the alkali atoms Cs and Fr have the lowest values (see values in Appendix 7.A.1). The noble gases have no EN values. Scheme 7.11 shows the EN trends in the periodic table using arrows to indicate the directions of increasing EN. It is seen that if we move in a period from left to right (e.g., from Li to F), the EN value increases. Similarly, if we move in a family from the bottom upward, the EN increases. Thus, the most electronegative atoms in the periodic table are N, Cl, O, and F.

7.5.2   Polarity Trends in Bonds

Let us look at the polarity of H—X bonds, where X is an atom in a given family. To do so, we must use the EN values of the atoms, which can be taken from the EN table in

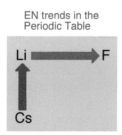

SCHEME 7.11   Electronegativity (EN) trends in the periodic table. The arrows indicate the directions in which EN increases in a period and in a family.

TABLE 7.1   H—X Bonds (X Varies down a Column), the Electronegativity (EN)^a of Their Constituents, and EN Differences

Bond/Property	H—F	H—Cl	H—Br	H—I
EN values	2.1/4.0	2.1/3.0	2.1/2.8	2.1/2.5
EN difference	1.9	0.9	0.7	0.4

^{a}EN values are from Appendix 7.A.1 and are given in the order of H/X.

Appendix 7.A.1. Table 7.1 shows the EN values for the bond partners in a sequence, for exmple, 2.1/4.0 for H/F in the H—F bond, etc.

It is seen that for all the bonds, X is more electronegative than H and hence the bonds are polarized in the sense of $H(\delta^+)$ and $X(\delta^-)$. However, as X changes down the column from F to I, the EN of X decreases and the EN difference also decreases, such that the partial charges get smaller as we go down the period from H—F to H—I.

Table 7.2 considers the polarity trends in H—X bonds, wherein X varies along a period. It is seen that as before, in all the bonds, X is more electronegative than H and hence the bonds are polarized in the sense of $H(\delta^+)$ and $X(\delta^-)$. As X changes from left to right in the period, the EN difference increases so that H—C is the least polar bond, while H—F is the most polar.

The same trends as in the above tables are expected for any bond type when varying one atom along a family and a period.

7.5.3   Molecular Polarity

All molecules which have bonds between different atoms will have polar bonds. But does this necessarily mean that the molecule as a whole will have a dipole? It turns out that this is not the case, and whether or not there is a molecular dipole depends entirely on the molecular geometry or 3D structure. Scheme 7.12 shows two such cases. In part (a), we show CO_2, where the two C=O bonds are definitely polar. However, since the dipole has a direction, we can see that the dipole arrows of the two bonds are directed in opposite directions, and hence, they cancel each other. So as a whole, CO_2 does not have a dipole even though both its bonds are polar.

Scheme 7.12b shows another molecule, CCl_4, which has a tetrahedral shape. Here too, the C—Cl bonds are polarized, but the dipole vectors of the bonds cancel each other such that the molecular dipole is zero. In fact, every molecule that looks the same from all directions (called symmetric) does not have a molecular dipole. You can try thinking about other such symmetric molecules.

TABLE 7.2   H—X Bonds (X Varies along a Period), the Electronegativity (EN)^a of Their Constituents, and EN Differences

Bond/Property	H—C	H—N	H—O	H—F
EN values	2.1/2.5	2.1/3.0	2.1/3.5	2.1/4.0
EN difference	0.4	0.9	1.4	1.9

^{a}EN values are from Appendix 7.A.1 and are given in the order of H/X.

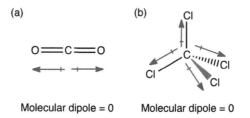

(a)   (b)

Molecular dipole = 0   Molecular dipole = 0

SCHEME 7.12   Individual bond dipoles in: (a) CO_2 and (b) CCl_4. Both molecules have zero molecular dipoles because the individual bond dipoles are oppositely directed and hence cancel one another.

7.5.4 Intermolecular Interactions and the Hydrogen Bond

Molecules that have molecular dipoles or polarized bonds can interact with each other in a favorable way when a positive pole in one molecule approaches a negative pole of the other. This will create an attractive interaction that will "glue" the two molecules and will lower the energy of the bimolecular aggregate. It may aggregate more molecules into shapes that enjoy as much as possible these interactions. These intermolecular interactions are relatively weak, about 1–10% of the strength of a covalent bond, but nevertheless, when there are many of them, they become very important. Have you ever thought how does the lizard (called a gecko) walk on glass and not fall? This is because of intermolecular interactions between the molecules making the skin of the gecko toe pads and the ones making the glass surface.

There are quite a few types of intermolecular forces (see Retouches section 7.R.5),[6] but we are going to focus here on a particular interaction, called the *hydrogen bond* (H-bond). The term "bond" is a misnomer, because this is not the click chemical bond that holds the atoms in a molecule. This is a much weaker interaction *that arises when a positively charged H(δ^+) in one molecule approaches the negatively charged X(δ^-) of another molecule, where X is one of the three most electronegative atoms, namely N, O, or F.*

Figure 7.12 shows three examples of H-bonding. Figure 7.12a shows the structure of solid H–F. It is seen that it is composed of zigzag chains of HF molecules, which are "glued" together by an interaction between H(δ^+) in one molecule and F(δ^-) in the other. As opposed to the covalent bond, which is drawn by a short full line, the dotted longer lines mark the H-bond. An important feature is the linear H-bond angle F–H••••F (180°). The structure zigzags because the H-bond is oriented toward the lone pairs of the accepting F.

Figure 7.12b shows four H-bonding interactions that a single water molecule can experience with others. First, note that in order to have a neutral H_2O molecule, the negative charge on the oxygen should equal the sum of the positive charges on the H's ($\delta^- = 2\delta^+$). Looking at the drawing, one can see that each molecule can be surrounded by four neighboring molecules; two of these accept H-bonds from the H's of the central molecule, and the other two donate H-bonds to the oxygen of the central molecule. All the H-bonds have O–H••••O angles of 180° each. But furthermore, the four H-bonds are distributed in space along the tetrahedral axes of the H_2O molecule;

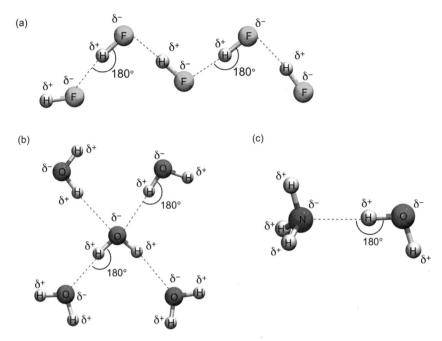

FIGURE 7.12 Hydrogen bonds (H-bonds) drawn as dotted lines (a) in HF, (b) in H_2O, and (c) between H_2O and NH_3. Note the H-bonds are not click bonds but rather are relatively weak interactions.

two are along the O—H bonds, and two are along the space occupying lone pairs of the O. Finally, Figure 7.12c shows an H-bond between two different molecules, ammonia and water.

7.5.5 Properties of Water

The properties of water are fashioned by H-bonding, that is, the dotted "bond" in O—H••••O. The accessibility of four H-bonds for each molecule makes the overall interaction quite significant. One sign of this strong interaction is the high boiling temperature (called the *boiling point*) of water. During boiling, individual H_2O molecules are released from the sticky interactions with other molecules and evaporate to give gaseous water vapors, wherein the molecules are far apart and virtually noninteracting. The stronger the interaction between the molecules, the more heat we have to invest in order to tear apart the molecules from their glued aggregates and have them escape into the vapor phase. The result is that the boiling point of water reaches 100°C (°C = degrees Celsius), which is a very high temperature. Just compare this to the situation in H_2S, which does not maintain those sticky H-bonds between the individual molecules. H_2S is already a gas at room temperature and its boiling point is −60°C (see Retouches section 7.R.5)!

Another feature of water is the beautiful airy structure of ice we saw in Figure 7.11. Usually, solids are more dense than liquids, but not in the case of solid water (ice) versus liquid water. Because of the H-bonds and their architectural elements (angles discussed above), the reclining chair aggregates of $(H_2O)_6$ pack to form ice, which has many empty spaces, many voids. Therefore, ice has a small density and it floats on water. Furthermore, when liquid water freezes and becomes ice, the volume expands. Try putting in the refrigerator a bottle full of water; it will explode because of the expansion of the ice. For the same reasons, when we "freeze to death," our cells, with their 70% water content, simply explode.

Liquid water has similar aggregates as ice, but they are neither uniform nor organized as in ice and hence the empty spaces are smaller. Water in the oceans also acts as the giant "air conditioning unit" on our Earth. During summer, water absorbs the heat radiated from the sun, and all this occurs without breaking all the H-bonds that hold a single molecule. The heat absorbed goes into vaporizing some of the water. As a result, the temperature of water does not rise steeply, and the environment is not heated excessively (as in the case where water is absent, e.g., in the Sahara). Also, in the winter, in places such as northeast America, water freezes, and all the heat stored in the motion of the molecules in the liquid state is freed to the environment and keeps the temperature from dropping too much (most North Americans would not agree with this understatement when winters get miserably cold. Still, without the water, the temperature drop would be deadly). Finally, since water is a polar liquid, it acts as a universal solvent, and it is the medium for *all* the reactions of living systems.

7.5.6   H-Bonds in Proteins

The properties of proteins are also dominated by their H-bond patterns, which are of the type N—H••••O=C (N—H••••N interactions and others contribute, too). There are two main architectural motifs that are shaped by H-bonds, which were already displayed in Figure 7.11b; one motif is called the α-helix, and the other is the sheet motif.

Figure 7.13 shows the α-helix (use a model: it will help). Here a protein chain coils by forming internal N—H••••O=C H-bonds. The α-helix is mechanically strong. It appears in our skin, nails, hair, and muscles. In the latter case, the protein is called collagen, which is made of three helices that are mutually embracing (see right-hand side drawing). The collagen is extremely strong, which is what our muscles and tendons can be. It is glued to the bones by interacting with the ions of the bone material. Cooking bones decomposes the collagen and produces jelly … Many of the cell receptors are made from bundles of α-helices, such as the serotonin receptor in Figure 5.15 in Lecture 5.

Figure 7.14 shows the sheet motif. Here polymer chains facing each other form many N—H••••O=C H-bonds, and this continues to form a two-dimensional (2D) sheet. In turn, the 2D sheets can pack by further H-bonding. The 2D sheets are strong against shearing, and the layering can vary the consistency. This motif is found in claws, armors (of reptiles), in silk where lesser layering give the beautiful airy look, and unfortunately also in brains suffering from Alzheimer's disease. Janusian is the chemical matter!

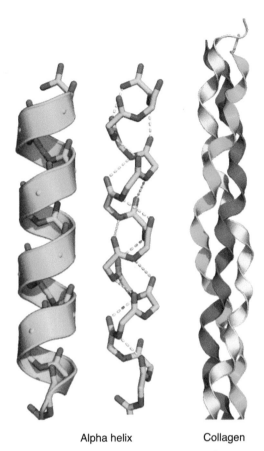

Alpha helix   Collagen

FIGURE 7.13 The architecture of an α-helical protein chain (drawn using 1OT0.pdb) and the prominent influence H-bonding interactions exert on this architecture (H-bonds are represented by the dashed yellow lines in the central picture). The hydrogen atoms are not shown. Shown on the right is the triple helix coil of the type appearing in collagen (drawn using 1BKV.pdb).

Finally, there is also a globular form of proteins, and one has been shown in Figure 7.11b. The globular forms contain α-helices, sheets, and protein strips, and these chunks are glued to form a globule by a mesh of H-bonding interactions. The globular proteins serve as "molecular havens," such as for active molecules that carry out certain chemical reactions, as in the case of enzymes (see Lecture 5). Some receptors like the estrogen receptor are globular.

7.6 H-BONDING AND OUR GENETIC CODE

Figure 7.15 shows a strand of DNA and near it the DNA structure[7] as it is stored in the nuclei of our cells. This is the famous double helix, which Watson and Crick

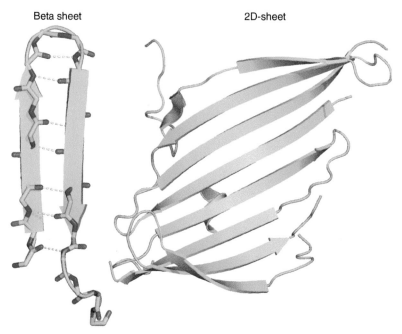

FIGURE 7.14 The architecture of protein sheets (drawn using 3UA0.pdb). Dashed yellowish lines connecting red and blue parts on the left are O••••HN H-bonds (the hydrogen atoms are not shown).

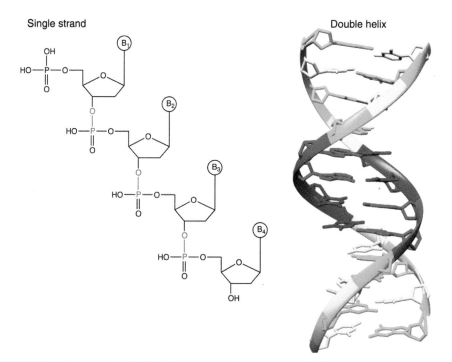

FIGURE 7.15 A strand of DNA (the bases are labeled generically as bases 1–4) on the left-hand side and the double helix structure of DNA (drawn using1FQ2.pdb) on the right-hand side. The bases in the double helix are color-coded and are paired by H-bonding.

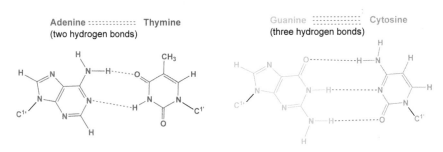

FIGURE 7.16   The base pairing combinations in DNA are A•••T and C•••G. The black lines indicate the site of attachment of the base to the carbon atom in the sugar ring.

came up with, as told in Watson's book, *The Double Helix*, which is also a fascinating story about the social aspects of science.

Look at the double helix and marvel at its architectural beauty.

The architecture of this molecule is governed by the N−H•••O=C and N−H•••N H-bonds between base pairs, and by the architectural element of these H-bonds, the 180° N−H•••O(N) angles. As shown in Figure 7.16, it turns out that bases of one strand pair up only with specific bases in the second strand. (Figure 7.16 gives the full names of the bases.)

Thus, we have exclusively A•••T and G•••C pairs, and this is called *base pairing*. Using this "faithful" base pairing, one strand recognizes the other. Furthermore, *this recognition enables one strand to act as a template for generating the other strand or part of it*. This happens during the cell's division, when the DNA has to be duplicated so that each cell will have all the genetic material. This is done first by opening the double helix. Subsequently, a specific enzyme progresses along each strand (see also Retouches section 7.R.6), adding nucleotides one by one, while conserving the A•••T and G•••C base pairings. In order to stitch each new addition to the growing strand, the enzyme removes a water molecule (recall: "clack-click"). In this manner, each strand forms its complementary strand and generates a full double helix, such that we end up having two double helices, hence self-replication!

The DNA is also a genetic template for making specific proteins, and this is the major purpose of DNA. By making specific proteins, DNA determines if you are blonde or have black curly hair, if you tend to have some illnesses, or if you have good speech language acquisition or not,[7, 8] etc. In this case too, H-bonding controls these traits. Thus, in order to guard the genetic information well encased and protected, the cellular DNA does not leave the nucleus. The double helix opens up at a site of a specific gene, which is copied to form a strand of RNA nucleotides with the help of the base-pairing mechanism (just that U replaces T in RNA). Like in the case of DNA replication, here too an enzyme moves along the DNA strand of interest. At each point, it recruits the necessary RNA nucleotide and then stitches the nucleotide onto the growing chain, to form an RNA strand. The DNA fragment from which the RNA is made is called a *gene*, and the formed RNA is a copy of the gene (see also Retouches section 7.R.6). This complementary gene copy then leaves the nucleus to move to the ribosome, which is a cell particle serving as a store of

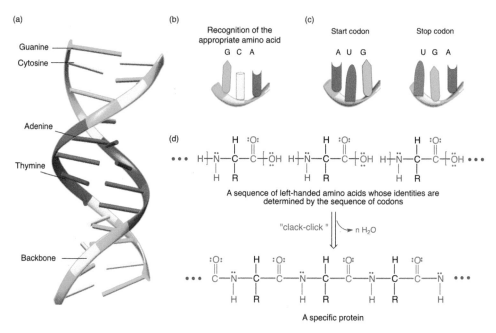

FIGURE 7.17 (a) The double helix. (b) A codon on a strand is a sequence of three bases, such as GCA. Each RNA codon recognizes one amino acid. (Since there are four bases, the number of codon combinations is 4^3, which equals 64 different codons. This number is more than sufficient to recognize the 20 natural amino acids.) A gene is a unit of DNA containing a sequence of codons that starts with a start codon and ends with a stop codon. It is used to direct the formation of RNA and defines the protein that will be made in protein synthesis. (c) The RNA start codon is AUG. One of the RNA stop codons is UGA. (d) Specific amino acids (with different identities specified by the R groups) are arranged according to instructions from a gene and are sequentially stitched together to form a specific protein.

enzymes and RNA nucleotides. At the ribosome, the genetic information on the RNA strand is translated into a protein (see Retouches section 7.R.6).

To comprehend this part a bit better, we have to understand what is a codon and what is a gene, by using Figure 7.17, which shows the double helix and three codons of the RNA copy. As shown in Figure 7.17b, a codon is made of *three consecutive nucleotides, which "recognize" one particular amino acid.* This "recognition" is in fact a complex multi-step process, whereby nucleotide carriers deliver the amino acids one by one to the RNA copy, and these amino acids are sequentially stitched together, "clack-click," to form the desired protein, as shown in Figure 7.17d (see more details in Retouches section 7.R.6). There are codons that signal where should the protein synthesis start; these are the start codons, such as AUG in Figure 7.17c. Other codons that *do not* recognize any amino acid, such as the codon UGA in Figure 7.17c, serve as *"stop" codons.* A gene is a sequence of codons that starts with a start codon and ends with a stop codon. In this manner, the gene will eventually create a protein of specific amino acid sequence and length.

Nucleotides are chiral because they contain chiral carbon centers. Therefore, the codons themselves are chiral, and as such, they recognize preferentially left-handed amino acids (although these details are not yet fully elucidated). In this manner, we end up with left-handed amino acids, which are sequentially stitched together, "clack-click," and a final protein molecule gets generated (Figure 7.17d), being made of specific and only left-handed amino acids. This is an ingenious genetic code that is based on H-bonding and a chiral recognition mechanism. There is much more about DNA that goes beyond the schematic description I outlined above. Some additional aspects of the DNA world can be found in Retouches section 7.R.6.

7.7 SUMMARY

This lecture was an intensive tour through the 3D structure of molecules and the architecture of matter.

The geometric and 3D structures of molecules follow a few rules, which are based on electron repulsion between electron pairs in the valence shell (and referred to as *VSEPR rules*), and which can be summarized as follows:

☞ Bond pairs and lone pairs around an atom will be distributed in space so as to maximize their distances and minimize the pair–pair repulsion. Multiple bonds of an atom count as a single spatial pair.

☞ The preferred molecular shape will depend on the number of spatial pairs that have spread apart around the atom on which they reside: (i) two pairs yield linear structures, (ii) three pairs yield planar structures with bond angles of 120°, (iii) four pairs lead to structures derived from a tetrahedron with bond angles of 109°, (iv) five pairs lead to structures derived from trigonal bipyramidal structures, with bond angles of 120° and 90°, and (v) six pairs lead to structures derived from an octahedron, with bond angles of 90°.

The architecture of matter is determined by weak interactions between molecules (called intermolecular interactions). The key factor that determines the strength of the interaction is an atomic property called *electronegativity* (EN). The EN is defined as follows:

☞ The EN of an atom is its ability to attract the bond-pair electrons from its bonding partner and acquire some negative charge at the expense of the bonding partner. The EN increases in a period from left to right, and in a family from the bottom up (see Scheme 7.11; numerical values are summarized in Figure 7.A.1).

☞ A bond between two atoms that differ in their EN values will be polarized, with a partial positive charge on the atom having the smaller EN value and a partial negative charge on the atom having the higher EN value.

Intermolecular interactions can occur between polarized bonds. The most important such interactions are called *hydrogen bonds* (H-bonds):

☞ An H-bond interaction occurs between an H that is attached to one N, O, or F atom and another F, O, or N atom in another molecule, such as $F-H \bullet\bullet\bullet OH_2$, $F-H\bullet\bullet\bullet FH$, $HO-H\bullet\bullet\bullet NH_3$, etc. The H-bond angle, such as the $F-H\bullet\bullet\bullet F$ angle, is $180°$, and as such it constitutes *an architectural element* in the structure of the macroscopic matter (the matter we see with our eyes).

☞ The H-bond interaction in water, $HOH\bullet\bullet\bullet OH_2$, fashions the properties of water (high boiling point, the anomaly of expansion during ice freezing, the role of water as an "air conditioner" on our Earth, etc.).

☞ The H-bond interactions in proteins, for example, $NH\bullet\bullet\bullet O{=}C$, fashion the properties of proteins (their mechanical strengths and appearances).

☞ The H-bond interactions, such as $NH\bullet\bullet\bullet O{=}C$, between DNA bases confer a structure of a double helix on the DNA. These H-bond interactions are specific ($A\bullet\bullet\bullet T$ and $C\bullet\bullet\bullet G$), and hence the interaction is called *base pairing.*

☞ Base pairing enables both DNA replication and genetic information transfer to RNA molecules that copy pieces (*genes*) of DNA. These RNA molecules then bring about the synthesis of proteins that are encoded by the DNA genes.

☞ Three consecutive nucleotides on a DNA gene define *a codon* that recognizes a single amino acid (see the Retouches). The nucleotides seem to selectively recognize left-handed amino acids. The exclusive left-handedness of the amino acids in biologically occurring proteins is due to chiral recognition governing the interactions between the genetic material and amino acids.

☞ We mentioned that enzymes polymerize DNA and RNA according to the base sequence information. These enzymes are themselves proteins, and DNA and RNA make proteins. It means that the genetic material and its product (the protein) are like a chicken and an egg. This is a poetically wonderful symbiosis!

In addition to the above H-bonding interactions, there are several other types of interactions. These are weaker interactions than H-bonds, but are nevertheless important, as discussed in the Retouches.

7.8 REFERENCES AND NOTE

[1] (a) W. H. Brock, *The Norton History of Chemistry*, W.W. Norton & Co., New York, 1992, Chapter 7; (b) See also: http://www.cosmolearning.com/video-lectures/vant-hoffs-tetrahedral-carbon-and-chirality-6672/

[2] P. J. Ramberg, *Bull. Hist. Chem.* **1994**, *15/16*, 45.

[3] (a) See E. A. P. Robinson, *Coord. Chem. Rev.* **2000**, *197*, 3; (b) R. J. Gillespie and R. S. Nyholm, *Quart. Revs. Chem. Soc.* **1957**, *11*, 339.

[4] S. Cotton, *Every Molecule Tells a Story*, CRC Press, New York, 2012, pp. 115–117.

[5] A. K. Stewart, *Science*, **2014**, *343*, 256.

[6] P. Ball, *Chemistry World*, 2014, *February Issue*, 50.

[7] S. Kean, *The Violinist's Thumb: And Other Lost Tales of Love, War, and Genius, as Written by Our Genetic Code*, Little, Brown & Co., New York, 2012.

[8] There are genes of language acquisition. See: P. Lieberman, *Science*, **2013**, *342*, 944.

7.A APPENDIX

After teaching this chapter, it is advisable to use a whole session of tutoring, which includes material from Lecture 6, and add to it tutoring on the 3D structure of molecules, electronegativity (EN), polarity, H-bonding, etc. Problems in the problem sets, in Sections **6.P** and **7.P**, can be used in the tutoring.

The Appendix provides the EN table and proposed demonstrations.

7.A.1 The Periodic Table of Electronegativity Values

Figure 7.A.1 shows the electronegativity (EN) values of atoms in the periodic table. The trends are also periodic.

7.A.2 Proposed Demonstrations for Lecture 7

The demonstrations for this lecture focus on H-bonding and polarity. One demonstration addresses chirality.

- **Water—a polar molecule**

In this spectacular demonstration, a light projected from a stroboscope is used to observe that water droplets coming from a separatory funnel are spherical, and that the direction of the stream of these droplets can be deflected by static electricity.

The demonstration involves an aqueous solution of fluorescein placed in a separatory funnel, under which a large beaker is placed on the floor. Then, the lights in the lecture hall are turned off, the stopcock of the funnel is opened just slightly to produce droplets, and the stroboscope light is turned on. As the green-glowing droplets fall, their shape is carefully examined; they are spherical.

Why are droplets spherical and not shaped differently? The water molecules are held by hydrogen bonds (H-bonds), and they prefer to surround themselves by other water molecules, so as to maximize the H-bonding interactions. Thus, a water molecule always prefers to be at the center of the drop rather than at the surface, where it will lack at least one H-bond (this phenomenon is also known as *surface tension*). Recalling geometry classes, the shape that has the smallest surface enclosing a given volume is a sphere. And this is the physical shape the droplets adopt, so as to minimize the loss of H-bonding.

The polarity of water molecules is demonstrated in the second part of the experiment by bending the stream of water droplets using static electricity. To do so, a

Electronegativities of the Elements

1998 Dr. Michael Blaber

1A	2A	3	4	5	6	7	8	9	10	11	12	3A	4A	5A	6A	7A
H 2.1																
Li 1.0	Be 1.5											B 2.0	C 2.5	N 3.0	O 3.5	F 4.0
Na 0.9	Mg 1.2											Al 1.5	Si 1.8	P 2.1	S 2.5	Cl 3.0
K 0.8	Ca 1.0	Sc 1.3	Ti 1.5	V 1.6	Cr 1.6	Mn 1.5	Fe 1.8	Co 1.9	Ni 1.9	Cu 1.9	Zn 1.6	Ga 1.6	Ge 1.8	As 2.0	Se 2.4	Br 2.8
Rb 0.8	Sr 1.0	Y 1.2	Zr 1.4	Nb 1.6	Mo 1.8	Tc 1.9	Ru 2.2	Rh 2.2	Pd 2.2	Ag 1.9	Cd 1.7	In 1.7	Sn 1.8	Sb 1.9	Te 2.1	I 2.5
Cs 0.7	Ba 0.9	La 1.0	Hf 1.3	Ta 1.5	W 1.7	Re 1.9	Os 2.2	Ir 2.2	Pt 2.2	Au 2.4	Hg 1.9	Tl 1.8	Pb 1.9	Bi 1.9	Po 2.0	At 2.2

Legend:
- 3.0-4.0
- 2.0-2.9
- 1.5-1.9
- <1.5

FIGURE 7.A.1 Atomic electronegativity (EN) values (the values are dimensionless numbers) in the periodic table. Noble gases are considered to have no EN.

balloon is rubbed against the demonstrator's (or the volunteered student's) hair and then brought closely to the droplet stream (without touching it). The water droplet seems to bend towards the balloon! Rubbing a balloon against one's hair produces a charge separation—the balloon becomes negatively charged and the hair becomes positively charged. The students may probably also notice how the hairs stand up in the air as each of them is repelled by the others. When the rubbed balloon approaches the water droplets, the positively charged hydrogen ends of the water molecules attract the negatively charged balloon. This attraction is sufficiently strong to pull in the water stream toward the balloon. Other materials get positively charged by rubbing (in a process called tribocharging); for instance, rubbing a glass against silk makes the former positively charged, and such positively charged materials also cause the stream of droplets to be attracted to them through the negatively charged ends of the water molecules. Had water molecules not been polar, the charged objects would have had no effect on the droplets. This demonstration was home-designed by Mr. C. DeLano.

- **The missing volume**

This demonstration shows that when two liquids are mixed together, the properties of the combined mixture can sometimes be quite different from the sum of the properties of its individual components. In this experiment, it is shown that mixing 100 ml of ethanol with 100 ml of water gives a solution with a volume of less than 200 ml. The reason for this anomaly is that, due to H-bonding, the water molecules are packed in relatively open structures, which form empty spaces (every water molecule is surrounded on average by only 3.5 water molecules in the liquid phase). The ethanol molecule is small enough to fit inside this void, so the addition of ethanol leads to a more closely packed liquid. The packing is aided by H-bonding to the ethanol.

To make the demonstration more visually vivid, a hydrophobic (a term meaning not dissolving in water) red dye, Sudan III, is first mixed with 100 ml ethanol and with 100 ml water separately in two graduated cylinders. The students note that the dye is easily dissolved in ethanol but is completely insoluble in water. Then, the two liquids are mixed together in a graduated cylinder and the lower part is examined. A volunteered student is asked to touch the cylinder vessel and to tell whether it is relatively warm/lukewarm/cold. The vessel feels warmer, which means that mixing water and ethanol is an exothermic (energy-releasing) process, in which new and stronger H-bonds between water and ethanol are formed.

In the second part of the experiment, the mixed solution is separated into its components, by the *salting out* effect, when 30 g of $MgSO_4$ are added to the ethanol–water mixture. Slowly, a transparent liquid starts forming at the bottom part of the vessel, leaving above it a red-colored liquid. Which liquid is the ethanol and which is the water? Recalling that the dye is soluble in ethanol and insoluble in water provides the answer: the lower transparent liquid is water and above it is the colored ethanol. As will be seen in the next lecture, salts such as $MgSO_4$ dissolve in water and form strong interactions with the water molecules. These interactions are strong enough to "push out" the ethanol molecules from the water voids. This demonstration was home-designed by Drs. Y. Aouat and R. Ben-Knaz.

- **Observation of chirality on an iPad**

In this demonstration, chiral and achiral transparent crystals are examined with polarized light emitted from an iPad or a tablet. The crystals are placed over the device's screen and viewed through a second polarized filter. Whereas achiral crystals seem transparent, much as they are under normal room light, the chiral crystals appear blue.[1A]

- **The buzzer experiment**

At this stage, it is advisable to repeat the buzzer experiment, which was described in Lecture 1, in order to better prepare the students for the next lecture, where the ionic bond is taught.

7.A.3 References for Appendix 7.A

[1A] P. M. Schwartz, D. M. Lepore, B. N. Morneau, and C. Barratt, *J. Chem. Edu.*, **2011**, *88*, 1692.

7.R RETOUCHES

7.R.1 Electron Pair Repulsion

The major repulsive mechanism between electron pairs around a given atom is a quantum mechanical (QM) effect, which we alluded to in Section 1.R.3 as *the exclusion principle*.[1R] According to this principle, if two electrons occupy the same space, they must possess different spin directions (Scheme 1.R.1). Indeed, all the electron pairs, be they bonds or lone pairs, are net spin-less, involving two opposite-spin electrons. Since there are only two directions of the spin property, this necessarily means that two electron pairs cannot occupy the same space. If they do, there will be two electrons of the one spin type (↑↑) and two others with the other spin type (↓↓). This is excluded by the principle; the exclusion actually means that this situation raises the energy very much, and hence, molecules avoid the situation as much as possible. The consequence of this is that *electron pairs around a given atom will be distanced maximally in space to lower the molecular energy*. This is the basis of the VSEPR rules.

7.R.2 Pictorial Description of Lone Pairs

As we already noted, according to the QM theory, electrons occupy *orbitals*, which can be described as spatial objects, with certain volumes around the atom. An orbital represents the volume that encompasses the maximum probability of finding the electron around the atom. The lone-pair orbitals of main elements have the lobe

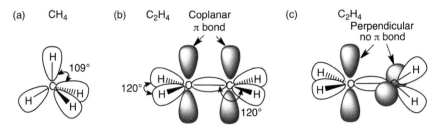

SCHEME 7.R.1   Lobes of bonding orbitals: (a) In methane, the lobes are oriented along the C—H bond axes. (b) In ethylene, one of the C—C bonds (the one with the shaded lobes, labeled as a π bond), lies above and below the molecular plane. (c) When the two CH_2 units of ethylene are perpendicular to each other, the π bond is broken.

shapes we drew in Scheme 7.4, which are referred to by chemists as "rabbit ears."[2R] The website en.wikipedia.org/wiki/Atomic_orbital includes many orbital drawings.

7.R.3   The Nature of the Double Bond

The orbitals of a given atom occupy different directions in space to abide by the exclusion rule. As such, they are oriented in the direction of the electron pairs around the atom. Furthermore, each bond orbital (orbitals can exist for bonds, as well as for atoms) is composed of an overlap of the orbitals of the bonding partners. For example, in the case of CH_4, the C—H bond orbitals are oriented along the C—H bonds and maintain tetrahedral angles, as in Scheme 7.R.1a. Since $H_2C=CH_2$ has two planar CH_2 units, three orbitals of each of these units are directed in a plane with 120° angles, and the fourth orbital is perpendicular to the CH_2 plane. As shown in Scheme 7.R.1b, these perpendicular orbitals (which are shaded) can "see" one another and overlap, *but only when the two CH_2 groups are in a single plane.* This component of the double bond is not oriented along the C—C axis, and is holding the two C atoms with overlap regions above and below the molecular plane. It is a weaker bond than the one along the axis, and is labeled by chemists using the Greek letter π (pai).[3R] When the two groups are in perpendicular planes, as in Scheme 7.R.1c, these two orbitals cannot overlap and the π bond is broken.

7.R.4   Conformations of C_2H_6

The conformations of ethane, C_2H_6, which are shown in Scheme 7.8, have C—H/C—H angles (angles between the two bonds) of 60° and 0°; the former is preferable. The term "preferable" means that this is the conformation that has the lower energy. To represent this aspect, we can draw in Scheme 7.R.2 an energy diagram of the conformers as we change the C—H/C—H angle from 0° to 60° in the two directions. To show the relative rotations of the two C—H bonds, we mark two of the H's in red. Thus, starting with an angle of 60° between the two red-colored H's, and rotating the CH_3 group in the back, the energy goes up and reaches a maximum when the angle reaches 0°. Continuing the rotation, the energy goes down and reaches another

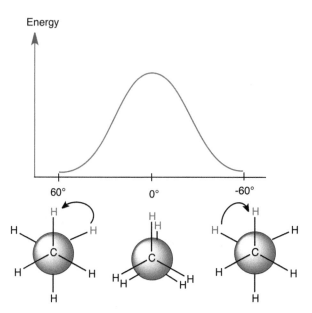

SCHEME 7.R.2   The energies of the conformations of ethane (C_2H_6) as the angle between the C—H bonds on adjacent carbons varies from 60° to −60° and passes through 0°. Two of the hydrogen atoms are marked in red to indicate the angle between the respective C—H bonds during the internal rotation of the CH_3 groups. The C—C bond is drawn as pointing into the page.

minimum when the angle reaches −60°. Further rotation will again eventually lead to 0°, then to 60°, and so on and so forth. It is seen that the unpreferred conformation with the parallel C—H/C—H bonds (angle of 0°) sits at the top of the energy curve, and hence it will "roll down" to its lower conformers.

7.R.5   Other Intermolecular Forces

Dipole-Dipole Interactions: All bonds between different atoms are polarized. Indeed, S—H bonds in H_2S are polarized. However, the HS—H••••SH_2 interaction is much weaker than the corresponding HO—H••••OH_2 interaction. There are two main reasons: (a) The S—H bonds are less polar, so the fractional charges on the H and S atoms are smaller than on H and O, and (b) as we shall learn later, the distance H••••S is longer than H••••O. Consequently, the HO—H••••OH_2 interaction is stickier and it prefers an O—H••••O angle of 180°, compared with the HS—H••••SH_2 interaction, which is weaker and has no angular preference. This is why chemists tend to use the name H-bond for interactions of the most electronegative atoms H••••X (X = F, O, and N), while other ones where X is less electronegative or belonging to higher periods are simply called dipole–dipole interactions. Dipole-dipole interactions occur also between molecules involving bonds where neither atom is a hydrogen atom, whenever there is an overall molecular dipole.

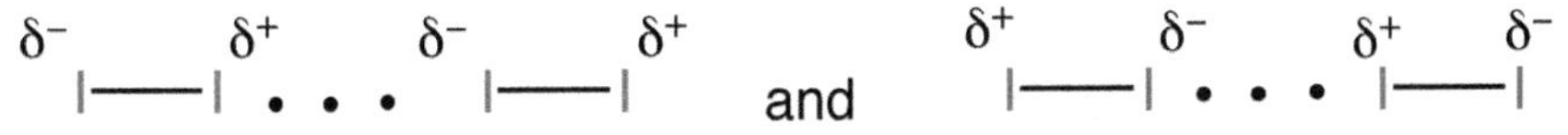

$$\overset{\delta^-}{|}\!\!-\!\!-\!\!\overset{\delta^+}{|} \cdots \overset{\delta^-}{|}\!\!-\!\!-\!\!\overset{\delta^+}{|} \quad \text{and} \quad \overset{\delta^+}{|}\!\!-\!\!-\!\!\overset{\delta^-}{|} \cdots \overset{\delta^+}{|}\!\!-\!\!-\!\!\overset{\delta^-}{|}$$

SCHEME 7.R.3   van der Waals interactions between two I_2 molecules (iodine atoms shown in red), which induce dipoles on each other. Both dipole orientations are shown, on the left and on the right.

van der Waals Interactions: It turns out that even bonds between identical atoms interact with one another by intermolecular interactions that are called van der Waals (vdW) interactions, after the name of the Dutch scientist who first hypothesized the phenomenon. Consider in Scheme 7.R.3 two molecules of I_2. When each is alone, then of course the bond is not polarized. But when they approach one another, they can induce polarity in each other's bond, in order to lower the energy. As seen in the scheme, the dipoles oscillate in the two molecules, so that on average the bond is not permanently polarized in any specific direction. The primary reason why the molecule I_2 exists at room temperature as a solid is these interactions between the induced and oscillating dipoles. These interactions are gaining increased interest in recent years, because when you have many bonds interacting, the sum of all the vdW interactions can become significant, for example, as between the layers of graphite.

7.R.6   More on DNA

The book of Sam Kean[7] is a good source for more information. The book is written in a fun style, which is amusing and enlightening. Other sources are listed in references for Retouches.

The story of DNA involves an intricately engineered flow of information, using both concepts from chemistry, such as hydrogen bonding and chirality, as well as biological machinery that facilitates reactions. The Central Dogma of Biology states that DNA nucleotide strands store genetic information and then the cell produces RNA nucleotide strands with complementary copies of this information. Next, RNA directs the synthesis of proteins.

Base pairing and the genetic code facilitate information transfer. As DNA undergoes copying to form new DNA or RNA, base pairing guarantees that nucleotides in one strand accurately specify the nucleotide sequence in the complementary strand, ensuring efficient information flow. The genetic code associates sequences of three nucleotides (called codons) with specific amino acids, so RNA sequences dictate amino acid sequences during protein synthesis.[4R]

Replication: Every cell of a living being contains a copy of the organism's DNA. In organisms that have a nucleus (a cellular compartment which stores and protects the DNA), the DNA is found there. Every time a cell divides, it must copy its DNA, so that each of the two resulting cells gets a copy. This copying process is termed replication.

During replication, the DNA helicase enzyme opens the double helix, exposing the bases. After grooming and relaxation of the strands, the DNA polymerase enzyme transforms each strand into a new double helix. Given the complementary and specific nature of base pairing, one DNA strand specifies which bases should be added during the formation of the new strand to generate thereby a new double helix. DNA polymerase moves along the unwound DNA, and it adds nucleotides one by one, stitching each new addition to the growing strand by removing a water molecule ("clack-click"). The charged site of the DNA polymerase enzyme accelerates this loss of water and the formation of the O–P bond. Each new nucleotide incorporated is complementary to the corresponding nucleotide in the strand copied.

DNA polymerase enzymes consider the shapes of the base pairings formed (like Cinderella's foot in the glass slipper, the new base pair must fit in the enzyme) as well as their hydrogen bonds, to conserve the A•••T and G•••C base pairings. With this and other quality control processes, the overall error rate is very small, for example, one mistake for every 1,000,000,000 to 10,000,000,000 nucleotides replicated. Replication culminates in two identical double helices, having started with one helix and added the complementary strand to each individual strand.[4R]

Transcription (Making RNA from DNA): DNA contains genes that encode proteins, starting with a start codon and ending with a stop codon. In a process called transcription, the cell uses DNA to create RNA that reflects the information contained in that DNA strand. Transcription occurs in the nucleus (when one exists in the cell). This copy of the DNA strand is called messenger RNA (mRNA), which eventually directs the synthesis of the protein the DNA encodes.

There are other types of RNA, which have different roles. Transfer RNA (tRNA) contains 73–93 nucleotides and recruits amino acids to participate in the protein synthesis process that the mRNA directs. Ribosomal RNA (rRNA) is part of the cell's machinery that enables the protein synthesis process to occur (by acting as a catalyst, see the discussion of protein synthesis further in this section). These three types of RNA molecules all contain nucleotides and constitute stepping stones along the path from DNA to protein.[4R]

Only some of the DNA undergoes transcription. To begin transcription, the DNA double helix opens. Only one of the two strands in the double helix is transcribed. An enzyme (RNA polymerase) moves along the DNA strand of interest. At each point, it recruits the necessary complementary RNA nucleotide and then stitches the nucleotide onto the growing RNA strand, using "clack-click" chemistry. The enzyme checks the hydrogen bonds formed between the DNA being copied and the forming RNA strand, as well as the geometries of the new DNA–RNA base pairs.[4R,5R] Thus, transcription forms an RNA strand complementary to the DNA strand that is being copied. Although Nature is a masterful, she is also a slightly messy writer of DNA: hence, the initially formed mRNA molecule is a rough draft, which is refined further after transcription; for example, one edit is that regions that do not code a protein are removed (by a process called splicing), etc.[4R]

Protein Synthesis: The processed mRNA leaves the nucleus and goes into the main area of the cell to associate with a cellular component called the ribosome. Built of proteins and RNA (rRNA), the ribosome is the site of protein synthesis. How do amino acids come to the ribosome?

The transfer RNA (tRNA) molecules introduced in the previous section accomplish this task. tRNA molecules first associate with amino acids, with the help of enzymes called aminoacyl tRNA synthetases. Then, tRNA molecules deliver these amino acids to the ribosome. Each such tRNA has an anticodon component composed of three nucleotides; the anticodon base pairs with the codon on the mRNA. There are 61 different tRNA molecules (one would think there are 4^3, with four bases and three codon sites, but one must subtract the three codons that do not code amino acids). tRNA molecules differ in both their anticodons and other parts of their nucleotide sequences. Each tRNA molecule corresponds to a particular amino acid (hence, many of the 20 naturally occurring amino acids have multiple tRNA molecules recognizing them).

Each amino acid has a different enzyme that recognizes that amino acid and its corresponding tRNA molecule(s). This enzyme first activates the appropriate amino acid and then releases it to the appropriate tRNA, by making an ester bond between the amino acid and tRNA molecules (an O–C bond, to form the O–C=O ester group encountered in Lecture 5). These loading enzymes use intricate intermolecular interactions (such as hydrogen bonds) to recognize the appropriate tRNA and amino acid. There are 20 variations on this theme, and how each enzyme recognizes its amino acid molecule and corresponding tRNA molecule(s) is a story in its own right.[6R] These confirmation mechanisms ensure that mistakes in loading amino acids on tRNA carriers occur only once for every 10,000 amino acids loaded.[4R, 7Ra]

To begin protein synthesis on the ribosome, the mRNA start codon AUG associates with a tRNA–amino acid complex. Then, another tRNA delivers another amino acid, and a C–N bond forms between the two amino acids, making a small peptide. The peptide then grows further and becomes a protein. This process is called translation, because it translates a nucleotide sequence into an amino acid sequence.

How does the ribosome know which amino acid to add? Each tRNA delivers a specific amino acid that is dictated by the original gene, and the tRNA–amino acid complexes come to the ribosome one by one. The tRNA accurately delivers the amino acid requested by the mRNA because the tRNA anticodon associates with the appropriate mRNA codon by base pairing. In this manner, tRNA brings the correct amino acid that matches the mRNA codon. The ribosome stitches the incoming amino acid onto the peptide, excising one water molecule in total ("clack-click"). Interestingly, the ribosomal RNA catalyzes this reaction. Then, the process repeats: the mRNA moves, the next codon requests another amino acid, and a tRNA delivers this amino acid. Accurate tRNA loading enzymes and hydrogen bonds between tRNA and mRNA ensure that amino acids arrive in the right order. The process continues until an mRNA stop codon (i.e., UGA) ends protein synthesis. *Escherichia coli* makes a 100-amino acid polypeptide in 5 seconds: think of how many proteins were made as you read this lecture![4R]

In nature, proteins have only left-handed amino acids (termed *homochirality*). This homochirality is established during tRNA loading and protein synthesis. The tRNA loading enzymes preferentially insert left-handed amino acids onto tRNA. Left-handed amino acids fit better into these enzymes than do right-handed amino acids. Also, nucleotides are chiral (see Problem 7.7). In studies of chiral recognition, in the loading enzyme, a tRNA model whose nucleotides had the chirality found in nature preferred to receive left-handed amino acids, and a model whose nucleotides had a different chirality preferred to receive right-handed amino acids. Chiral preferences also occur on the ribosome, where left-handed amino acids fit better. Two left-handed amino acids interact more favorably and form a bond more easily in the ribosome than do a left-handed amino acid and a right-handed amino acid. Biology is exquisitely sensitive to molecules' handedness.[4R,7Ra]

More details can be found in Ref. [7R].

All Kinds of Genes: Among the interesting genes are the ones called "HOX" and "SHH" genes. The HOX genes are responsible for making the signaling proteins that are in control of the shape of the body, and thus they are responsible for the fact that humans have the shape of humans and not of flies, etc. The SHH genes are responsible for creating signaling proteins that take care of our lateral symmetry, including the right and left hemispheres of the brain, the spinal cord (central nervous system), eyes, limbs, and many other parts of the body.

Mitochondrial DNA (mtDNA): Although most of the human DNA resides in the nuclei of the cells, some DNA is present in the organelle called the mitochondrion. The mitochondrion is the site where the energy stored in foods gets converted to a form that the cells can use (a molecule looking like a nucleotide with three consecutive phosphate units and an adenine base). The mtDNA is inherited from the mother, thus enabling scientists to trace the entire human race to "a single mother" living in Africa some 170,000 years ago. The mtDNA undergoes self-replication but remains residing in the mitochondrion. Some of the mtDNA genes have been transferred during evolution to the nuclear DNA.

Gene Regulation: The DNA story has many, many more facets. The long double helices in the nucleus of the cell are wrapped like a yo-yo string around spools of proteins called histones. A gene is activated into action only when the histone undergoes a chemical modification by attaching an acetyl group ($H_3C-C=O$) that causes the DNA to unwind and expose the requisite gene (for copying). Another means of regulation is by attaching a CH_3 group to the cytosine (C) base. This causes the respective gene to be silenced. The distribution of silenced genes in the cell is different for different cells, and this allows the cells to become specific (e.g. heart cells, brains cells, etc.) For example, cells destined to become skin cells must have turned off all the genes that produce liver enzymes, neurotransmitters, ... and so on.

A variety of small molecules called *transcription factors*, or put simply copying factors, achieve finer control of genes. Vitamin A (which is found in carrots and has resemblance to the retinal molecule) is such a transcription factor. It causes the copying of genes that are responsible for growth and for maturation of cells, especially skin cells.

Gene "Jumping": Although we inherit the DNA of our parents, it turns out that the DNA has a degree of freedom to change when the strands clap and exchange or copy pieces of DNA into one another. This phenomenon of "jumping genes," which was discovered by Barbara McClintock in the 1940s, creates DNA variation. In corn, where this was discovered (for this discovery, McClintock was finally awarded the Nobel Prize in 1983), the phenomenon of jumping genes creates fast mutations and generates new forms and colors of corn. In humans, it is more restrained and not well understood. Nonetheless, such jumping genes are postulated to have been a major force in human evolution, to be important for brain plasticity, and so on (see Ref. 7 in the text).

The DNA story is a vibrant science in development!

7.R.7   References for Retouches

[1R] S. S. Zumdahl and S. A. Zumdahl, *Chemistry*, 9th Ed., Cengage Learning, Brooks College, 2010. Chapter 7, p. 314. Available at: www.cengage.com/global

[2R] Ibid, Chapter 8, p. 314.

[3R] Ibid, Chapter 9, pp. 419–20.

[4R] D. L. Nelson and M. M. Cox, *Lehninger Principles of Biochemistry*, 6th Ed., W. H. Freeman & Co., New York, 2013, pp. 1009–1153.

[5R] (a) M. W. Kellinger, S. Ulrich, J. Chong, E. T. Kool, and D. Wang, *J. Am. Chem. Soc.,* **2012,** *134,* 8231; (b) S. Zhang and D. Wang, *Israel J. Chem.,* **2013,** *53,* 442.

[6R] R. Giegé, M. Sissler, and C. Florentz, *Nucleic Acids Res.,* **1998,** *26,* 5017.

[7R] (a) For general reading, see: S. D. Banik and N. Nandi, Chirality and protein biosynthesis. In P. Cintas, Ed., *Biochirality: Origins, Evolution and Molecular Recognition* (Topics in Current Chemistry, Vol. 333), Springer, Heidelberg, 2013, pp. 255–305; (b) J. L. Marx, *A Revolution in Biotechnology*, Cambridge University Press, Cambridge, 1989, pp. 4–10.

7.P   PROBLEM SET

7.1 Construct the molecule made from Be and Cl. (a) Specify its geometry and bond angle. (b) In the solid, the molecule forms "infinite" chains. What is the geometry around the Be centers in these chains?

7.2 Construct the following molecules, draw their 3D structures, name the 3D structure (e.g., octahedron, or derived from an octahedron, etc.), and indicate the bond angles: (a) B and as many H's as required. (b) The molecule in (a) after accepting an H^-.

7.3 Do as requested in Problem 7.2 for the following molecules, which obey the octet rule: (a) S and as many F atoms as needed. (b) P and as many Br atoms as needed.

7.4 Do as requested in Problem 7.3 for the following molecules, which possess an even number of electrons but without the octet rule limitation: (a) one S atom and as many F atoms as needed (there are multiple answers). (b) P and as many Br atoms as needed. Do not consider electron-deficient molecules.

7.5 Draw 3D structures for all the molecules that can be generated by the replacement of two H's in C_2H_4 with methyl groups. Characterize the 3D structures, indicate all unique bond angles (using idealized angles), and try to name the different molecules.

7.6 Draw and characterize the 3D structures of H_2SO_4, SO_2, and SO_3, and indicate all unique angles.

7.7 Find all the chiral (handed) carbon centers in a nucleotide of DNA and mark them by asterisks on the drawing. You can use generic labels for the phosphate (as P) and the base (as B).

7.8 Pick an amino acid of your choice and draw its two enantiomers. Try marking the sense of rotation from the NH_2 group first to the COOH group and then to the R group (as done in the text, place the H away from you). To determine your assignment, consider using plastic models of the molecule.

7.9 C_4H_4 is a cage molecule. Convert it to a chiral one by changing the C atoms to other atoms.

7.10 The C_4H_4 molecule in (Problem 7.9) has a planar isomer. Draw it and replace its H's by four different single connector atoms or groups. Is this molecule handed?

7.11 Acetic acid is known to form pairs of molecules even in the vapor state. Draw this pair, characterize the intermolecular interactions, and indicate all the geometric details.

7.12 Construct all the EH_4 molecules where E is an atom belonging to the 4A family (considering from C to Pb) and rank the relative polarity of the E–H bonds in these molecules. How do the corresponding molecular dipoles vary?

7.13 Draw benzene, characterize its 3D structure, and specify the unique bond angles. Now, replace two H's by two Cl's and draw only the isomer that has no molecular dipole.

7.14 Draw a chain of HF molecules. What kind of interaction holds the molecules? What are the two unique angles in the chain? Explain your reasons.

7.15 Draw the most stable conformation of C_4H_{10}.

7.16 Draw three isomers of tartaric acid, showing the two representing enantiomers and the one where the chirality cancels out. (bonus question).

THE IONIC BOND AND IONIC MATTER

8.1 CONVERSATION ON CONTENTS OF LECTURE 8

Sason: Welcome back, Racheli!

Racheli: Thank you, Sason. But are you sure you can handle two interviewers once more?

Sason: I am somewhat tensed by these prospects, but I shall try ...

Racheli: OK, let me begin then. After two demonstrations of the "buzzing crystal" of the common salt NaCl versus the non-buzzing sugar crystal, are you *finally* going to teach about ionic bonds?

Sason: Yes, this is my intention ...

Usha: Why did you delay it, Sir?

Sason: We just learned about polar bonds. So this is the right time to teach about the ionic bond, which manifests itself in extremely polar situations like Na^+ and Cl^- in the common salt. The activation of the electric buzzer occurs because the ions Na^+ and Cl^- move in the electric field and complete thereby the electric circuit.

Racheli: Are there other proofs that a material like NaCl is intrinsically ionic?

Usha: Of course, Racheli! Haven't you heard about *colligative properties* of ionic versus covalent solutions?

Sason: Now, now, Usha. I am trying to avoid using strange terms and you are bringing one. Let me at least explain it.

Chemistry as a Game of Molecular Construction: The Bond-Click Way, First Edition. Sason Shaik.

The term *colligative properties* refers to the effect of dissolving compounds in a solution (such as a water solution) on the properties of this solution. It turns out that the water solution boils at a higher temperature than 100°C and freezes below 0°C, and so on.

Racheli: What does this have to do at all with the distinction between ionic and covalent materials?

Sason: Elementary, my dear Racheli! Imagine that we are warming pure water in a teapot to bring it to a boil. When the water is pure, warming it will cause water molecules to leave the liquid and go into the vapor phase after detaching all the H-bonds holding each such escaping molecule. You need to warm the pure water up to 100°C.

Imagine now that we dissolved an Avogadro's number of molecules in the water, and we again warm it up to boiling. When the solution contains so many other molecules of the dissolved material, using a tongue-in-the-cheek expression, these molecules act as a blanket that retards the escape of the water molecules from the solution to the vapor. Therefore, you would need to heat the solution above 100°C to bring it to a boil. Just for the sake of argument, let's say that the boiling point will be 101°C.

Racheli: But you did not answer my question, Sason.

Sason: Be patient … Imagine now that we dissolved an Avogadro's number of NaCl molecules in water. If NaCl is indeed an ionic material, it will separate in solution into the Na^+ and Cl^- ions. So how many dissolved particles does this solution have?

Usha: Two times Avogadro's number, Sir!

Sason: This is correct. Because of the dissociation to ions, the effect of the dissolved NaCl will be twice as large compared with the effect of a nondissociating covalent material. Hence, the NaCl solution will have a higher boiling point than the sugar solution.

Usha: The ratio of the magnitudes by which the boiling points are elevated will be the ratio of the number of dissolved particles in the two solutions. Hence, let's suppose that the sugar solution boils at 101°C, 1°C above the boiling temperature of pure water; this means that the NaCl solution will boil at 102°C, 2°C above the natural boiling point.

Sason: So, if we boil the solution and find what Usha just described, we may conclude that the NaCl bond is ionic, and upon dissolution, the Na^+ and Cl^- ions separate and move independently.

Usha: Do not forget, Sir. NaCl solution will conduct electricity, as opposed to a solution of sugar. We therefore have two more indications of the bond ionicity in NaCl.

Racheli: But Sason, Usha, you missed the gist of my question. Some covalent compounds like acids and bases also dissociate to ions in solution. Would you still say they have ionic bonds?

Sason: But as usual, Racheli, you were jumping too far ahead with this question.

You are right, of course, that when we put an acid like HCl in water, each HCl molecule is attacked by a water molecule that grabs H^+ from the HCl and cleaves the bond to two ions, H_3O^+ and Cl^-. So, even though the H–Cl bond is covalent (with some polarity), its reaction with water produces ions, and hence, if we rely only on colligative properties or on the conductivity of the water solution, we shall err in our deduction about the intrinsic nature of the H–Cl bond.

Racheli: This is all fine, Sason. I trust, you can prove that H–Cl is covalent, but how would you reason the ionicity in limestone, $CaCO_3$, which does not dissolve in water?

Sason: Heavens, Racheli! Just imagine that water could dissolve limestone. Then all the beautiful houses in Jerusalem would have dissolved every winter! I hate to think about that, but even the Wailing Wall in Jerusalem would have been in solution by now ...

Racheli: So what about the ionicity of limestone?

Sason: $CaCO_3$ is made of Ca^{2+} and $CO_3{}^{2-}$ ions. As I plan to teach, when the ionic charge is high, water cannot dissolve anymore the ionic material. Luckily, there are other measurements of *ionicity*, like conductivity of molten salts and measurements of the dipole moment of salt molecules in the gas phase. When we apply these methods to Na–Cl, we find that the bond is intrinsically ionic.

Usha: But, Sir, you forgot two additional methods. For example, theory is sufficiently accurate today to tell you that NaCl is made from two ions held together by the attraction of the opposite charges. You and your collaborators performed such calculations several times.

A second method that is suitable for the insoluble ionic material looks at the structure of the solid material ...

Sason: Looking at solids is a great way! In solid sugar, you would see discrete sugar molecules. In solid NaCl, you do not see molecules but an infinite array of Na and Cl species packed together at equal Na–Cl distances in all directions. This indicates that instead of covalent molecules with short bonds and separated by long distances between the molecules, you see close packing of ions. All the ionic materials, even those that do not dissolve in water, obey this rule, and we can deduce the ionic nature of their bonds from the structure of the solid.

We have plenty of ways to characterize the ionic bond ...

Racheli: Tell me, Sason, how did all these chemists come up with the ideas of these bond types? And why were they concerned about these issues at all?

Sason: This is a wonderful chapter of science, Racheli. Much of it has to do again with the great Gilbert Newton Lewis.[1–3]

Usha: Sir, I like stories. Please tell us!

Sason: The story goes back to the end of the nineteenth century and at the turn of the twentieth century, when prominent European chemists like Ostwald, van 't Hoff, Arrhenius, and Nernst (Figure 8.1) started developing an understanding about the behavior of materials in solution, and they established a new school, which they called *physical chemistry*. They went against the tide and revolutionized chemistry.

FIGURE 8.1   Photographs, from left to right, of W. Ostwald, J. H. van 't Hoff, S. Arrhenius, W. Nernst, and J. J. Thomson. (Ostwald's photograph is taken from *Popular Science Monthly*, Vol. 67; van 't Hoff's photograph is taken from Les Prix Nobel published in 1904; Arrhenius's photograph is taken from German Wikipedia, with credit to Matthias Bock; Nernst's photograph is used with credit to the Smithsonian Libraries; and J. J. Thomson's photograph is from *The Great War*, edited by H. W. Wilson and J. A. Hammerton, 1915, Amalgamated Press, London.) All images are in the public domain.

Oh, these were great days for chemistry in the public eye! There is a famous novel, *Arrowsmith,* which was written in 1925 by the American playwright, Sinclair Lewis, about a young rebel scientist who goes his own way to make a breakthrough, and who is reminiscent of the young physical chemists who revolutionized chemistry.[3]

Racheli: Sason, the students will think you are talking about the chemist Lewis who wrote a novel …

Usha: Racheli, Sir likes to tell stories within stories … A story is not a scientific piece …

Sason: You are a good audience, Usha. And you Racheli, there's no one like you to keep me on track … Let me get back to the main story.

In those days, batteries were already known, and conductivity measurements could be done; one of the major interests of the physical chemists was the conductivity of materials in water solution. They discovered that some materials like NaCl conducted electricity very well, while others like acetic acid (CH_3COOH) exhibited only weak conductivity in solution. They called these material *electrolytes* because they conducted electricity.

The physical chemists understood that the materials produced ions upon dissolution, but they did not really understand why and what is the difference between weak electrolytes like CH_3COOH and strong ones like NaCl.

Usha: How did Lewis get into this European scene at all?

Sason: Well, as it usually happens in science, his PhD adviser at Harvard, Theodore Richards, who had himself studied with Ostwald and Nernst, advised Lewis to go to Germany and have an internship with the two great scientists. Germany was then the leading country in science.

Racheli: And …

Sason: So, after his time in Europe, Lewis returned to the US with this background of questions about the behavior of weak and strong electrolytes in solutions. While back at Harvard, he and Richards did not get along anymore. Both were not very easy characters.[2]

Lewis was subsequently offered to join the new group of Arthur Noyes at MIT. In the court of King Arthur, as Noyes was referred to,[3] Lewis's creativity soared and he began thinking about the chemical bond in terms of the newly discovered elementary particle, the electron ...

Usha: ... which was discovered by the physicist J. J. Thomson (Figure 8.1), who also had a general theory of the chemical bond in terms of ions.[1]

Sason: Chemists who knew molecules like sugar or benzene, which are not ionic, dismissed Thomson's idea of chemical bonding. Bonding had clearly to go beyond just ionic matter.

Racheli: Sason, you are digressing and keeping everyone in suspense. So what happened when Lewis started thinking...?

Sason: Hey, let me finish the story. Stories are convoluted ... Lewis's famous 1916 paper is a fascinating piece of literature,[1] since unlike the way we write papers today, his paper looks like a story of how the idea actually evolved during the writing until it converged to the picture we have today about bonding.

Lewis was interested in understanding the difference between weak and strong electrolytes in solution. However, like the Biblical story of King Saul, who initially went seeking for his father's donkeys, and then ended up finding a Kingdom, so did Lewis, who wanted to understand the behavior of electrolytes in solution and found the electron-pair bond as an intrinsic property of molecules. In so doing, Lewis changed chemistry. The historian Servos[3] likens this situation to the early voyagers to America who sailed to seek the Spice Islands and discovered a new continent.

Racheli: We are still in suspense ...

Sason: Just relax. I am getting to the point ... About halfway through his paper, *Lewis suddenly postulates that pairing of electrons forms a bond.* There was no precedent for this idea anywhere, because it defied the laws of physics.[1] It takes some courage to do that ...

Lewis then argued that when the electron pair is shared equally or almost so, the bond is covalent or polar covalent, and when one of the bonding partners appropriates the electron pair for itself, the bond is ionic and contains a pair of ions that are attracted to one another. In this manner, *Lewis unified the chemical bond and stretched its territory all over the chemical knowledge of the day.*

Racheli: Lewis did not receive the Nobel Prize for his contribution. Why?

Sason: His candidature was submitted numerous times. Scientific rewarding tends to be once in a while political. You can read all about this in Coffey's book.[2] Lewis's story is full of drama, and even his death is shrouded in mystery.

Usha: If you have finished your story, Sir, let me remark that it is not easy to teach an ionic bond using your "bond-click" approach. So how do you plan to teach that NaCl is actually Na^+Cl^- and not, for example, $Na^{7+}Cl^{7-}$ or $Na_2^+ Cl^-$, etc.?

Sason: Fear not, Usha! I am going to use a simple and fun method, which I call "click-clack," and which follows the Lewis logic.

First, I will "click" the available valence electrons to make covalent bonds, and then "clack," I shall shift the electron pair to the more electronegative atom and create a pair of ions. I will then show how this method enables constructing ionic materials of the correct stoichiometry and charges.

Racheli: The ionic bond seems to be quite dull. OK, you make it and then what? You cannot design a beautiful chemical synthesis to make new molecules. It seems that the bond is a dead end.

Sason: Well, Racheli, the ionic bond has some beautiful architecture too, and it lends itself to design, such as the very useful ionic liquids. I shall also illustrate the ingenious usage of ionic matter in biological systems. We are the salt of the earth …

Usha: I have a feeling, Sir, that you will have to teach at least a small section on acids and bases, since these molecules dissociate to ions in water.

Sason: Certainly … Let us proceed now with the lecture. This reminds me that a good way to start the lecture is to let NaCl activate the buzzer …

8.2 IONIC BONDS VERSUS COVALENT BONDS

We have mentioned all along that in addition to all the beautiful molecules we constructed up to now using shared electron pairs (covalent bonds), there are materials wherein the electron pair in the bond is completely "owned" by one of the bonding partners, and as such the bond involves two oppositely charged ions. This ionic constitution of the bond can be detected in a variety of ways. For example, if you connect a crystal of common salt, NaCl, to a buzzer and wet the crystal a bit, the buzzer will whistle due to the movement of the ions and the creation of an electric current. A crystal of sugar that is made exclusively of covalent bonds will not whistle even if you drown the sugar in water. Similarly, a solution of an ionic material like NaCl will conduct electricity, while a solution of a covalent material like acetone or sugar will not.

There is no life without ionic matter. If you recall, our neurons communicate by neurotransmitters and by currents of ions, which flow through proteins called receptors. In this manner, we can move, think, feel, and be what we are. It is important to get familiar with the ionic bond.

8.2.1 The Formation of Ionic Bonds. How and When?

An ionic bond is formed as a result of extreme polarity of the electron-pair bond in a situation where one of the bonding partners is much greedier for electrons than the other. Scheme 8.1 shows this in a pictorial manner, by looking at the changes in the bond character along a scale of electronegativity difference between the constituent

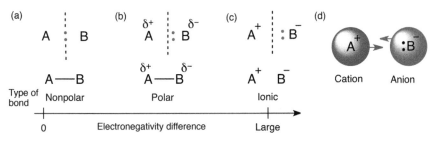

SCHEME 8.1   A schematic description of how the polarity of a bond A—B changes as the electronegativity difference between A and B increases. The vertical dashed line between A and B marks the midpoint of the bond: (a) When the electronegativity values are the same or similar, the electron pair (marked in red) is located on average in the midpoint between A and B, and the bond is nonpolar. (b) As B becomes more electronegative than A, the electron pair shifts closer to the electronegative fragment B, creating a polar bond ($A^{\delta+}$—$B^{\delta-}$). (c) Upon further increase in the electronegativity difference, the electron pair moves completely to the electronegative fragment B, thus forming an ionic bond, A^+ $:B^-$. This ionic bond is different from click bonds and, thus, is not depicted with a line as the click bonds in (a) and (b) are. (d) In ionic bonds, the two ions, the cation (the positive ion) and the anion (the negative ion), are being pulled together by the attraction between the opposite charges; the arrows pushing the two ions together indicate the attractive force.

fragments A and B. Thus, when the electronegativities of the two fragments are close or identical (Scheme 8.1a), the electron pair is located on average in the bond midpoint, and A—B is a nonpolar bond. As B starts becoming more electronegative than A, in Scheme 8.1b, the electron pair starts shifting its position toward B, thereby generating a polar bond with small charges ($\delta^{+/-}$) on the fragments. As the electronegativity difference further increases, at some point, B appropriates the electron pair for itself, thus forming an ionic bond $A^+:B^-$ (Scheme 8.1c). In the ionic bond, the positive ion is called a *cation* and the negative one is called an *anion*, and the two ions are pulled together by the attraction between the opposite charges (Scheme 8.1d).

It is not a simple matter to give a definite value of the electronegativity difference where the bond shifts character from being polar covalent to becoming ionic. However, chemists have developed a practical guideline that defines when we actually observe ionic bonds. In principle, the ionic bonds are formed from combinations of the least electronegative atoms with those that are the most electronegative. Scheme 8.2 shows these combinations. Thus, the least electronegative atoms in the periodic table are the metals of groups 1A—3A (with the exception of Be and B in groups 2A and 3A, respectively), *which will form cations*. On the other hand, the late nonmetals of groups 5A—7A are the most electronegative ones and *will form anions*. The combinations of atoms of groups 1A—3A with those of groups 5A—7A form ionic bonds. Transition metals also form cations, which in combination with atoms from groups 5A—7A will form ionic bonds (see Lecture 9).

The classical example of an ionic bond is Na^+Cl^-. Figure 8.2 shows pictorially the formation of Na^+Cl^- from its elements. Na is a metal, which is a grayish solid,

Cations The metals = The least electronegative				Anions The late nonmetals = The most electronegative		
1A	2A	3A		5A	6A	7A
Li	—	—		N	O	F
Na	Mg	Al		P	S	Cl
K	Ca	Ga		:	Se	Br
Rb	Sr	:		:	:	I

Other cations
transition metals
e.g.: Mn, Fe, Co, Ni

SCHEME 8.2   The common cation and anion constituents of ionic bonds are the metals of groups 1A—3A, which form cations, and the late nonmetals of groups 5A—7A, which form anions. Transition metals also form cations.

usually kept under kerosene oil because the metal goes wild if exposed to humidity or to oxygen, bringing about vigorous reactions, heat and fire. Chlorine, Cl_2, is a yellow-greenish gas, which is extremely toxic, and it was used in World War I as a chemical weapon. On the other hand, it can be used as a water disinfectant and hence a "provider" of life-giving water. So the gas chlorine is another manifestation of the chemical Janus. But I am digressing. Getting back to our reaction in Figure 8.2, we can see that as a grayish metal and a green gas meet, lo and behold, they both disappear and give rise to the neat-looking white crystals of the salt Na^+Cl^-, which we can taste and use in our food. This is the chemical magic at its best: two materials disappear and a new one, a completely different one, appears instead.

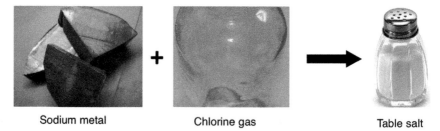

Sodium metal   Chlorine gas   Table salt

FIGURE 8.2   The chemical "magic": formation of the salt Na^+Cl^- (table salt) from a grayish metal (Na) and a yellow-greenish gas (Cl_2). These photographs were obtained from http://en. wikipedia.org/wiki/File:Na_%28Sodium%29.jpg, http://commons.wikimedia.org/wiki/File: Chlorine2.jpg (public domain image), and http://en.wikipedia.org/wiki/File:Salt_shaker_on _white_background.jpg.

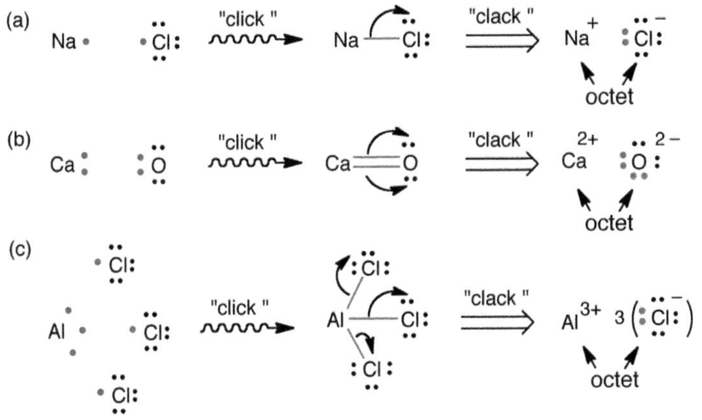

SCHEME 8.3 Construction of ionic molecules by "click" followed by "clack" starting from neutral atoms: (a) NaCl, (b) CaO, and (c) AlCl$_3$. The electron pairs that are connected by the "click" and are then transferred by the "clack" are marked in red. The clack is indicated by the curved arrow that shifts an electron pair.

8.2.2 Construction of Ionic Bonds by "Click-Clack"

How do we construct ionic molecules of the correct stoichiometry and ionic charges? An effective method is by first "clicking" the connectivities of the atoms, and then "clacking" (breaking) the covalent bonds and transferring the electron pairs to the more electronegative atoms. Scheme 8.3 shows three such examples of bonds between metal atoms and more electronegative atoms, which qualify as forming ionic bonds according to the guidelines in Scheme 8.2.

Thus, as shown in Scheme 8.3a, since Na and Cl are single connectors, they will "click" and form a single bond. Since the electronegativity difference of the Na and Cl qualifies this bond to be ionic (see Scheme 8.2), we "clack" and transfer the electron pair to the Cl. In this manner, we construct Na$^+$Cl$^-$.

Note that the ions have unit charges and the stoichiometry is 1:1. The reason for that is that by gaining the electron pair to itself, the Cl$^-$ anion has attained an octet. What about Na$^+$? The atom Na has 11 electrons, which are arranged as 2 electrons in the deeper shell, 8 in the second shell, and only 1 in the frontier shell. Thus, by loss of an electron, the resulting ion Na$^+$ has attained an octet in the exposed second electronic shell, which now becomes the frontier shell. Any other stoichiometry or different charges would have violated the rule of Nirvana and so these do not occur.

Scheme 8.3b shows a more challenging case, of Ca and O, which according to Scheme 8.2 qualify to form ionic bonding. Thus, since Ca and O are double connectors, "click" and they form a double bond. Subsequently, by "clacking," we transfer the two electron pairs to oxygen and generate the ionic molecule Ca^{2+}O^{2-}. The ions are now doubly charged, but the stoichiometry is still 1:1. As we reasoned in the above example, here the two doubly charged ions are in their states of Nirvana, and the ionic bond is kosher.

(a) Complex anions   (b) Complex cations

PO_4^{3-}, SO_4^{2-}, CO_3^{2-} NH_4^+

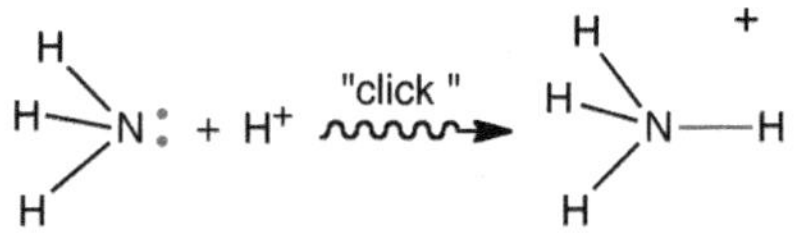

SCHEME 8.4   Complex ions: (a) The anions PO_4^{3-}, SO_4^{2-}, and CO_3^{2-}. (b) The complex cation NH_4^+ and its construction by clicking ammonia and a proton, where ammonia donates both electrons to form a bond.

Scheme 8.3c further increases the complexity, by considering a molecule made of Al and Cl. According to Scheme 8.2, these atoms will combine to generate ionic bonds. Now, since Al is a triple connector, it will require three Cl atoms, each being a single connector. By "clicking," we form $AlCl_3$ with three Al–Cl electron pairs, and by "clacking," we transfer the pairs to the Cl atoms and generate the ionic molecule Al^{3+} $(Cl^-)_3$. The molecule has three ionic bonds made from Al^{3+} holding three Cl^- anions. With this stoichiometry and charges, the two ions attain Nirvana (see Retouches section 8.R.1).

8.2.3   Ionic Molecules Containing Complex Ions

The rules of electronegativity difference and ionic Nirvana apply also to more complex ions, some of which are shown in Scheme 8.4. Thus, Scheme 8.4a shows three anions with their charges. In a minute, we are going to derive these charges using our "click-clack" trick, but for the moment, just consider their chemical constitution. All these ions contain several electronegative oxygen atoms attached to centers like P, S, and C, and therefore, they follow the requirement that the anion should be highly electronegative. If you recall the acids H_3PO_4, H_2SO_4, and H_2CO_3, you can immediately identify that PO_4, SO_4, and CO_3 are fragments of these acids.

Scheme 8.4b contains a complex cation made from nitrogen that has four covalent bonds to H. Recalling that the molecule NH_3 has three N–H bonds, but it also possesses a lone pair on the nitrogen, we can immediately determine that this cation is generated when ammonia donates this lone pair to the electron-deficient proton H^+, thus leading to NH_4^+, known as the ammonium ion. NH_4^+ has achieved an octet for N and a duet for H, and hence the cation as a whole obeys the rule of Nirvana. You can easily extend this to other ammonium ions by attaching a positive fragment to the lone pair of amines. You can further apply this idea to the generation of analogous complex anions, like BF_4^-, which is common in ionic chemistry.

Let us construct now some ionic molecules from these complex ions. Scheme 8.5 is instructive since it shows why should PO_4 have a triple negative charge. Thus, let us start with the fragment PO_4 and figure out the molecule it will form with Na. Since the neutral PO_4 fragment is a triple connector (you can also practice constructing it from H_3PO_4, which we did in Lecture 6), we need three Na atoms, and "click," we

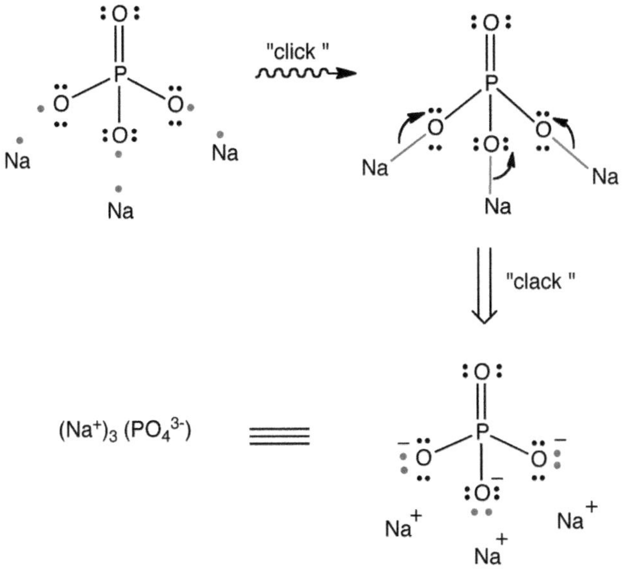

SCHEME 8.5 Constructing the ionic molecule $(Na^+)_3PO_4^{3-}$ from neutral Na and the modular neutral fragment PO_4 with the click-clack approach.

form three O–Na bonds. Subsequently, we clack and transfer the electron pairs to the oxygen atoms, thus forming the ionic molecule $(Na^+)_3PO_4^{3-}$.

In a similar manner, one can deduce the double-negative charges in SO_4^{2-} and CO_3^{2-}. For example, constructing $(Na^+)_2CO_3^{2-}$ (used as washing soda) is a good exercise. Once we deduce the charges of all the complex anions, we can combine them with NH_4^+ to form new ionic molecules. Scheme 8.6 shows two of these molecules. The first couples the ammonium ion with Cl^-, and since both ions are singly charged, the molecule is $NH_4^+Cl^-$. This material is called ammonium chloride, and it is used as a fertilizer (it was made synthetically by the Arab alchemists in the tenth century). It also has a strong smell of ammonia due to its decomposition to NH_3 + HCl. In my childhood, when I accompanied my late father to the synagogue on Yom Kippur, I recall that fasting people in the synagogue used to smell $NH_4^+Cl^-$ vials, to keep themselves awake whenever hunger caused them some drowsiness during the prayer.

The second ionic material in Scheme 8.6 couples NH_4^+ with sulfate, SO_4^{2-}. Since the anion is doubly charged, we need two ammonium ions to create a neutral

$NH_4^+ Cl^-$ Ammonium chloride $(NH_4^+)_2 SO_4^{2-}$ Ammonium sulfate
    NH_4Cl $(NH_4)_2SO_4$

SCHEME 8.6 The ionic molecules made from NH_4^+ combined with Cl^- and with SO_4^{2-}. Note: The different stoichiometry is determined by the different charges of the two anions. The formulae below the names of these materials constitute the implicit way chemists write ionic materials by deleting the charges of the constituent ions.

molecule, with two ionic bonds, hence $(NH_4^+)_2SO_4^{2-}$, which is called ammonium sulfate. This material also serves as fertilizer. NH_4^+-based molecules are fertilizers because they are good sources of nitrogen.

Note that underneath the names in Scheme 8.6, I rewrote the formulae of the two materials while omitting the charges. This is not because I dislike ionic bonds. This is simply the economical way chemists write these materials, because supposedly, every chemist knows that NH_4Cl stands for $NH_4^+Cl^-$, and $(NH_4)_2SO_4$ for $(NH_4^+)_2SO_4^{2-}$, and so on. In fact, chemists use this economical way, which omits the ionic charges, to write all the ionic molecules in an implicit manner.

8.2.4 Why Are Ionic Materials Generally Solids?

Up to this point, we constructed isolated ionic molecules. However, if the common salt Na^+Cl^- were indeed made of isolated molecules, we would have sprayed our morning omelets using cylinders of NaCl gas. But NaCl is solid, and its name "rock salt" testifies to its solidity. Scheme 8.7 provides a glimpse of why ionic compounds tend to aggregate and form solids.

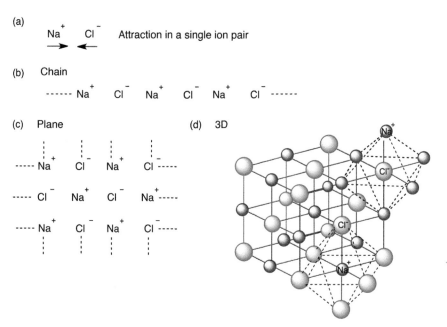

SCHEME 8.7   Stepwise aggregation of Na^+Cl^- "molecules," from a single ion pair (in a) to a chain where each ion is attracted to two adjacent oppositely charged ones (b), onward to a plane where each ion is attracted to four oppositely charged neighbors (c), and finally, to a three-dimensional (3D) solid, where the ions are packed into a regular arrangement (d). In this 3D solid, each ion has six oppositely charged nearest neighbors, as shown by the octahedra drawn around the Na^+ and Cl^- ions on the cube's edge and corner, respectively. The purple spheres represent Na^+, while the green spheres represent Cl^-.

Consider in Scheme 8.7a a single ion pair, Na^+Cl^-. As we have already argued, the opposite charges pull the two ions together by electrostatic forces (see Retouches sections 8.R.1 and 8.R.2), and this lowers the energy of the pair of ions. So, if a single attraction is favorable, why not consider a chain of Na^+Cl^- molecules, as in Scheme 8.7b? Now each ion sees two oppositely charged ions, one to its left, the other to its right. The ionic attraction is now stronger than in an isolated molecule, and the energy further drops.

Since a chain is more favorable than an isolated molecule, why not aggregate in a layer, as in Scheme 8.7c? Here, each ion is attracted by four oppositely charged ions, and therefore a layer is better than a chain. Continuing the same vein of reasoning, since a layer is better than a chain, how about packing an infinite number of planes one on top of the other? Indeed, as shown in Scheme 8.7d, in a three-dimensional (3D) structure, each ion sees, in addition to the four oppositely charged ions in the same layer, two more, one from the layer above and the other from the layer below. The improved attraction and the consequent energy lowering are the driving forces for the aggregation to form a solid material.

Figure 8.3a shows the 3D structure of Na^+Cl^-. It is seen that all the ions are arranged in an alternating $+/-$ pattern in all the three dimensions. In this energetically favorable arrangement, one cannot anymore distinguish individual Na^+Cl^- molecules! This is unlike the solid structure of covalent molecules, where the covalently bonded molecules retain their individuality. An example of a covalent solid is the solid structure of the Cl_2 molecule (shown in Figure 8.3b), where the individual covalently bonded molecules are clearly conserved and they maintain longer distances among themselves.

There are other architectures of solid ionic compounds, and they all share the above common feature of close packing of oppositely charged ions, in order to optimize the attraction between the cations and the anions.

(a)   (b)

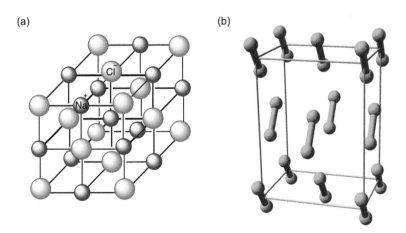

FIGURE 8.3   (a) The structure of solid Na^+Cl^-. (b) The structure of solid Cl_2. Note the short distance within each Cl_2 molecule and the long distances between molecules. (The image presenting the structure of solid Cl_2 is a public domain image obtained from http://commons.wikimedia.org/wiki/File:Chlorine-unit-cell-3D-balls.png.)

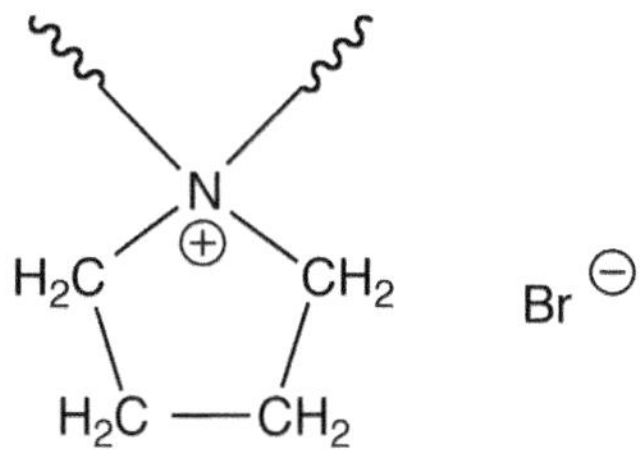

SCHEME 8.8 A cation that is protected by a ring and two carbon chains (shown by wiggly lines emanating from N). This protection of the positively charged site prevents the anion Br^- from packing well with the cation. Together, they form an ionic liquid.

8.2.5 Ionic Liquids?

The ability of chemists to manipulate matter and craft new forms is ingenious. Even in the ionic materials, where variations do not seem to be many, chemists found ways to make new ionic materials that cannot pack the ions too well. An example is shown in Scheme 8.8, where the cation is an analog of NH_4^+, but now the four bonds to nitrogen contain chains of CH_2 and CH_3 fragments (please verify that the nitrogen and its attached carbons are in Nirvana). In this manner, the positive charge of the cation is masked and protected by the chain such that the Br^- anion has little incentive or ability to pack close to the positive charge. The result is that the material cannot form a solid, and it exists as a liquid, an ionic liquid. There are many more ionic liquids available.

Ionic liquids have found numerous applications in green chemistry, in batteries, and in catalysis,[4] and it is thought that they will become useful in toxic waste removal. It turns out that the ant species called the tawny crazy ants (*Nylanderia fulva*) are the only ant species that can resist the amine-based venom of the awesome fire ants (*Solenopsis invicta*). The crazy ants do so by "washing" the venom of the fire ants with formic acid (which is their bodily fluid) and generating thereby a viscous ionic liquid, which is detoxified.[5] The element of design of ionic liquids is so evident. The imagination can soar in crafting new cations and anions. Try to design ones by yourself ...

8.2.6 Solubility and Insolubility of Ionic Materials

Anyone who has ever put some NaCl in water noticed that the solid dissolves almost instantly, as illustrated in Figure 8.4a. This often happens when the ions in the material are singly charged, as in Na^+Cl^-. Still, we have just argued that the solid is formed because it optimizes the number of attractions between opposite charges (Scheme 8.7) and lowers the energy. In other words, the ions in the solid are held by several strong ionic bonds. So why does Na^+Cl^- dissolve so easily?

Figure 8.4b show that when Na^+Cl^- dissolves in water, each ion is surrounded by a few water molecules that stabilize the ions. The cation Na^+ is surrounded by an average number of six water molecules, which stabilize it through their negatively polarized partial charges on the oxygen atoms. Similarly, the Cl^- anion is surrounded

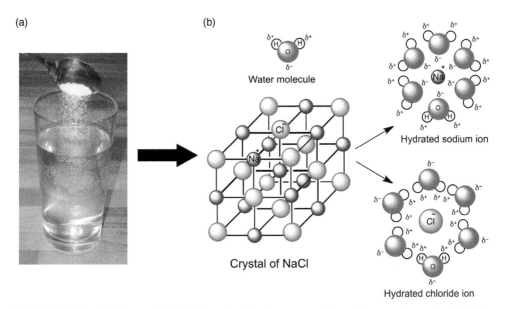

FIGURE 8.4   (a) The solid Na^+Cl^- salt dissolves in a glass of water instantly. (b) The decomposition of the ordered structure of the ions as they disperse throughout the water solution and interact with water molecules.

by a similar number of molecules of water which interact with it via strong hydrogen bonds. In this manner, the water molecules stabilize the ions and compensate them for the lost attraction they had in the solid before it dissolved. This is the reason for the instant dissolution of the ions. In many cases, the dissolution is accompanied by heating of the water in the glass. This heating means that the bonding of the dissolved ions is stronger than the ionic bonds in the solid, and hence the energy of the dissolved ions is lower compared with their state in the solid. This excess energy is dissipated as heat; this is called an exothermic process, that is, a heat-evolving process (see Retouches section 8.R.3 for the opposite case).

If Na^+Cl^- dissolves so easily in water, why is it that ionic solids like $(Ca^{2+})_3[PO_4{}^{3-}]_2$ or $Ca^{2+}CO_3{}^{2-}$ do not dissolve in water? Actually, the dissolution process is complex. It depends on the charges of the ions, on the structure of the solid, and on the water environment around the ions during the dissolution process. Therefore, an explanation for this phenomenon cannot be too simple. In part, the reluctance of an ionic solid to dissolve reflects the fact that the doubly and triply charged ions are attracted to one another very strongly and that the surrounding water molecules during solvation cannot compensate anymore for the loss of these strong attractions in the solid. This is because of the nature of the attraction between charges, which depends on the product of the two charges (see Retouches sections 8.R.1 and 8.R.2). Thus, for example, the attraction in $Ca^{2+}CO_3{}^{2-}$ is approximately four times stronger than in Na^+Cl^-, and hence $Ca^{2+}CO_3{}^{2-}$ does not dissolve. Similarly, $(Ca^{2+})_3[PO_4{}^{3-}]_2$ and $Al^{3+}PO_4{}^{3-}$ do not dissolve in water. However, as the ionic charge is but one of the factors in the propensity for dissolution, we can find ionic materials with multiple charges that dissolve in water, like $Ca^{2+}(Cl^-)_2$ that dissolves in water despite the dipositive charge on Ca^{2+}.

8.3   THE USE OF IONIC MATTER IN LIVING ORGANISMS

Living systems have evolved to make use of ionic matter in some ingenious ways. The soluble ionic material is used as a means of electric communication between our brain and the body, while the insoluble ionic material creates our skeleton and our teeth.

8.3.1   Soluble Ionic Material Takes Care of Biological Communication

The soluble ionic material creates free ions, which are mobile and can move under the influence of voltage variation (electric fields). Living organisms have made use of this mobility and harnessed ions like Na^+, K^+, and Ca^{2+} in neurons (nerve cells). We discussed already in Lecture 5 the receptor (bundle of proteins) that opens a pore in response to the binding of serotonin. Once the pore is opened, K^+ ions flow through the receptor and cause communication between billions of neurons. In the case of serotonin, this results in our being psychologically balanced.

Figure 8.5 shows a similar mechanism of ion flow that is used for muscular contraction (flex your muscles and you will be activating this mechanism). As can be seen from the figure, in addition to acetylcholine neurotransmission, the simple action

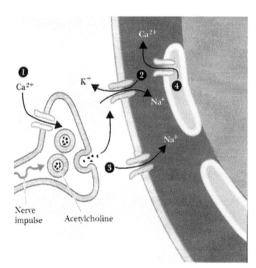

FIGURE 8.5   Activation of a neuromuscular junction that causes muscular contraction. The sequence of the ions' movements is indicated by the numbers near each ion. A flow of Ca^{2+} into the neuron (1) causes the release of the neurotransmitter acetylcholine, which is taken up by receptors on the muscle cell. This acetylcholine uptake in turn creates Na^+ and K^+ ion currents (2, 3), followed by Ca^{2+} release from one of the muscle cell's compartments into the muscle cell (4). The Ca^{2+} release subsequently induces the motion of proteins within the muscle cell, giving rise to the muscles' contraction. Adapted from Figure 8.23 in M. K. Campbell, *Biochemistry*, 2nd Ed., 1995, Saunders College Publishers.

of muscle contraction involves four different events of ion flow from the neuron to the muscle cell and inside the latter cell. What induces the muscle contraction is the flow of Ca^{2+} ions from a compartment of the muscle cell into another part of the cell (4 in the figure). The actual mechanism of the contraction is quite complex and involves a variety of Ca^{2+}-binding proteins. Being a doubly charged ion, Ca^{2+} attracts the polar C=O bonds of the protein, which claw the ion like in a cat's grip. Long-term memory involves an inflow of Ca^{2+} ions into neurons. Blood flow (and hence also male erection) is controlled by the outflow of Ca^{2+} ions from cells. Take these processes away and we are "dead meat."

8.3.2   The Insoluble Ionic Material Makes Our Skeleton and Teeth

The insoluble ionic material in living systems is mostly hydroxyapatite, which has the complex formula, $Ca_{10}(PO_4)_6(OH)_2$, and it involves a mixture of two anions, PO_4^{3-} and OH^-. All our bones and teeth (enamel) are based on hydroxyapatite. However, hydroxyapatite by itself would be extremely brittle. Such a property is no good for a skeleton that should keep us upright and enable us to walk and carry our weight, and at the same time be able to absorb hits and blows without easily breaking. And our teeth … They sustain enormous pressure during chewing and grinding. $Ca_{10}(PO_4)_6(OH)_2$ does not have these properties by itself. Therefore, it finds a suitable companion, the protein called collagen, which is made of a bundle of three α-helix proteins, and which is therefore flexible and very strong. The mixture of hard hydroxyapatite and flexible and strong collagen forms our skeleton and teeth. Figure 8.6 shows the hydroxyapatite in a bone. Without collagen, our bones and

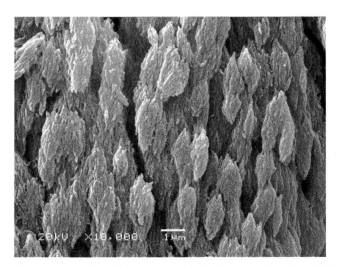

FIGURE 8.6   A constituent of our bone material is hydroxyapatite, $Ca_{10}(PO_4)_6(OH)_2$. Note that it is made of small crystals. In the living body, these small crystals are dispersed in a matrix of collagen. This scanning electron microscopy image of a rat bone is in the public domain and was obtained from http://commons.wikimedia.org/wiki/File:Bertazzo_S_SEM _deproteined_bone_-_cranium_rat_-_x10k.jpg.

teeth would be pulverized easily, and without hydroxyapatite, we would be gel-like creatures. We are indeed the salt of the earth.

8.4   COVALENT MOLECULES THAT FORM IONS IN SOLUTION: ACIDS AND BASES

At the outset of this lecture, we mentioned that there are molecules which are composed of covalent bonds, and nevertheless, in solution they produce ions because of a reaction they undergo with water molecules. Chief among these molecules are the ones we call acids and bases. Let me dedicate a few words to these molecules and to the ones they form (salts) when they react with one another. This topic belongs to the dawning of chemistry.[6]

Acids and bases have been known for millennia. For example, H_2SO_4 was known by the Sumerians, and most likely the acids HCl and HNO_3 were also known since their mixture (called *aqua regia*—the "royal water") was used to dissolve gold. Alchemists discovered that the stomachs of humans and animals contain acid (HCl), and some of them like van Helmont, who we mentioned before, thought that acids are "the force of life." Bases have been known for a long time, and they are even mentioned in the Bible (as "neter") as a means of laundering.

When alchemists noticed that acids (and bases) change the colors of fruits and vegetables (e.g., grapes, beets, cabbage ...), they started using water solutions of extracts made from flowers, fruits, and vegetables as indicators to identify the presence of acids and bases. These "detective" materials are called, to this very day, *indicators*, and they are used to indicate to us when the reaction between an acid and a base is over. In fact, the organization of chemistry into a materialistic science with clear rules owes much to the chemistry of acids and bases. This goes back to the eigtheenth century when Lavoisier discovered the gas oxygen. He called this gas by the name oxygen since he thought that it is an essential ingredient of acids, as oxy = acid and gen = generator. So, in a single articulation, he has linked his discovery of one gas to a large body of knowledge, the chemistry of acids and bases, which was the most developed branch of chemical knowledge at his time.[6]

8.4.1   Acids in Water: A Proton Transfer Reaction from the Acid to Water

Acids are generally molecules wherein a hydrogen atom is covalently bonded to a highly electronegative group. Figure 8.7 shows some acids which we have encountered during the various lectures. In each of the acids, there is one H or more connected to O or to Cl. For example, sulfuric acid has two such hydrogen atoms, and so does carbonic acid, while phosphoric acid has three such hydrogen atoms, and the rest of the acids in the figure have only one H connected in this fashion.

When an acid is put into water, what transpires is a chemical reaction that is called a *proton transfer* between the acid and the water. Scheme 8.9a shows this reaction between a water molecule and a generic acid, H—X acid. The reaction generates two

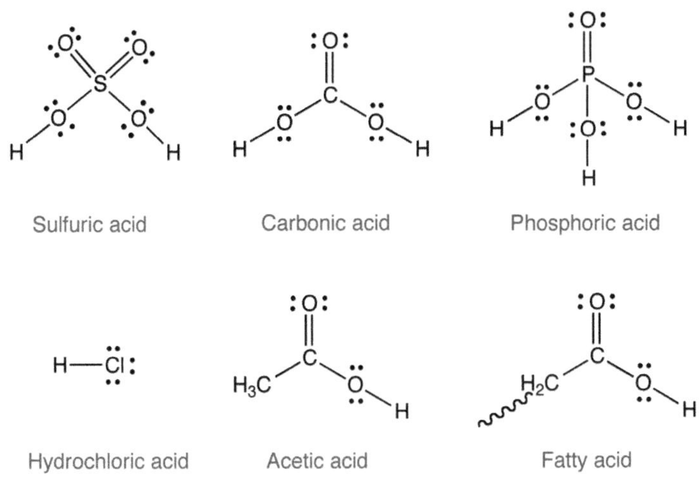

Sulfuric acid   Carbonic acid   Phosphoric acid

Hydrochloric acid   Acetic acid   Fatty acid

FIGURE 8.7   Some acids and their names.

ions. The positive ion is H_3O^+, which is called a *hydronium ion*, and the negative one is an anion, $:X^-$. Scheme 8.9b shows two specific anions.

The rationale for this reaction is the fact that the proton in the acid (H—X) has a partial positive charge, while the oxygen in the water molecule has a partial negative charge. As such, there is an attraction between the water molecule and the acid, and as the two molecules collide, one of the lone pairs of water "attacks" the H—X and makes a bond to the H. Since the H accepted an electron pair from the water, it must let go of the electron pair of the H—X bond that is taken by the departing X, or else the H and X will violate their states of Nirvana. This electron flow is described in Scheme 8.9a using two curved arrows: a lone pair moves from O to make an O—H bond, while the bond pair H—X moves to X, which leaves as $:X^-$. Note that the two ions each have an octet, and usage of *the curved arrow* shows that the octet is kept throughout the proton transfer reaction. The curved arrow is a powerful and useful symbol used by chemists to describe chemical reactions.[7]

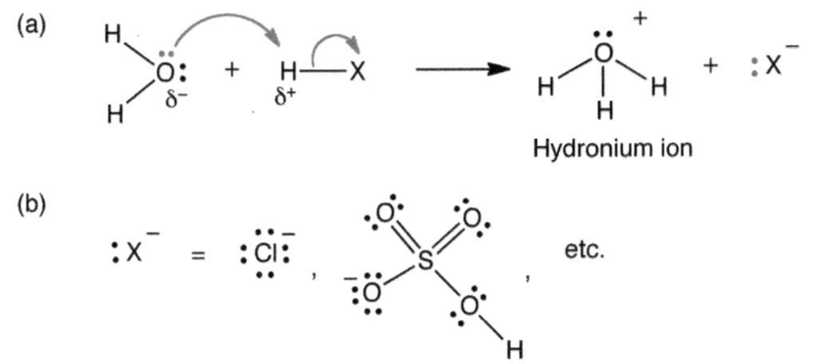

Hydronium ion

SCHEME 8.9   (a) A proton transfer reaction between a water molecule and a generic acid molecule, H—X, leading to a cation, H_3O^+, and an anion, $:X^-$. The H_3O^+ cation is called the *hydronium ion*. The electron flow in the reaction is shown using curved arrows. (b) Two examples of $:X^-$ anions which can be formed by this reaction.

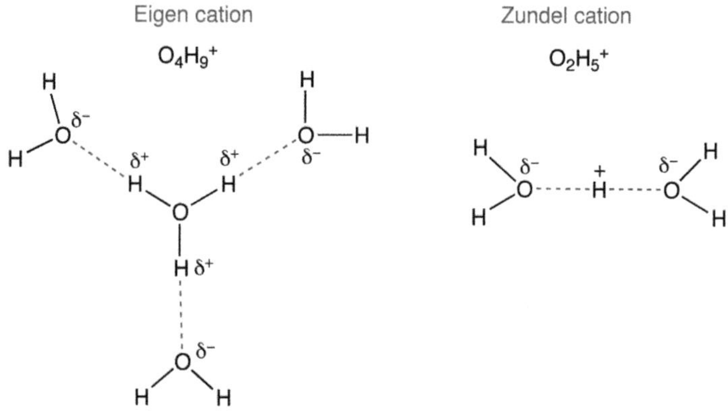

FIGURE 8.8   The hydronium ion in water is stabilized by hydrogen bonds with water molecules and may exist as $O_4H_9^+$ and/or $O_2H_5^+$.

As we explained before, the driving force for this breakage of the H—X bond of the acid, to generate ions in solution, is the fact that these ions are stabilized by the water molecules. Thus, for example, the anion :X⁻ may be surrounded by 4–6 water molecules, as we have already described for :Cl⁻ in Figure 8.4. The hydronium ion is also not free and is surrounded by additional water molecules. As shown in Figure 8.8, currently, we think that there are two states of this ion,[8] both named for the chemists who postulated them. One, $O_4H_9^+$, is called the *Eigen cation*, and it involves three water molecules stabilizing the hydronium ion by hydrogen bonds. The other one, the *Zundel cation*, $O_2H_5^+$, involves just one additional water molecule. The actual situation may well involve both as well as other structures, since the hydronium ion is very mobile in water.

As we mentioned in the introduction to this lecture, the presence of ions in a water solution of an acid can be probed by testing the conductivity of the solution. In so doing, the pioneering physical chemists found that some acids almost completely dissociated in water (e.g. HCl, H_2SO_4), while others only dissociated in part (e.g. CH_3COOH). The former types are called *strong acids*, while the latter types are called *weak acids*. As the ionization process depends on at least three different factors, the explanation of why a certain acid is strong while another is weak is not straightforward, and I am going to waive it. Instead, let us talk about how chemists define the strength of an acid.

Scheme 8.10 provides this information. As shown in part (a), a strong acid, like HCl or H_2SO_4, dissociates almost completely when it dissolves in water, and hence the resulting solution has plenty of hydronium ions running to and fro through the solution. Chemists define the number of hydronium ions per a given volume of water (say 1 liter) as concentration (see Retouches section 8.R.4). So, in the case of a strong acid, the concentration of hydronium ions is high, and since transfer of H^+ to the water achieves the generation of hydronium ions, we simply say that a strong acid is typified by a high H^+ concentration in water. In contrast, a weak acid like CH_3COOH dissociates only in part; namely, not each molecule dissociates, and

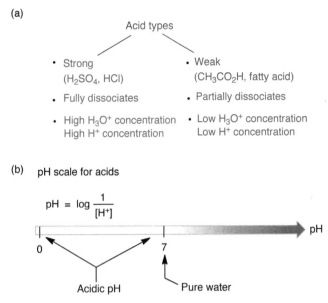

SCHEME 8.10   (a) The classification of acids as strong and weak is based on the concentration of H_3O^+ ions (represented simply as H^+) in solutions of these acids. (b) The pH scale (as defined mathematically) measures an acid's strength: the pH of pure water is 7, while acidic solutions have pH values smaller than 7.

therefore CH_3COOH and its weak acid congeners are typified by a relatively low H^+ concentration in water.

The physical chemists even devised a quantitative measure of acids' strengths. As shown in Scheme 8.10b, this scale is called pH, and it is defined mathematically as the logarithm of 1 divided by the concentration of H^+; the concentration is symbolized as H^+ in square brackets, that is, $[H^+]$. There are various means of measuring the pH for a given solution of an acid. The simplest method is called a *litmus test*, which uses a paper soaked with a dye extracted from plants. This paper serves as an indicator that changes color when put in acid (blue turns to red). There are, of course, more sophisticated methods based on conductivity measurements, using *pH meters*.[9] Some more information about pH and the associated calculations is provided in Retouches section 8.R.4, while here we simply use the scale qualitatively. Thus, the pH of pure water is 7, and all solutions of acids have pHs smaller than 7; the stronger the acid, the smaller the pH. For example, when we dissolve an Avogadro's number of HCl molecules in 1 liter of water, almost all the HCl molecules dissociate, and the pH will be zero, while if we do the same with the weaker acetic acid, the pH will be around 2.4, which means that a small minority of the acid molecules will dissociate to yield H^+ ions.

The take-home lesson about acids can be summarized as follows:

☞ When an acid dissolves in water, the water molecules attack the acid molecules and abstract from them a proton, H^+. When the acid is strong, almost all the

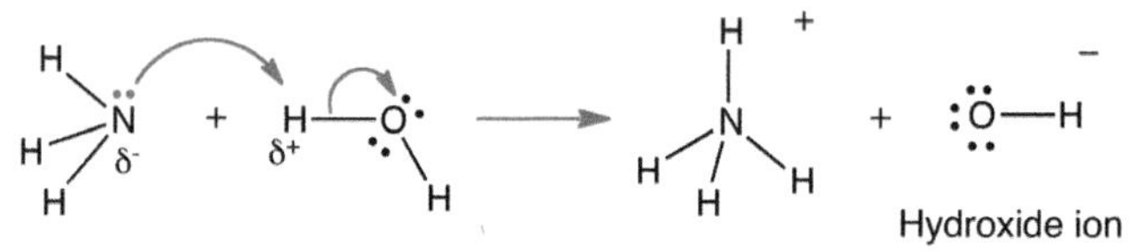

Hydroxide ion

SCHEME 8.11   A proton transfer reaction between the base ammonia and a water molecule, leading to the ammonium cation (NH_4^+) and the hydroxide anion (OH^-). Here, water behaves as an acid toward ammonia.

acid molecules donate their protons to water, and when an acid is weak, only a small fraction of the molecules do that. The quantity pH expresses the acid's strength; the smaller the pH, the stronger the acid.

8.4.2   Bases in Water: A Proton Transfer Reaction from Water to the Base

When a base is put into water, an opposite proton transfer reaction transpires, whereby the base abstracts a proton from a molecule of water. Bases are generally molecules bearing atoms that have lone pairs, like ammonia and amines, as well as the DNA and RNA bases. Scheme 8.11 shows the proton transfer reaction when ammonia dissolves in water. Thus, using the curved arrow method, we can see that upon collision of the two molecules in an orientation that causes attractions between the partial charges on N and the H of water, the lone pair of the ammonia makes a bond to the H, which in turn must give up the electron pair in the H—OH bond, and this lone pair ends up on the departing hydroxide anion, OH^-. Once again, the products are two ions, the ammonium cation and the hydroxide anion.

Compared with pure water, in the presence of ammonia, we increased the number of OH^- anions in the solution. As we mentioned, all the amines are bases. But generally, amines are not strong bases. The strong ones are ionic materials that contain the hydroxide anion like K^+OH^- or Na^+OH^- (known by the name "caustic soda"). These ionic materials fully dissociate in water and release essentially all their OH^- ions. Thus, a base's strength can be gauged by the concentration of OH^- ions. It turns out that the concentrations of H^+ and OH^- ions in water are dependent on each other, such that *the product of the two concentrations at a given temperature has a constant value* of 10^{-14} (0.00000000000001) under all conditions (see Retouches section 8.R.4). In pure water, where the pH is 7, there are equal concentrations of the two ions; in an acidic solution, the H^+ concentration exceeds the concentration of OH^-, while in a basic solution, the opposite is true, and the H^+ concentration is lower than that of OH^-.

Because of this mutual dependence of the H^+ and OH^- concentrations in water, it is possible to use the pH scale for both acidic and basic conditions. This scale is shown in Scheme 8.12. In the middle at pH 7, we have the situation in pure water (where the concentrations of the ions are identical). At pH values smaller than 7, we are in the acidic regime, which is typified by high H^+ concentrations; meanwhile, at pH values larger than 7, we are in the basic regime, which is typified by low H^+ concentrations (and of course high OH^- concentrations).

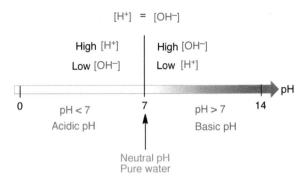

SCHEME 8.12 The common pH scale for acidic and basic water solutions: At the pH of pure water, which is 7, we have identical concentrations of H^+ and OH^-. At pH values smaller than 7, we are in the acidic regime, typified by high H^+ and low OH^- concentrations; meanwhile, at pH values higher than 7, we are in the basic regime, typified by low H^+ and high OH^- concentrations.

8.4.3 A Proton Transfer Reaction from Acids to Bases

The take-home lessons from the behaviors of acids and bases can be summarized as follows:

☞ Acids provide protons, H^+, in proton transfer reactions in water, while bases take up protons in proton transfer reactions in water.

☞ When both acids and bases are present in the same solution, they will transfer protons between themselves.

Indeed, acids and bases react with one another by proton transfer. Scheme 8.13 shows by way of example the process occurring between NH_3, which is a base, and the acid HCl. Using the curved arrow method (see Retouches section 8.R.5), we can see what we saw before: the lone pair of ammonia makes a bond with the H-end of the acid and causes thereby breakage of the H–Cl bond and departure of Cl^-. In fact, what we formed is the ionic material $NH_4^+Cl^-$, which we constructed before.

An interesting case is an amino acid, since it has on the same molecule a basic amino group and an acidic end. As shown in Scheme 8.14, the result is that the basic end will abstract the proton from the acidic end, and this process will generate the double ion form (called by chemists a *zwitterion*, which is a German word for "a

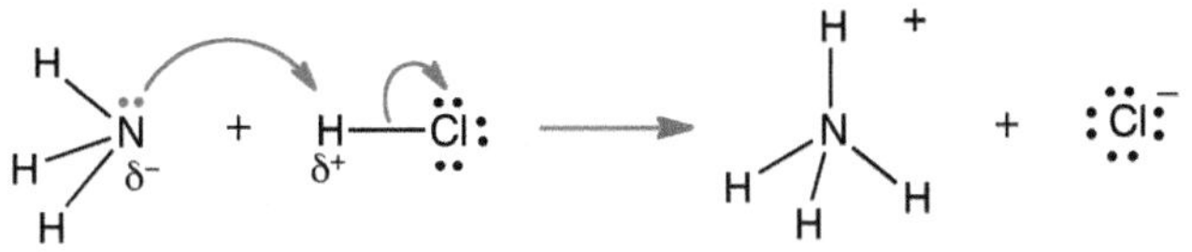

SCHEME 8.13 Using the curved arrow method to represent the flow of electrons in a proton transfer reaction between NH_3 and HCl.

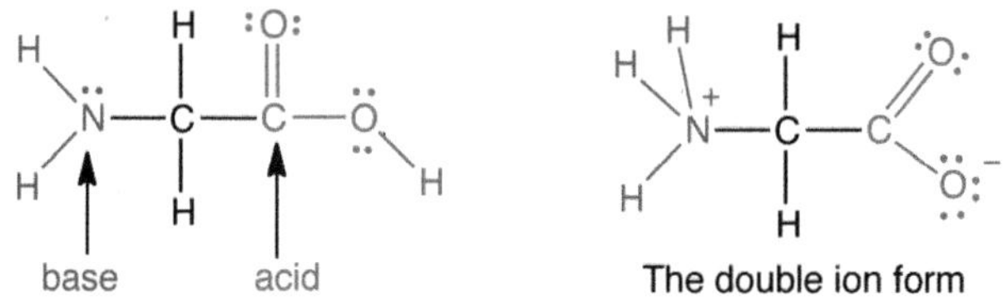

base   acid   The double ion form

SCHEME 8.14   The structure of an amino acid without charged components, on the left, and its structure in the double ion (zwitterion) state, on the right.

double ion"). Note that because the COOH group is a weak acid and the NH_2 group is a weak base, the amino acid will exist as a mixture of the two forms.

The historical name of the reaction between an acid and a base is a *neutralization reaction*. This term means that we are adding to the acid just sufficient base such that each acid molecule transfers a proton to a base molecule, and there remains no excess of either the acid or the base. That the reaction is indeed over can be ascertained by inserting a few drops of some indicator (the one I have in mind is called phenolphthalein, which is colorless in acid and pink in base [see Retouches section 8.R.4]). Imagine now that I placed an unknown quantity of HCl in water, and I ask you to tell me the quantity of acid I used. What you can do is add a drop of the indicator to the acid, and then add a base of known concentration, say NaOH, in a dropwise manner to the solution. At some point, the base would neutralize all the HCl molecules. Then, when you add one more drop of the base, tiny as you wish, the solution will turn pink all of a sudden. The magic of chemistry at its best! Based on the quantity of the base needed to reach this *endpoint*, you can determine the unknown quantity of the acid in solution.

8.4.4   A Few Facts About Our Acids and Bases

Healthy blood is very weakly basic, with a pH range of 7.35–7.45. If the pH drops below this range of values, the result is a condition called acidosis. Severe acidosis (pH < 7.0) leads to depression of the central nervous system, which starts malfunctioning (leading to disorientation and, if the condition is very severe, possibly a comatose state). If the pH rises above 7.45, the condition is called alkalosis. This condition causes hypertension and muscle spasms, and it can even lead to death. As a result, the pH value of our blood is highly regulated by *buffers*, which are molecules that consume any extra protons or hydroxide ions and keep the pH constant or variant in a narrow interval.

Our stomach produces large quantities of the strong acid HCl, which is required to break down the food we eat, such as to decompose the proteins to amino acids. So acidic is the stomach that if you were to dip your finger into it, the acid will "chew" the finger quite fast. How then is the stomach itself being protected against being self-digested by its own acid content? It turns out that the stomach is protected by a few mechanisms. One is a layer of mucosa (mucous membrane) that covers the inner linings of the stomach and acts as a gel that cannot be penetrated by acids, and the other is the secretion of bicarbonate by cells in the mucosa.

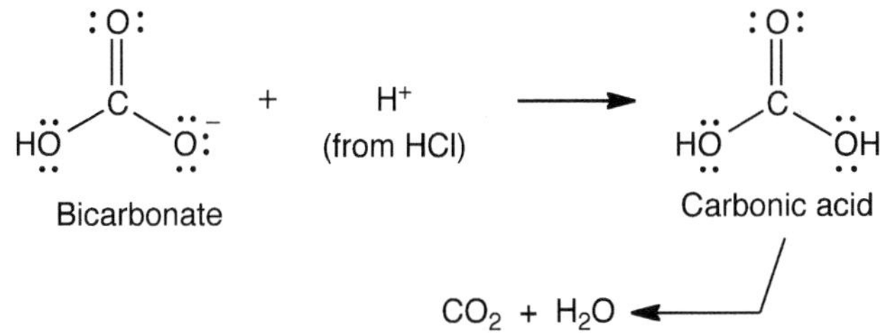

SCHEME 8.15 Our stomach is protected from being digested by its own HCl content. This protection involves generation of bicarbonate (HCO_3^-), which takes up a proton and generates carbonic acid (H_2CO_3). Carbonic acid subsequently falls apart to CO_2 and H_2O.

The action of the bicarbonate is seen in Scheme 8.15. Bicarbonate is the anion of carbonic acid (Figure 8.7) after the acid releases one of its protons. Carbonic acid is a relatively weak acid, and bicarbonate is a relatively strong base. So in the presence of the strong acid HCl, the bicarbonate anion accepts a proton and forms carbonic acid. Carbonic acid tends to break down in acidic environments to form H_2O and the gas CO_2, the latter of which happily bubbles off. In this manner, the stomach neutralizes the excess acid and protects itself against digestion. Many of us are familiar with the antacid relief of tablets, some of which may contain bicarbonate.

A daily usage of bicarbonate is as baking soda, which reacts with an acidic ingredient in dough and produces carbonic acid, which decomposes and gives off CO_2, which in turn causes the dough to leaven.

8.5 SUMMARY

In this lecture, we learned some new principles:

☞ Ionic bonds are formed by combinations of the least electronegative atoms—the members of groups 1A–3A (except for H, Be, and B) and the transition metals, which form cations (positive ions)—with the most electronegative atoms, the members of groups 5A–7A, which form anions. A typical example is Na^+Cl^-. There are also ionic materials made from complex ions, such as $NH_4^+Cl^-$ or $Mg^{2+}SO_4^{2-}$.

☞ Ionic bonds involve opposite ions glued together by electric attraction. The principle of construction of ionic bonds uses the "click-clack" method. Initially, we click the valence electrons based on the connectivity of the two atoms; for example, for the single connectors Na and Cl, we can form one "click" bond. Subsequently, we "clack" and transfer the so-generated electron pair(s) to the more electronegative atom. In this manner, the two ions maintain their states of Nirvana and generate ionic bonds that obey the octet rule.

☞ Ionic bonds tend to aggregate to form solid 3D structures that contain many charge alternating ions, in a structure which maximizes the attraction of the opposite charges. A result of this close packing of ions is that no individual ionic molecules can be recognized in ionic solids. For example, in $(Na^+Cl^-)_{solid}$, there are no individual Na^+Cl^- pairs.

☞ Ions that are protected by bulky groups do not aggregate and instead form ionic liquids.

☞ Generally, ionic materials in which the ions are singly charged, such as Na^+Cl^-, dissolve in water, but those with higher charges, such as $[Ca^{2+}]_3[PO_4^{3-}]_2$, do not.

☞ Living systems utilize soluble ionic matter, such as Na^+ and K^+ (in combination with neurotransmitters) as a means of *communication between the brain and the body* and as nanomotors for movement and flexing muscles (Ca^{2+}).

☞ Insoluble ionic material, in the form of hydroxyapatite, $Ca_{10}(PO_4)_6(OH)_2$, creates our skeleton and our teeth by weaving in the protein collagen, which provides enormous strength and flexibility to these parts of the body.

☞ Acids are covalent molecules that create ions in water as a result of a chemical reaction with the water. The acid donates proton(s), H^+, to water. Bases are molecules that take up protons from water molecules in water solution. The pH scale of water solutions defines the properties of acidity and basicity. Pure water has a pH equal to 7, and an acidic solution has a pH smaller than 7, while a basic solution has a pH larger than 7.

☞ Proton transfer reactions are important in living systems, as pH regulators (e.g., the bicarbonate reaction, which protects our stomachs) and as a means of catalyzing some reactions. Living cells have special receptors (bundles of proteins) that transfer H^+ selectively.[10]

☞ Acids and bases transfer protons between themselves. The H^+ transfer reaction can be described using *the curved arrow method* (for more, see Retouches section 8.R.5), which represents the electron flow during the reaction.

8.6   REFERENCES AND NOTES

[1] S. Shaik, *J. Comput. Chem.* **2007**, *2*, 51.

[2] P. Coffey, *Cathedrals of Science*, Oxford University Press, Oxford, England, 2008.

[3] J. W. Servos, *Physical Chemistry from Ostwald to Pauling*, Princeton University Press, Princeton, NJ, 1990, p. 111, pp. 133–135.

[4] E. Salminen, P. Virtanen, and J.-Y. Mikkola, *Front. Chem.*, **2014**, *2*, 1.

[5] E. Stove, *Chemistry World*, August 1, 2014.

[6] R. Siegfried, *From Elements to Atoms: A History of Chemical Composition*, American Philosophical Society, Philadelphia, PA, 2002.

[7] On the history of arrow usage in chemistry, see: S. Alvarez, *Angew. Chem. Int. Ed.* **2012**, *51*, 590.

[8] E. Codorniu-Hernández and P. G. Kusalik, *Proc. Nat. Acad. Sci. USA.* **2013**, *110*, 13697.

[9] S. S. Zumdahl and S. A. Zumdahl, *Chemistry*, 9th Ed., Brooks College, Cengage Learning, 2010, Chapter 14, pp. 660–664. Available at: www.cengage.com/global

[10] T. E. DeCoursey and J. Hosler, *J. R. Soc. Interface*, **2014**, *11*, 20130799.

8.A   APPENDIX

8.A.1   Proposed Demonstrations for Lecture 8

• **Water—a good solvent for salts**

This demonstration compares the dissolution of cobalt chloride in water versus in ethanol. As both solvents are polar and capable of forming hydrogen bonds, they are able to dissolve cobalt chloride, but to different degrees. The experiment begins by observing the color of an anhydrous sample of $CoCl_2$, which is sky blue.[1A] The students are told that the cobalt's environment influences its color. A small amount of the powder is dissolved in water and in ethanol. Whereas the color of cobalt chloride in ethanol remains blue, the aqueous solution turns pink to dark purple.[1A] The water molecules are able to separate Co^{2+} from Cl^- and stabilize these ions by surrounding them, as we showed in Figure 8.4. The less polar ethanol molecules do not separate these ions completely, and the chloride ions are still close to Co^{2+}, as reflected in the blue color of the solution.

• **Colligative properties**

The following demonstration shows one of the properties of a solution, called colligative properties, which can be observed when a nonvolatile compound dissolves in a solvent, such as water. The normal boiling point of water under standard atmospheric pressure (1 atmosphere [a unit of pressure]) is 100°C. At high altitudes, the atmospheric pressure is lower, and the boiling point is lower. Let us say that we perform the experiment at standard pressure (although in Jerusalem, where the demonstrator lives, the pressure is lower, and the boiling temperature is also lower).

– **Boiling point elevation:**

This experiment involves three beakers, each of which is filled with 500 ml of distilled water. We add 142 g of NaCl to the first beaker, 835 g of sugar (sucrose) to the second, and nothing to the third. The demonstrator explains to the students that the two solids are composed of the same number of "molecules." The three beakers are then heated until the solutions boil. The boiling temperature is measured by a thermometer. The boiling point of the pure water is, as expected, 100°C, the NaCl solution boils at 105.0°C, and the sugary solution boils at 102.5°C. In order to understand the phenomenon of the boiling point elevation, one should first explain what happens at the boiling point, as we did in Section 8.1. The difference between the NaCl and sugar solutions is that NaCl dissociates in water to form two ions, Na^+ and

Cl⁻, whereas the sugar dissolves as complete molecules. Hence, the NaCl solution contains twice as many particles, which retard the escape of water molecules from the liquid, compared with the sugar solution, and therefore the boiling temperature of the NaCl solution is raised by twice as much compared with the raise in the boiling point of the sugar solution. (One may also use a more technical explanation of boiling as the balance between the atmospheric pressure versus the vapor pressure above the solution. A liquid boils when these pressures are equal. When the water contains a dissolved compound, this dilutes the water to such an extent that the vapor pressure of the solution is lowered, and its boiling point becomes higher compared with that of pure water. The NaCl solution has twice as many particles, and hence, its dilution effect is twice as large as that of the sugar solution. The boiling point increases of the two solutions are proportional to the numbers of particles dissolved.)

- **Determining acidity/basicity with red cabbage juice**

Leaves, stems, roots, and flowers of many plants, as well as fruits and vegetables, contain pigments that change their color depending on the acidity/basicity of their environment. The pigment of red cabbage belongs to this class of molecules. In this demonstration, an extract of red cabbage is used as a pH indicator to determine the acidity/basicity of various household solutions such as: lemon juice, baking soda (sodium bicarbonate), household ammonia, Tums pills (calcium carbonate), vinegar (acetic acid), caustic soda (sodium hydroxide), etc. The procedure for extracting the pH indicator pigment from red cabbage is explained by Fortman and Stubbs (1992).[2A]

- **The stoichiometry of the reaction between vinegar and baking soda**

Following the above experiment, the students learn that vinegar is an acid and that baking soda is somewhat basic. When these are put together, they react violently and produce a fizzing carbon dioxide gas and water. This demonstration uses this acid–base reaction to also show the basic concept of stoichiometry. In the experiment,[3A] increasing amounts of baking soda are added to a given volume of vinegar, and the amounts of carbon dioxide gas produced indicate the extents to which the reactions occurred. As the amounts of the added baking soda increase, growing amounts of carbon dioxide are produced until a limit is reached, where the added baking soda exceeds the amount of vinegar and no additional gas is produced. The experiment is carried out in plastic bottles stoppered with balloons. The sizes of the balloons are observed and used to quantify the amounts of gas produced.

8.A.2   References for Appendix 8.A

[1A] http://en.wikipedia.org/wiki/Cobalt%28II%29_chloride
[2A] J. J. Fortman and K. M. Stubbs, *J. Chem. Educ.*, **1992**, *69*, 66.
[3A] Editorial staff, *Journal of Chemical Education*, **1997**, *74*, 1328A.

8.R   RETOUCHES

8.R.1   Energetic Aspects of Ionic Bonding

It is important to stress that the formation of an ionic bond involves two opposing energetic aspects. Creation of the two ions independently of each other requires an investment of energy; that is, this raises the energy of the system relative to that of the atoms.

For example, consider the generation of Na^+ and Cl^- from their respective atoms. We can generate the ions in two steps:

$$Na \rightarrow Na^+ + e^-$$

$$Cl + e^- \rightarrow Cl^-$$

In the first step, we remove an electron (e^-) from an Na atom and create Na^+. In the second step, we allow Cl to take up this electron and create Cl^-. If we now sum up these two chemical equations, we get the equation for the ions' formation:

$$Na + Cl \rightarrow Na^+ + Cl^-$$

(Note that chemical equations behave like algebraic equations, so that the electron that appears on the two sides of the arrow cancels out upon summation).

The energy of the formation of the ions is then the difference of two energies: one is the energy invested in creating Na^+ from the atom, called the ionization energy (IE), and the other is the energy released as the electron is accepted by Cl to form the anion Cl^-, called the attachment energy (AE). (Note that the attachment energy is also sometimes called the electron affinity). These energies have been measured by chemists and are freely available on the website of the NIST, which is an acronym for the National Institute of Standards and Technology (just Google "NIST database" and you shall find these energies and many others). It turns out that the energy invested, IE, is larger than the energy released, AE, and therefore the entire process requires an overall investment of energy.

Clearly, if we considered only the formation of the ions, we would not have had any ionic bonds. *The driving force for the formation of this bond is the attractive pull of the two oppositely charged ions.* This is called the Coulomb energy, after the name of the French physicist Charles-Augustin de Coulomb, who established in 1785 the quantitative law of electricity. The law states that two opposite charges Q_1 and Q_2 will apply a force on each other such that the *force* is proportional to the product of the charges, $Q_1 \times Q_2$, and inversely proportional to the square of the distance between them $(1/R_{12}^2)$. If the charges are opposite, one negative, the other positive, this will be an attractive force; meanwhile, if the charges are of the same sign (either positive or negative), they will repel one another. *Attraction between opposite charges results in energy lowering*, which is proportional to $Q_1 \times Q_2$ and inversely proportional to the distance between the ions $(1/R_{12})$.

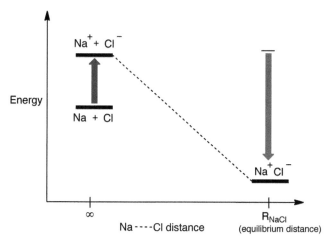

SCHEME 8.R.1   Energy diagram for the neutral atoms and the ions of NaCl, drawn as a function of the distance between Na and Cl. The broad arrows indicate when the energy is raised (an arrow pointing up) and when it is lowered (an arrow pointing down). The brown arrow, which points up, represents the energetic investment required to form ions from neutral sodium and chlorine atoms when the two atoms are infinitely far apart. The red arrow, which points down, signifies the energetic lowering resulting from the attraction of the charges being brought from infinity (∞) to a distance R_{NaCl} (this special distance, R_{NaCl}, is the most favorable one for lowering the energy by optimizing the attraction between the two ions).

Thus, using as an example Na^+ and Cl^-, this *lowering of the energy* is proportional to the $Q_{Na} \times Q_{Cl}$ product (which is -1) divided by the distance R_{NaCl}. Thus, the sign of the energy due to the attraction is negative, *which means that the attraction lowers the energy of the ions as they are brought together* to a certain equilibrium distance specific to NaCl, R_{NaCl}. The calculation is very simple (see later), and the results are qualitatively shown in Scheme 8.R.1.

It is seen that the generation of the ions requires an investment of energy. But when the ions are brought close together (to the short distance marked as R_{NaCl}), the attraction of the ions lowers the energy so much that the ion pair Na^+Cl^- attains a lower energy than that of the two atoms (compare the positions of the isolated atoms and the bonded ionic compound in the energy diagram in Scheme 8.R.1). This in turn means that the ion pair is bonded relative to the two atoms. Thus, the ionic bond behaves according to the rule of successful financial transactions; you invest something in order to get back a greater profit. The "profit" of the ionic bond is the Coulomb attraction of the ions.

The Origins of the Law of Nirvana in Ionic Bonds: So why do we have Na^+ Cl^- and not, for example, Na^{2+} Cl^{2-}, which will certainly render four times as much lowering of the energy? This is because the creation of Na^{2+} requires ejection of the second electron from the exposed valence shell, which has achieved an octet as Na^+, and simultaneously creating Cl^{2-}, which exceeds the octet. This would be a huge investment, which would not be overcome by the Coulomb attraction between the ions. This is the reason for our statement in the text that the charges of the two ions are determined by their need to achieve Nirvana.

Why Do Ionic Bonds with Multiple Ionic Charges Exist? This is a good place to comment on ionic molecules with ionic charges that exceed unity, such as $Al^{3+}PO_4{}^{3-}$. Here the profit is huge, because the product of the ionic charges is -9, namely nine times as large as that in Na^+Cl^-. However, the three electrons that we have to eject from the atom, to generate Al^{3+}, and attach to the PO_4 fragment, to generate $PO_4{}^{3-}$, make the investment huge. Therefore, the profit due to the Coulomb attraction may not overcome the investment. However, recalling that when the ions aggregate in a solid, each cation feels an attraction to several anions, and vice versa, we can understand that an ionic solid can have a larger profit that exceeds the investment when it has multiply charged ions, as in the example of $Al^{3+}PO_4{}^{3-}$. Still however, the ions must obey the Law of Nirvana, as is the case for Al^{3+} and $PO_4{}^{3-}$. For cases without Nirvana, as with Na^{2+} and Cl^{2-}, the investment will be so huge that even the profit in the solid would not overcome it.

8.R.2   Energy Units and Bond Energy Calculation for Ionic Bonds

We have repeatedly used the term "energy," but we never said anything quantitative about this term. Of course, the way the quantitative scale of energy was shaped in science is philosophically and historically very interesting, but let me resist the temptation to digress and limit myself to the quantity itself.

Energy has many different units depending on whether one measures, for example, mechanical energy, heat energy, electric energy, radiation energy, magnetic energy, etc. Translating the energy quantity from one unit to another involves some constant numbers, which have been evaluated by careful experiments, and are precisely known today.

Since we are all interested in diets, the most popular unit is a calorie. What is a calorie? Suppose we have a pot with 1 liter of water (1,000 g), and we want to heat it. We will use for this the heat energy from a gas burner. How much energy do we have to invest? It turns out that raising the temperature of the water in the pot by just 1°C requires us to invest 1,000 calories. A total of 1,000 calories is called 1 kilocalorie, which we label as 1 kcal. Bringing this liter to its boiling temperature (from room temperature, e.g., 20°C) will take about 5 minutes, and we shall invest about 80 kcal, that is, 16 kilocalories (kcal) per minute.

To get a further idea how much is 1 kcal of energy, consider that you are using a treadmill and you are gently jogging on it at a speed of 6 kilometers per hour. If you keep jogging for 40 minutes, you will spend only about 250 kcal, slightly over 6 kcal per minute of jogging. Considering that our daily diet is about 2,000 kcal, then trying to lose this amount of energy you would have to jog for more than 5 hours (at the gentle treadmill pace). Also, after a long run, many of the calories burned will be consumed by eating a bowl of cereal, a smoothie, eggs, bacon, toast, and maybe a few well-earned biscuits. It is not easy to lose weight by jogging. Right?

To calculate the energy of attraction between Na^+ and Cl^- at a distance R_{NaCl} given in units of Å (which we defined in the Retouches in Lecture 1; $1 \text{ Å} = 0.00000001$ cm), we use the following expression:

$$E_{Na^+Cl^-} = 332 \left[Q_{Na^+} \times Q_{Cl^-} \right] / R_{NaCl}$$

The constant 332 is what we alluded to as a translator from units of electric energy to units of heat energy (in kcal/mol). Thus, this expression yields energy in kilocalories per mole of ions (i.e., Avogadro's number of the ions), and it is written as kcal/mol. Since the charges of the ions are $+1$ and -1, respectively, and since $R_{NaCl} = 2.79$ Å, we get $E_{Na^+Cl^-} = -119$ kcal/mol. This is a huge quantity. If we look on the NIST website, we find that $IE_{Na} = 119$ kcal/mol and $EA_{Cl} = -84$ kcal/mol. So the formation of the two ions costs 35 kcal/mol ($119 - 84 = 35$). Therefore, the balance of the investment and profit is -84 kcal/mol. If we refer back to Scheme 8.R.1, then we can say that the energy of the ion pair Na^+Cl^- is 84 kcal/mol lower than the energy of the atoms $Na + Cl$. Just think how long you have to spend on the treadmill to get rid of so much energy, provided of course you will not eat anything during that day …

Energy of Solid Ionic Material: We argued that aggregation of the ions is beneficial since this amplifies the number of attractions each ion feels. Consider Scheme 8.R.2 where we place two ion pairs in a charge-alternating manner.

You can see that, in this system, there are three attractions between Na^+ and Cl^- that are separated by 2.79 Å and one long-range attraction with a distance of 8.37 Å. On the other hand, there are also repulsions between the identical ions, and there are two such repulsions with distances of 5.58 Å. For each such interaction, we can use the equation above, but now for general charges Q_1 and Q_2, which can be opposite or identical ($E_{Q_2Q_1} = 332 [Q_1 \times Q_2]/R_{12}$), and we can sum up all the attractions and repulsions. Doing that, we shall find that the total energy of the two moles of ion pair units is $E_{2[Na^+ Cl^-]} = -278$ kcal. Since we want to know the energy for one ion pair, then we divide by 2 and get $E_{Na^+ Cl^-} = -139$ kcal/mol. If you compare this result to the one we calculate for just an isolated ion pair, -119 kcal/mol, it is clear that aggregating the ions increases the attractive interactions, in this case, by 20 kcal/mol.

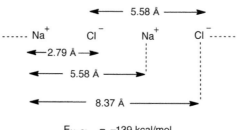

$E_{NaCl} = -139$ kcal/mol

SCHEME 8.R.2 Two adjacent $Na^+ Cl^-$ pairs shown along with the distances between the oppositely charged and identically charged ions.

The ratio between the energies for one unit in an aggregated state, as shown in Scheme 8.R.2, and in the isolated state, as shown in Scheme 8.R.1, is $-139/-119 = 1.17$. Such a ratio is called the Madelung constant (M), after the name of the physicist who first demonstrated the use of such calculations. So, if I use an infinite chain of ion pair units, the ratio will be $M = 1.39$, and for the full 3D $[Na^+ Cl^-]_{solid}$, we get $M = 1.75$. Thus, you can clearly see that it is indeed beneficial for ionic molecules to pack in solid lattices.

8.R.3   Dissolution of Ionic Solids in Water

Solvation energies of ions with unit charges are of the order of -100 kcal/mol, and therefore, the combined solvation energies can lower the energy of solid ionic materials by ca. 200 kcal/mol, which usually exceeds the Coulomb attractions in the solid (which are lost during dissolution) for monocharged ions. In such a case, the solvation of the solid will be exothermic, and the excess energy be released as heat.

The Resistance of Ionic Materials with Multiply Charged Ions to Dissolve: $Ca^{2+}CO_3^{2-}$ does not dissolve in water. The reason is rather simple and can be seen if we use the equation $E_{Q^2Q^1} = 332 [Q_1 \times Q_2]/R_{12}$ for the attraction of the ions. Even if we assume that the distance R_{12} is identical to the one in Na^+Cl^- (it is shorter in fact), the attraction energy for $Ca^{2+}CO_3^{2-}$ will be huge, -476 kcal/mol, and in the solid (assuming the same structure as in Na^+Cl^- and hence also the same $M = 1.75$ value), this quantity will become -833 kcal/mol. Now the solvation energy of the ions will not compensate for the loss of the attraction in the solid, and so the solid will resist dissolution.

What Happens When the Solvation Energy Is Only Slightly Less Favorable than the Ionic Attraction? As we mentioned, there are cases, like $[NH_4^+Cl^-]_{solid,}$ where the dissolution is accompanied by the cooling of the water in the flask. The cooling of the solution means that the ionic attraction in the solid is stronger than the interaction of ions with the water molecules, and hence the energy difference is absorbed from the surroundings, and this is the reason for the observed cooling. This is called an endothermic (energy-absorbing) process. This occurrence of an endothermic process sounds a bit odd, since such a process should not be spontaneous. So what could be the driving force for such a process?

There is a branch of chemistry called *thermodynamics,* and one of its important laws is called the Second Law,[1R] which states that all spontaneous processes in nature will be accompanied by an increase in disorder in the system of interest and its surroundings (the law refers to the entire universe, but using the word "universe" for a spoon of salt in a glass of water is a bit intimidating or overstated). Thermodynamics calls this "disorder parameter" by the name *entropy.* It turns out that the driving force for an endothermic dissolution process is entropy. To have some appreciation of the meaning of "disorder"/entropy, let us think about a drop of ink (it is made of dye molecules which have no strong interactions among themselves or with the water molecules). When you drip this ink drop into a glass of water, you see gradually how the ink spreads and is immersed in the entire volume of the water, such that, after a time, you can barely see a trace of the ink; all its dye molecules are wandering without any order among the many, many water molecules. So we went from a state of order where there was a drop made of a bunch of identical dye molecules grouped together and, separately, there were water molecules arranged together by hydrogen bonds, to a much less organized state, where the dye molecules are completely dispersed and are wandering in many possible ways among the water molecules. So, this dispersion of the ink molecules in the water is driven by an increase of the "disorder"/entropy in both the ink and the water.

Thus, when no energetic effect (of breaking bonds and making new ones, as in the dissolution of an ionic material in water) is involved, the most probable situation we anticipate to find is that which has maximum entropy/disorder.

Getting back to the dissolution of an ionic solid in water, it is clear that going from the nicely ordered ionic solid to separate ions in solution, arranged in many different and random patterns, increases the entropy/disorder, and hence, solvation will be preferred unless the attraction in the solid is much stronger than the attractions felt by each ion from the water molecules that surround it. When the interaction energies in the solid and the solution are not too different, then endothermic dissolution can nevertheless happen, and it will be driven by the increase of entropy/disorder. This is the case of the dissolution of $[NH_4{}^+Cl^-]_{solid}$ in water.

8.R.4   Concentration, the pH Scale, and Indicators

Concentration Units: The concentrations used by chemists are measured in molar (M) units. Let us take, for example, a solution of 1M NaCl in water. Saying this means that we weighed 1 mole of NaCl (which is the combined atomic weights of Na and Cl, namely 58.44 g/mol) and dissolved it in 1 liter of water. We write concentrations with square brackets, so the concentration of our solution would be written as follows: [NaCl] = 1M. A liter of water weighs 1,000 g, and since 1 mole of water in gram units is 18.015 (let us round it to 18), then 1 liter of water contains 55.5 moles. Therefore, in a solution of 1M NaCl or the same concentration of any other compound, we have for each molecule of the dissolved material 55.5 molecules of water, or better said, there are 110 water molecules for each two molecules of the dissolved material.

The pH Scale: In water, there are hydronium (H_3O^+) and hydroxide (OH^-) ions. These ions are formed when one water molecule acts as a base and abstracts a proton from a second water molecule that serves as an acid, as illustrated using the *curved arrow method* in Scheme 8.R.3.

Of course, this process is dynamic, and a molecule that acted as a base in one direction acts as an acid in the reverse direction and vice versa. So the process goes to and fro, and there is a dynamic equilibrium between the forward and the reverse reactions. Equilibrium is another important issue in chemistry (see Zumdahl and Zumdahl, 2010).[9] Although the equilibrium is dynamic (both processes occur all the time), the concentrations of the various species do not change with time.

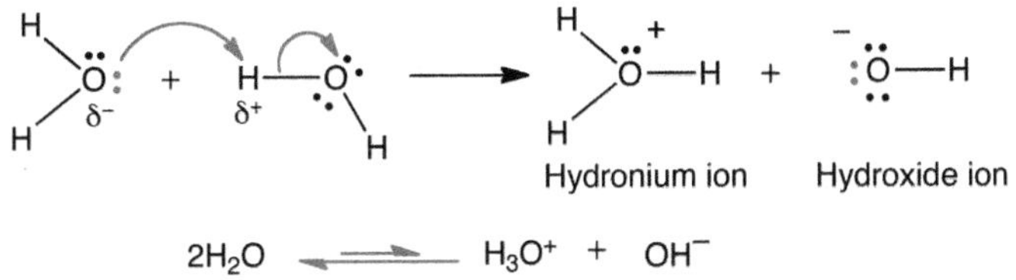

Hydronium ion   Hydroxide ion

$$2H_2O \rightleftharpoons H_3O^+ + OH^-$$

SCHEME 8.R.3   Using the curved arrow method to describe the reaction that forms H_3O^+ and OH^- in water. As indicated by the two arrows at the bottom of the drawing, the process is reversible and goes in both directions, what chemists call a dynamic equilibrium. The longer backwards arrow indicates that, at any given time, the concentration of discrete water molecules is larger than that of the ions.

In the case of the acid–base equilibrium in Scheme 8.R.3, the product of the concentrations of the ions behaves as a constant such that at 25°C, it has the following value:

$$[H_3O^+] \times [OH^-] = [H^+] \times [OH^-] = 10^{-14}$$

In pure water, the concentrations are identical, namely $[H^+] = [OH^-] = 10^{-7}$ M. Since pH is defined as follows: $pH = \log(1/[H^+])$, where log is a logarithm, the pH of pure water is $\log(1/10^{-7}) = 7$. Note that 10^{-7} M is a tiny little concentration, where for every one ionized molecule, there are many tens of millions intact water molecules. This is why we drew in Scheme 8.R.3 the forward arrow as being shorter than the reverse arrow, which in the symbolic language of chemistry means that most of the water is not ionized.

Imagine now that you prepared a solution of HCl with a concentration of 10^{-3} M and wanted to know its pH. Since HCl is a strong acid, it will dissociate completely to form ions, and therefore $[H^+] = 10^{-3}$ M. Using the pH expression, $pH = \log(1/[H^+])$, we get that $pH = \log(1/10^{-3}) = \log(10^3) = 3$. If you prepared a solution of 10^{-3} M NaOH, which is a strong base, then the concentration of hydroxide ions is $[OH^-] = 10^{-3}$ M. Since the product $[H^+] \times [OH^-]$ is a constant and equal to 10^{-14}, if $[OH^-] = 10^{-3}$, then to satisfy this constant, the proton concentration is $[H^+] = 10^{-11}$ M, and the respective pH is 11. This is a basic pH.

Indicators: Indicators are molecules that react with acids and bases and give products that have different colors, which depend on the pH. The phenolphthalein shown in Figure 8.R.1 is such a molecule. At pH < 7, it has one form, which is colorless in water, and at pH > 7, it assumes another molecular form, which has a bright red color. During acid–base neutralization, it is added to the acid solution and the base is added in a dropwise fashion; when the base consumes all the protons, the next drop raises the pH and the color changes all of a sudden. This is how we know that the reaction has reached its endpoint. This is also a method to determine how much of an unknown quantity of acid is in solution.

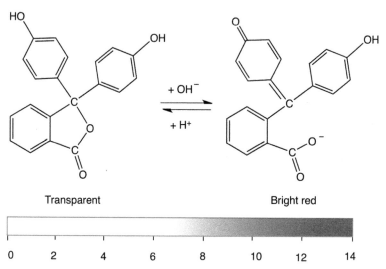

Transparent   Bright red

```
0    2    4    6    8    10   12   14
```

FIGURE 8.R.1   The molecular forms and colors of phenolphthalein in solutions with pH < 7 and pH > 7. The scale at the bottom represents different pH values.

8.R.5   Symbolic Representations of Chemical Reactions Using Curved Arrows

As we mentioned in the text, chemists use a symbolic way of representing chemical reactions. This symbolic element is a *curved arrow*, which serves as an indicator of the flow of electrons and reorganization of bonds during the reaction. It is very helpful in order to classify reactions, remember many of them, and design new ones by analogy. Scheme 8.R.4 shows a few elementary reactions with curved arrows.

Part (a) in the scheme shows an electron transfer reaction, using Na and Cl atoms to yield the ionic molecule Na^+Cl^-. To signify that this is a single electron transfer, we use a single-headed arrow to move the electron. Part (b) shows an H-atom abstraction from a molecule R'—H by a radical R$^•$. It is seen that here too we use single-headed arrows to move electrons, one from the radical and another from the H—R' bond, in order to create the new electron-pair bond for the R—H molecule, and meanwhile the

(a)   Electron transfer

Na $^•$ $^•$ $\overset{..}{\underset{..}{Cl}}$:   ⟶ Na^+ : $\overset{..}{\underset{..}{Cl}}$:$^-$

(b)   Hydrogen transfer

R $^•$ H—R'   ⟶   R—H   + $^•$R'

(c)   Proton transfer

B :   H—A   ⟶   B—H$^+$ +   :A$^-$

SCHEME 8.R.4   Representations of elementary reactions using curved arrows: (a) electron transfer, (b) H atom transfer, and (c) proton transfer.

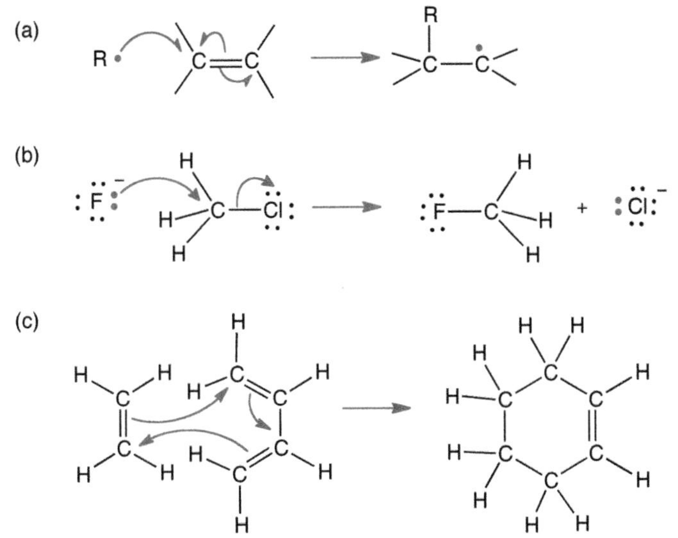

SCHEME 8.R.5   Representations of some reactions involving bond breaking and bond making using curved arrows: (a) a radical's addition to a double bond, (b) an ionic displacement reaction, and (c) a cycloaddition reaction.

third electron remains with the departing $R' \bullet$ radical. Finally, in part (c), we show a proton transfer reaction between a base, B:, and an acid, H–A. Here we use regular arrows to move electron pairs. One arrow shifts the electron pair from the base to form a bond with the H of the acid, and the electron pair that was initially in the H–A bond shifts to the A, which becomes A^-, as the second arrow indicates.

These elementary processes show themes that manifest themselves in other reactions, some of which are shown in Scheme 8.R.5.

Thus, the reaction in Scheme 8.R.5a is an analog of the H-atom transfer reaction in Scheme 8.R.4b, where in both cases, we use single-headed arrows to move electrons and reorganize the bonds. In Scheme 8.R.5b, we see an ionic displacement reaction, where F^- attacks the C–Cl bond, forms a new F–C bond, and displaces Cl^-. This is a famous mechanism in chemistry and a very common one in nature (e.g., DNA mutations occur when the bases use their lone pairs to attack molecules with C–halogen bonds and undergo methylation). It can be seen that here we use regular arrows to shift the electrons and reorganize the bonds, much as we did in the reaction in Scheme 8.R.4c. Finally, in Scheme 8.R.5c, we show a reaction called a *cycloaddition*, named as such because the electron pairs flow in a cycle and cause the two molecules to add to one another by making two new C–C bonds. This is another famous reaction in chemistry, called the Diels–Alder reaction, named after the two discoverers.

Inspection of the two last schemes shows, in fact, that chemical reactions involve bond reorganization; old bonds are broken and replaced with new ones. This is the reason why almost all chemical reactions have energy barriers that separate the reactants from the products.[2R] These barriers give all molecules lifetimes; the larger the barriers for the reactions of a molecule, the longer the lifetime of the molecule.

8.R.6   References for Retouches

[1R] P. Atkins, *What Is Chemistry?* Oxford University Press, Oxford, United Kingdom, 2013, pp. 32–45.
[2R] D. Usharani, W. Z. Lai, C. Li, H. Chen, D. Danovich, and S. Shaik, *Chem. Soc. Rev.*, **2014**, *43*, 4968.

8.P   PROBLEM SET

8.1 Potassium cyanide is a deadly poison made from K and CN fragments, and it is ionic. (a) Construct it using the "click-clack" method. (b) Show how it forms when the HCN gas is bubbled into a solution of KOH in water.

8.2 Construct all the possible ionic materials which can be made from the following fragments: Na, Mg, Al, O, Cl, SO_4, PO_4. Arrange your answers in the form of a table. Here and in subsequent problems in this problem set, treat all ions as being in Nirvana when determining their charges.

8.3 Construct all the ionic materials that can be made from the following fragments: BF_4, $N(CH_3)_4$, NH_4, CH_3O, Br.

8.4 Construct the molecule made from one N and as many Li fragments as needed using the "click-clack" method.

8.5 Construct the molecule made from the fragments K and ClO_4 (no O–O bonds here) using the "click-clack" method.

8.6 Devise an ionic liquid based on what you studied.

8.7 Quicklime is a white solid made from Ca and O. It has the same structure as solid NaCl. Draw the Ca and its close neighbors in the solid. In a separate structure, draw the O and its close neighbors in the solid.

8.8 Write the chemical reaction that occurs when you mix NaOH and HCl.

8.9 Adding HCl to Na_2CO_3 leads to a vigorous reaction, which forms bubbles of gas. Write down the chemical reaction and balance it.

8.10 When you mix solutions of Na_2CO_3 and $CaCl_2$ you get a white precipitate that does not dissolve in water. Write the equation for this chemical reaction.

8.11 Characterize the pH as equal to, larger than, or smaller than 7, of (a) a DNA base, (b) HCOOH, (c) ammonia, and (d) CO_2 in water.

LECTURE 9

BONDING IN TRANSITION METALS, SPECTROSCOPY, AND MOLECULAR DIMENSIONS

9.1 CONVERSATION ON CONTENTS OF LECTURE 9

Racheli: Sason, in previous courses, you avoided even a single mention of bonding in transition metals. Why?

Sason: I did not feel then that students of the humanities or the social sciences ought to know *everything* about chemistry …

Usha: But then, Sir, what made you change your mind for the book?

Sason: One of the referees of the book wrote that she/he could not imagine chemistry without transition metal–containing molecules. I thought to rise up to the challenge and include the topic as a *bonus* lecture.

Racheli: This is a surprise, Sason. I was not aware that you are afraid of referees. Is this really the reason you decided to teach bonding in transition metal–containing molecules?

Sason: You know me too well Racheli … The referee's comment hit on one of my softer spots. These molecules, which are also called *complexes*, are so pretty! Hence, triggered by the referee's comment, I decided to include them in the bonus material. Just look at Figure 9.1!

Magnus' green salt (**1**) was discovered in 1827 by Heinrich Gustav Magnus, who worked with the great Berzelius in Sweden.[1] Glancing at it shows a chain of platinum atoms wearing necklaces of four chlorines and four ammonia molecules in an alternating manner. A fascinatingly eclectic structure!

Chemistry as a Game of Molecular Construction: The Bond-Click Way, First Edition. Sason Shaik.
© 2016 John Wiley & Sons, Inc. Published 2016 by John Wiley & Sons, Inc.

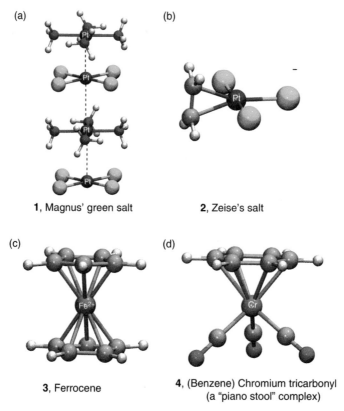

FIGURE 9.1   Four iconic transition metal complexes: (a) Magnus' green salt (**1**), (b) Zeise's salt (**2**), (c) ferrocene (**3**), and (d) the "piano stool" complex (benzene) chromium tricarbonyl (**4**). Color coding: Cl (bright green), N (blue), C (green), O (red), and H (gray).

Zeise's salt (**2**), made in 1827 by the Danish chemist William Christopher Zeise, is an unusual-looking anion, wherein a planar charged $PtCl_3$ moiety is linked to an ethylene molecule "hanging in the air" near the platinum.[1]

Of course, these structures were not known then, but as the technique called X-ray crystallography developed, the structures, as I drew them in Figure 9.1, were solved accurately. Aren't these beautiful?

Usha: You are jumping to mention X-ray crystallography, Sir, before telling us about **3** and **4**, which are landmark molecules. They are very important. Compound **3** is called ferrocene. It has a beautiful orange color. Its structure looks like a sandwich where two anionic $C_5H_5^-$ rings flank an Fe^{2+} ion. Ferrocene was made in 1951 accidentally, but its structure was solved a year later by two groups, led by Geoffrey Wilkinson and Ernst Otto Fischer, who were awarded the Nobel Prize in 1973.

Sason: These molecules also marked the revival of the field of inorganic chemistry and the dawn of organometallic chemistry, which unified organic and inorganic molecules. And by the way, complex **4** (Figure 9.1) contains a benzene molecule hovering above a chromium atom that is bonded to three CO molecules. It was made by Fischer and is called a "piano stool." These hovering halos created a lot of fascination in chemistry because it was not clear how are the "halos bonded to the metals." So rich! So intriguing!

Racheli: Aren't you basking in your aesthetic pleasures, Sason? Is this a sufficient reason to teach a relatively complicated topic?

Sason: You got me here, Racheli! I like pretty molecules ...

Racheli: But why teach a complicated topic?

Sason: Not only pretty ... they are also hugely important for chemistry and mankind.

Many of the good catalysts are based on transition metal compounds. The Ziegler–Natta catalyst makes polymers. Enzymes like hemoglobin and cytochrome P450 (CYP 450), which we discussed in Lecture 5, contain iron–oxygen bonds held by an organic molecule called a porphyrin. CYP 450 is by itself "an entire laboratory" for catalyzing vital reactions. Take hemoglobin away and we'll suffocate in a split second. Other enzymes with metal–oxo units prevent DNA damage by removing unnecessary chemical fragments that were attached to DNA bases, for example, by reacting with oxidized lipids (from the cell membrane). Just imagine our DNA being constantly modified without having repair mechanisms to restore it ...

Do not forget the future renewable and sustainable energy source, the water oxidation reaction (conversion of water to $O_2 + H_2$), which is catalyzed by transition metal complexes. So, yes we are dealing with enormously important molecules that ought to be taught.

Usha: But what are *you* going to teach, Sir, about transition metal complexes, considering that the topic is rather complex and extremely broad?

Sason: My intentions are very focused. I will teach the Law of Nirvana for transition metals, and I will then show how many other atoms/groups must a transition metal bind to achieve electronic Nirvana.

Racheli: Sason, remember that the transition metal community calls the groups that are bonded to the transition metal by the name *ligands* (Latin, *ligare* = to bind). This name was used for the first time in 1916 by the German chemist Alfred Stock.[2]

Usha: Will you not teach about the three-dimensional (3D) structures of these complexes?

Sason: Sure! What we learned in Lecture 7 will allow me to discuss the 3D structures of some of these complexes.

Racheli: How did the understanding of complexes come about? Who was largely responsible for it? Was it Berzelius, who also coined the term *catalysis*?[3]

Sason: This is a good question, my impatient interviewer. The issue of 3D structures brings me to the father of transition metal chemistry, Alfred Werner, whose photo is shown in Figure 9.2. In 1893, Werner published his seminal paper, "A contribution to the constitution of inorganic compounds,"[2] which eventually won him the Nobel Prize in Chemistry in 1913. In this paper, he laid down the basis for

FIGURE 9.2   Photo of Alfred Werner. Reproduced here with permission from Universität Zürich.

understanding and predicting the structures of transition metal compounds that had a given number of ligands (3, 4, 6, and 8 ligands, but particularly 6).

Werner was also a gifted teacher, and soon enough his classroom was full and teeming with students who came to hear him. His research group was international, as is science, and he had many female students. Quite unusual!

Usha: But, Sir, metals like iron, copper, gold, silver, etc. have been known for millennia, and many great chemists (even some great alchemists) were preparing transition metal compounds. So what was so special about Werner's coordination theory that earned him such acclaim?

Sason: In some ways, the story of Werner is similar to van 't Hoff's. It is a story of constructing a 3D structural theory that predicts correctly key chemical features such as the isomerism of transition metal compounds.

For example, consider the cobalt complex that was made by many and was described in the Berzelius notation as $CoCl_3 \cdot 6NH_3$, in Figure 9.3a. The Berzelius notation did not explain anything about this compound other than indicating its stoichiometry. Werner's coordination theory gave these complexes a defined structure for the first time.

By measuring conductivity and the percentage weight of chlorine in the complexes, Werner was able to determine that this compound was made from the cation $Co(NH_3)_6^{3+}$ and three Cl^- anions. Furthermore, in a stroke of genius, he

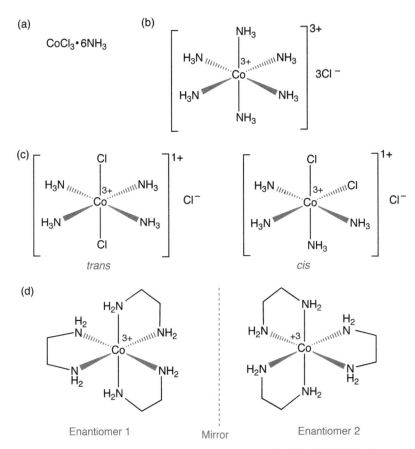

FIGURE 9.3 Werner's coordination complexes: (a) $CoCl_3 \cdot 6NH_3$ in the notation of Berzelius. (b) The same complex depicted with insights gained from Werner's coordination theory. (c) The structure of the $CoCl_3 \cdot 4NH_3$ complex, based on Werner's theory. Note the *cis* and *trans* isomers. (d) The complex $Co(NH_2CH_2CH_2NH_2)_3{}^{3+}$ using Werner's theory (the 3 Cl^- anions are omitted), written in the implicit representation without explicitly showing the C and H atoms between the nitrogen atoms. Note that this complex ion has a propeller shape, and it is hence chiral (the two mirror images cannot be superimposed).

also proposed that the six ammonia ligands form a coordination sphere around the Co^{3+} ion and are arranged in an octahedral 3D structure as shown in Figure 9.3b. Just compare the information in part (a) versus (b) of the figure and you will understand the intellectual leap made by Werner. If you look at figure 9.3c, you will immediately see that his scheme predicts new isomers, and hence new chemistry …

Usha: Sir, I am looking at Figure 9.3c and still do not see the leap. One could have equally assumed that the complexes are planar and still come up with *cis* and *trans* isomers. So?

Racheli: Let me reinforce Usha's dilemma. I also fail to see the leap.

Sason: Werner was lucky that you were not part of his Nobel committee. If you look at Figure 9.3d, you will immediately see his greatness …

Racheli: Yes, Usha, Sason is right. In fact, if you look at Figure 9.3d, you will see that when the ligand is ethylene diamine ($NH_2CH_2CH_2NH_2$), Werner's octahedral coordination would predict a chiral compound that has a propeller shape. The octahedral structure is critical for such a prediction.

Sason: This prediction of a chiral complex before it was actually made was one of Werner's decisive victories. And with evidence of this sort, his theory was accepted and became a useful driver for the development of coordination chemistry.

Usha: Transition metal complexes are also beautifully colorful. For example, $CoCl_3 \cdot 6NH_3$ is yellow, $CoCl_3 \cdot 5NH_3$ is purple, and *trans*-$CoCl_3 \cdot 4NH_3$ is green, while the *cis* isomer is violet. Will you try to teach the reasons for these very different colors?

Sason: The color is connected to spectroscopy …

Racheli: Spectroscopy is even a greater challenge for students. Are you really going to teach spectroscopy? You then need to also teach quantum mechanics (QM) theory…

Sason: I cannot teach much if anything of this material, Racheli and Usha. My focused intention is to convey the message of the immense importance of spectroscopy as the Rosetta stone of chemical matter.

Usha: But, Racheli, I think we should ask Sir to tell us something about the history of spectroscopy. Can you, Sir?

Sason: In spectroscopy, we irradiate the molecule with light and we "listen" to what the molecule has to tell us as it absorbs some of the light, diffracts or refracts it, or as it emits some of the light it absorbed. Spectroscopy is about light and its interaction with matter. The outcoming information from this interaction reveals the structures of molecules and their constituent bonds, groups, and atoms.

Just consider how easy it is to tell whether we have a C–H bond by infrared (IR) spectroscopy, or see if there is an unpaired electron by electron paramagnetic resonance (EPR) spectroscopy, or determine the molecular weight of a huge protein by mass spectrometry. What is the distance between the transition metal and the center of the rings in **3** and **4** in Figure 9.1? How do we know that the complexes in Figure 9.3 above are octahedral? These questions can be answered by X-ray crystallography. Try to imagine chemistry without spectroscopy …

Racheli: You are digressing, Sason … Usha just wanted to hear about the history of spectroscopy …

Sason: The most famous written evidence of spectroscopy is the story of the Biblical Noah, who saw the rainbow in the sky and took it as a sign of the covenant between the Creator and Mankind (Genesis 9:8–17).

White light is actually made from a bundle of colored components, which together blend into a "white" color. A rainbow occurs when sunlight passes via droplets of water, which refract (change direction) and reflect the light, such that its colored components separate according to their different wavelengths and thereby form a spectrum (Latin *spectre* = image, apparition), which is the band of beautiful colors we see sometimes after the rain or if we visit the Iguassu Falls.

Usha: Isn't this going a bit too far back, Sir? All I asked for is some scientific history …

Sason: Any history is history … The scientific history of spectroscopy is a vast field, so let me try to provide a bird's eye view. Here are some facts I picked up by reading the literature.[4,5]

Newton (Figure 9.4a) in fact created the science of spectroscopy by "repeating Noah's experiment" in 1666 and separating light into its components by passing it through a glass prism, and then by reflecting those components back into the prism, he showed that they mixed again to form "white" light.

This was followed by contributions of Joseph von Fraunhofer, who used a telescope and thereby created the first spectrometer, as a tool for seeing the different components of radiation. In 1868, Ångström (Figure 9.4b) measured 1,000 Fraunhofer lines and expressed their wavelength using the unit 0.0000000001 meters (10^{-10} m), which is now called 1 Å after his name, and serves to gauge the dimension of atomic and molecular size.

Racheli: But how has spectroscopy become a tool for analyzing matter?

Sason: This is due to Kirchhoff and Bunsen (Figure 9.4c).

Kirchhoff was the first to understand that the emitted and absorbed lights are intrinsic properties of materials, and that by absorbing and emitting light, matter actually "speaks" to us … He formulated a theory for this.

Together with Bunsen (who developed the famous burner carrying his name), Kirchhoff exploited the theory for chemical analysis. Together, they discovered two new elements, namely cesium (Cs) and rubidium (Rb), by the color of their flames, which is a form of spectroscopy using the naked eye.

A few years later, helium (He) was discovered and recognized by Frankland as a new gas in the atmosphere.

Usha: But, Sir, many others contributed, such as Balmer and Rydberg, who used the electric arc lamp and quantified the spectrum of the H atom.

Sason: True, the findings of Rydberg and Balmer formed a basis for the development of the quantum theory of light by Planck and Einstein, who viewed light as a stream of energy bundles called *photons* (see Retouches section 9.R.5).

Usha: This idea of a photon is important because Bohr (Figure 9.4d) used it later to develop the first quantum atomic model, which included absorption and emission of photons by H atoms with jumps of an electron between orbits. This paved the way for the later development of QM as a theory of matter.

Racheli: As chemists, we must not forget the important contribution of Beer, who recognized that spectroscopy is sensitive to concentration. This made spectroscopy an analytical tool for determining minute quantities of a desired molecule in a mixture of many different molecules.

Sason: Both of you seem to know the facts and you are still asking me to tell you the stories …

FIGURE 9.4 A few of the many contributors to spectroscopy: (a) Isaac Newton, (b) Anders Ångström, (c) Gustav Robert Kirchhoff (left) and Robert Wilhelm Eberhard von Bunsen, (d) Niels Bohr, (e) Joseph John Thomson, (f) Felix Bloch (left) and Edward Mills Purcell, (g) Richard R. Ernst, (h) Dorothy Crowfoot Hodgkin (second row right) and Ada Yonath, (i) From left to right, Wilhelm Conrad Röntgen, Max von Laue, William Henry Bragg, and William Lawrence Bragg. Richard Ernst's and Ada Yonath's photos are used with their kind permissions. Felix Bloch's and Edward Mills Purcell's photos are from: *Nobel Lectures, Physics 1942–1962*, Elsevier Publishing Company, Amsterdam, 1964. The portrait of Dorothy Crowfoot Hodgkin is by Walter Stoneman and is copyrighted by Godfrey Argent Studio. William Henry Bragg's portrait comes from the IUCr (International Union of Crystallography) photo gallery, http://www.iucr.org/gallery/individual-no-hires?show= 26565. The source for William Lawrence Bragg's image is: Lipson, H. S. (1972). William Lawrence Bragg, 1890–1971. Acta Cryst. A28, 225. This image is used with the copyright permission of the IUCr. Meanwhile, the following scientists' photographs are in the public domain and were obtained from these sources: Isaac Newton (http://commons.wikimedia.org/ wiki/File:GodfreyKneller-IsaacNewton-1689.jpg), Anders Ångström (http://commons.wiki media.org/wiki/File:Anders_%C3%85ngstr%C3%B6m_painting.jpg), Gustav Robert Kirch-hoff (http://commons.wikimedia.org/wiki/File:Gustav_Robert_Kirchhoff.jpg), Robert Wil-helm Eberhard von Bunsen (http://commons.wikimedia.org/wiki/File:Robert_Bunsen_portrait .jpg), Niels Bohr (http://commons.wikimedia.org/wiki/File:Niels_Bohr_Date_Unverified_LOC .jpg), Joseph John Thomson (http://commons.wikimedia.org/wiki/File:Jj-thomson2.jpg), and Wilhelm Conrad Röntgen (http://commons.wikimedia.org/wiki/File:Roentgen2.jpg). Max von Laue's image is attributed to Bundesarchiv, Bild 183-U0205-502 / CC-BY-SA and is licensed under the Creative Commons Attribution-Share Alike 3.0 Germany license.

Racheli: We do not really know all the stories, Sason.

Sason: Anyway, in 1947, after the discovery of ultraviolet (above violet) rays by Ritter, the first machine that recorded UV-VIS (VIS = visible) spectra became commercially available (it was called CARY 11), and the recording of UV-VIS spectra of molecules became routine, and such spectra proved to be a fingerprint of the electronic structure of molecules.

Usha: What about the development of infrared spectroscopy, Sir?

Sason: The astronomer Herschel discovered in 1800 infrared (IR) light (below red), and through a similar chain of discoveries, IR spectroscopy also culminated in 1930 in the PerkinElmer machine, which recorded IR spectra of molecules on a routine basis. IR spectroscopy proved to provide a fingerprint of the bonds in a molecule.

Racheli: What about mass spectrometry?

Sason: This is a relatively late development, mostly due to the work of J. J. Thomson (Figure 9.4e) in 1907. He was by then already a Nobel laureate for his discovery and characterization of the electron as a fundamental particle of matter. His technique to analyze ions involved subjecting them to a combination of electric and magnetic fields, which deflected the ions in proportion to the ratio of their mass to their charge. This allowed Thomson to determine the mass of positive ions. In this manner, he tested Frederick Soddy's hypothesis about the existence of isotopes, and Thomson showed that neon (Ne) has two isotopes, which differ in their masses. Today, mass spectrometry has become indispensable for determining the presence of very small amounts of material, for accurately determining molecular masses, and for analyzing some structural features of huge molecules, based on the observed molecular pieces as the original molecular cation falls apart.

Racheli: These were physicists who invented nuclear magnetic resonance (NMR) spectroscopy. Always physicists. Right?

Sason: This is correct. Physicists invent techniques, and chemists know how to utilize them for probing chemical matter. In this manner, the two sciences enrich one another.

Anyway, about NMR: Wolfgang Pauli showed in 1924 that since the proton has two different spins (like the electron), if the atom is placed in a magnetic field, this makes these spin states energetically different. One can then induce transitions from one spin state to the other if the atom is irradiated by and absorbs a photon of a specific radio frequency.

In 1946, Bloch and Purcell (Figure 9.4f) carried out the first NMR experiment, for which they were awarded the Nobel Prize in 1952. Later, Dickinson, Yu, and Proctor showed that the frequency of the radio waves from which the nuclei absorb energy and the shape of the absorption peak both depend on the chemical environment of the nucleus absorbing this energy. This has eventually become the basis for magnetic resonance imaging (MRI).

Usha: I did not want to interrupt you, Sir, but your sentence, "… [they] showed that the frequency of the radio waves from which the nuclei absorb energy…," would sound a bit opaque to the students. Can you say it in a different way?

Sason: Well, let's see … It turns out that QM theory predicts that the energy levels of atoms and molecules maintain gaps. In order to change an energy level, the atom/molecule absorbs radiation of a specific frequency. This frequency is determined by the energy of the photon (the particle of light) that matches the energy difference (see Retouches section 9.R.5) and induces a transition. Niels Bohr (Figure 9.4d) was the first to theorize this mechanism for the UV-VIS absorption of the H atom.

In NMR, the energy levels of the nuclear spin in a magnetic field maintain a gap, and this gap is sensitive to the chemical environment of the nucleus, because all the other nuclei, which are by themselves tiny magnets, modify slightly this energy gap. For example, a hydrogen atom nucleus in a molecule like $CH_3CH_2CH_3$ will absorb energy at different radio frequencies for the CH_2 and CH_3 groups, and the shapes of the absorption peaks will be different and will reflect the effect of all the neighboring hydrogen nuclei, which are themselves little magnets.

Recheli: So why did chemists pick up this NMR technique?

Sason: Already Bloch realized the potential of his NMR technique for chemistry. But it took a chemist like Shooley, who was the first to appreciate the huge potential of NMR for probing molecular structure. He joined the company Varian which built the first NMR spectrometer. And soon after, the technique was taken over by the community of chemists, who found it to be immensely useful for identification of molecular structure. The chemist who developed the technique immensely and received the Nobel Prize in 1991 was Richard R. Ernst (Figure 9.4g).

A similar technique is EPR, which deals with the spin of the electron. This technique is very useful for free radicals and for systems with unpaired electrons, like many of the transition metal enzymes I mentioned previously.

Usha: In my humble opinion, the most important spectroscopy is X-ray diffraction spectroscopy. What is your opinion, Sir?

Racheli: Sason, do not forget that this technique yielded a few years ago a Nobel Prize for one of your colleagues, Ada Yonath (Figure 9.4h).

Sason: Well, Usha, it is better not to rank these techniques by importance. You are right that X-ray diffraction gives structures, with bond lengths and bond angles, and hence it is very important. Structure is the heart of chemistry.

Racheli, if we are mentioning the Nobel Prize to Yonath, we should not forget Dorothy Hodgkin (Figure 9.4h), one of the pioneers of X-ray crystallography.

Anyway, the discoverer of X-rays emanating from radioactive compounds was Röntgen, who received the Nobel Prize in 1901 (Figure 9.4i). Max von Laue discovered that molecular crystals diffract X-rays (Figure 9.4i), yielding very ordered diffraction patterns. He reasoned that the 3D structure of the crystal was ordered and periodic (its structure repeats itself an infinite number of times). He got the Nobel Prize in 1914. The Braggs (Figure 9.4i), father and son, worked out the conditions for X-ray diffraction, what is now known as Bragg's Law. Bragg junior used this to solve for the first time the structure of the Na^+Cl^- crystal. He is mentioned in the book *The Double Helix* as the head of the laboratory where Crick and Watson worked. The Braggs shared the Nobel Prize in 1915.

Usha: Please do not forget that the first ones to apply the technique to solve the structures of hemoglobin and myoglobin were Max Perutz and John Kendrew. They also got the Nobel Prize, and they are also mentioned in *The Double Helix*.

Racheli: And let's not forget Watson and Crick, who used the X-ray information generated by Rosalind Franklin to determine the structure of DNA. What a great technique!

Sason: Thank you, Usha and Racheli, for these comments.

Now that the interview is over, let me do some teaching …

9.2   THE 18-ELECTRON RULE FOR TRANSITION METAL BONDING

Transition metals form fascinating molecules because their valence shell can accommodate more electrons than that of main group elements can, and therefore they bind to ligands that possess many available electrons like benzene or to many different ligands having available lone pairs.

Let us start then with counting electrons in transition metals. Scheme 9.1a displays the transition metals, which start from period 4. In each period in the scheme, we start with two main group elements, which belong to the families 1A and 2A, such as K and Ca, and we proceed with 10 transition metals that belong to families marked as groups 3 to 12. As we saw for main elements, here too, the family number is also the number of electrons in the valence shell of a given atom. Thus, moving along

(a)   The periods of transition metals

1A	2A	3	4	5	6	7	8	9	10	11	12
K	Ca	Sc	Ti	V	Cr	Mn	Fe	Co	Ni	Cu	Zn
Rb	Sr	Y	Zr	Nb	Mo	Tc	Ru	Rh	Pd	Ag	Cd
Cs	Ba	La	Hf	Ta	W	Re	Os	Ir	Pt	Au	Hg

(b)

Fe (8e) $\Longrightarrow$ Fe^{2+} (6e), Fe^{4+} (4e)

Co (9e) $\Longrightarrow$ Co^{3+} (6e)

Cu (11e) $\Longrightarrow$ Cu^{1+} (10e)

(c)

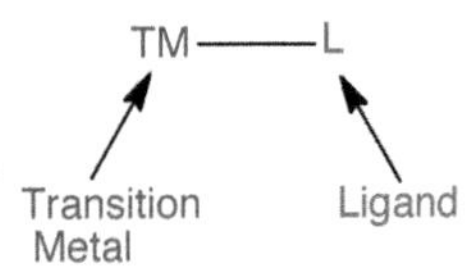

TM——L

Transition Metal   Ligand

SCHEME 9.1   (a) The periods of transition metals and their family numbers (the first two atoms in each period are main group elements). (b) Some neutral transition metals and transition metal cations with their valence electron counts given in parentheses. (c) The transition metal (TM) is bonded to a ligand (L), which is a modular fragment that consists of a molecule or an atom.

each period, the number of valence electrons increases successively from 1 to 12. For example, in the first period in the scheme, K has 1 valence electron (1e), while Zn has 12 (12e).

However, many complexes are formed by transition metals' cations, and we ought to learn how to count the valence electrons for these species as well. Scheme 9.1b shows a few cations with their number of valence electrons within parentheses. Thus, iron appears frequently as Fe^{2+}, and since the atom has eight electrons, then the corresponding cation has only 6e. In metalloenzymes like CYP 450, iron appears as Fe^{4+}, which has four valence electrons. Cobalt with 9e forms complexes frequently as Co^{3+}, which has six valence electrons, and copper as Cu^{1+} with 10 valence electrons, etc.

What is the Law of Nirvana for transition metal elements? In main elements, atoms reach Nirvana when they are surrounded with eight electrons by forming bonds with other elements. The octet rule is a very strong one, but even in this case, we saw that as we move down the periods, the atoms (like S, P, etc.) are able to attain different states of Nirvana and form electron-rich molecules, like SF_6 with 12 electrons around sulfur. The reason is that for atoms belonging to higher periods, beyond the second one, the valence shell can accommodate more than eight electrons, so these atoms have more choices of Nirvana states. For transition metals (TMs), the valence shell can accommodate up to 18 electrons. As such, instead of octet, in transition metal compounds, it is the 18-electron (18e) rule that determines the bonding capability of a given transition metal with ligands. Scheme 9.1c shows the general terminology we shall be using from this point onward.

Since the number of electrons required to achieve Nirvana is so large for transition metal species, the stability differences of other electron counts are not too forbidding. Consequently, the 18e rule is softer than the octet rule, and we may expect to find relatively persistent radical complexes with 17 electrons, and complexes with 16e or even less.[6] These cases are in fact extremely interesting because the electron-deficient complexes can serve as catalysts that activate other molecules. Despite all these qualifications, the Law of Nirvana for transition metal compounds is a very useful guide for constructing transition metal complexes and for considering their reactivity (propensity to react) and properties. Let us see how the rule is applied along with the "click" bond method.

9.2.1 An Example of a Transition Metal Complex That Obeys the 18e Rule

The 18e bar makes all transition metals and their cations severely electron-deficient and lacking a substantial number of electrons required to achieve Nirvana. The missing electrons must be supplied therefore by the ligands. To refresh our memory about the binding mode of electron-deficient species, let us repeat the binding of BH_3 with NH_3 (ammonia), which we did in Lecture 3 (Scheme 3.1). This is depicted here in Scheme 9.2a, which shows that BH_3 is missing one electron pair to reach an octet, while ammonia has a lone pair that can be used for this purpose. Hence, both species are single connectors. "Click," and the lone pair is employed in making a new B–N bond. Now both B and N have obtained an octet and have thereby reached Nirvana.

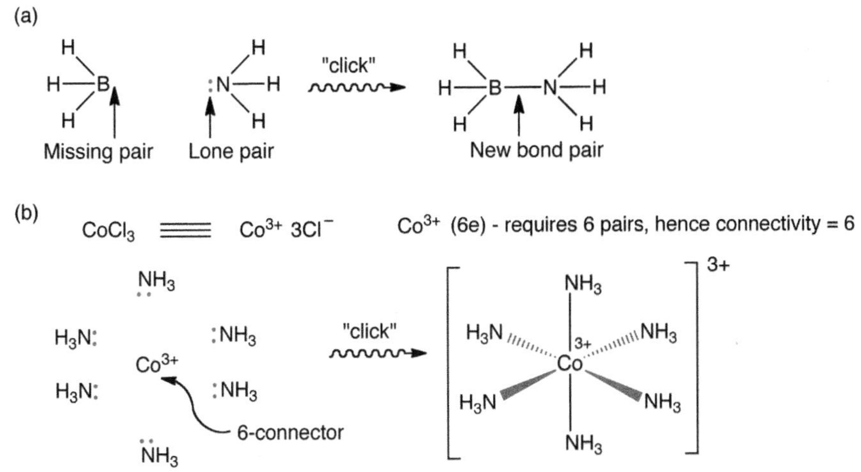

(a)

(b)

SCHEME 9.2   Binding of electron-deficient species to molecules having lone pairs: (a) BH_3 with NH_3. (b) Co^{3+} has 6e, and since it lacks six electron pairs (considering the state of Nirvana of 18 electrons), it is a six-fold connector transition mental ion, which binds six NH_3 molecules.

Let us now turn to Scheme 9.2b. Consider the cobalt complex that was made by the combination of $CoCl_3$ and ammonia. The use of $CoCl_3$ means that we have an ionic compound $Co^{3+}(Cl^-)_3$. Therefore, we must consider the binding capability of Co^{3+} with ammonia. Since Co^{3+} has only six valence electrons, it lacks six electron pairs to reach Nirvana, and hence it will be a six-fold connector and will form bonds with six ammonia molecules, each of which has a lone pair that acts as a single connector. The respective species are arranged in Scheme 9.2b, and "click," they form the complex ion $Co(NH_3)_6{}^{3+}$, which possesses six Co–N bonds, where both Co and N reached Nirvana.

9.2.2   Electron Counts of Ligand Contributions

In order to construct a variety of complexes, we must look at the binding capabilities of different ligands, as shown in Scheme 9.3. As a reminder, ligands labeled generally as L are molecules or ions that are bonded to the transition metal. To unify the considerations, the ligands are considered as species that have already satisfied their octet, and like ammonia, they possess pairs of electrons that can participate in further binding without affecting the octet of the ligand.

Scheme 9.3a shows typical two-electron (2e) ligands. Some of them are neutral molecules like ammonia, water, phosphines (PR_3), and ethylene. Others are anions like chloride or other halides and cyanide (CN^-). As shown by $CH_3{}^-$, even alkyl anions can serve as ligands. All the 2e-ligands are *single connectors*.

Scheme 9.3b shows three 4e-binders that serve as *double connectors*. A typical one is a diamine molecule like $H_2N(CH_2)_2NH_2$ that uses its two amino centers to form two bonds to a TM. Chemists call such ligands *bidentate*, namely a ligand that has two "teeth" to "bite" on the transition metal. Another type of bidentate ligand is the one

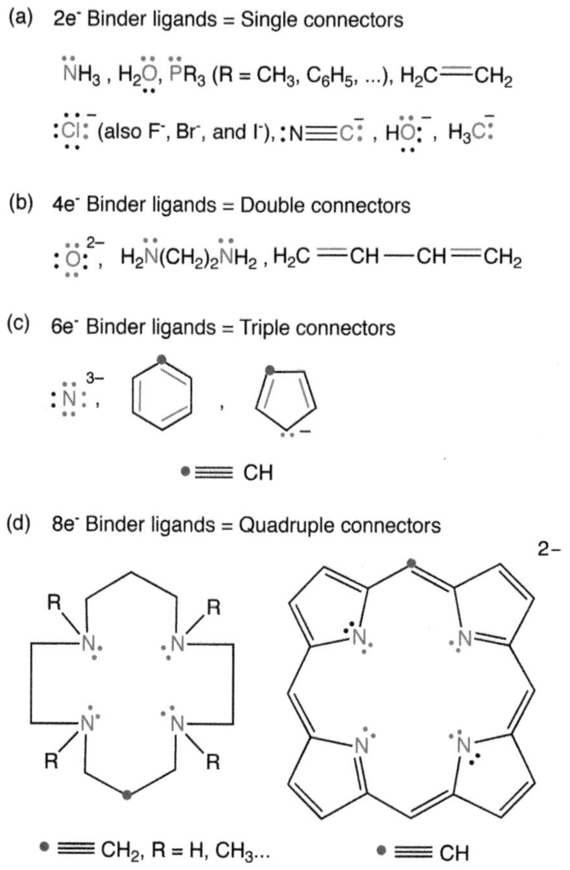

(a) 2e⁻ Binder ligands = Single connectors

$\ddot{N}H_3$, $H_2\ddot{O}$, $\ddot{P}R_3$ (R = CH_3, C_6H_5, ...), $H_2C{=}CH_2$

$:\!\ddot{\underset{\cdot\cdot}{C}l}\!:$ (also F⁻, Br⁻, and I⁻), $:N{\equiv}\ddot{C}:^-$, $H\ddot{O}:^-$, $H_3\ddot{C}:^-$

(b) 4e⁻ Binder ligands = Double connectors

$:\!\ddot{\underset{\cdot\cdot}{O}}\!:^{2-}$, $H_2\ddot{N}(CH_2)_2\ddot{N}H_2$, $H_2C{=}CH{-}CH{=}CH_2$

(c) 6e⁻ Binder ligands = Triple connectors

$:\!\ddot{\underset{\cdot\cdot}{N}}\!:^{3-}$,

● ≡ CH

(d) 8e⁻ Binder ligands = Quadruple connectors

● ≡ CH₂, R = H, CH₃...   ● ≡ CH

SCHEME 9.3 The ligands (L), which bind to transition metals (TM), and the number of electrons (marked in red) they contribute to bonding to the TM: (a) 2e-ligands, (b) 4e-ligands, (c) 6e-ligands, and (d) 8e-ligands. The ligand atom that will attach to the transition metal and its respective electron pair or pairs participating in bonding to the TM are marked in red. When a double bond is the ligand, only its electron pair, not the atoms, is marked in red. Note that, in some cases, the ligand is not connected to the transition metal using only one specific atom/group, but is bonded using a few of its atoms or groups. Note that the terms three-fold connector and triple connector are synonymous. The same applied to four-fold connector, 4-connector, and quadruple connector. The same applies to all other schemes in the lecture.

that possesses two double bonds, and this type is represented by $H_2C{=}CH{-}CH{=}CH_2$. But as we already know about chemistry, there are many possible variations on this theme, like two double bonds in a ring and so on. Finally, I show the ligand O^{2-}, which uses two of its lone pairs to form a double bond to a TM.

In Scheme 9.3c, we show three-fold connector ligands, which are 6e-donors. Among them are N^{3-}, the benzene molecule, and the cyclopentadienyl anion ($C_5H_5^-$); the latter two use their six mobile ring electrons to form three bonds to a TM.

Finally, Scheme 9.3d shows four-fold connector ligands. The first one is a cyclic molecule that has CH_2 groups and four amino (N–H, N–CH_3, ...) groups, and the second is an anionic porphyrin ligand, which is common in enzymes like hemoglobin and CYP 450.

9.3 CONSTRUCTION OF TRANSITION METAL COMPLEXES THAT OBEY THE 18e RULE

Once we know the electron counts and connectivities of various transition metals and ligands, we can proceed to systematically construct a few beautiful complexes.

Let me start with two of the complexes we mentioned in the conversation, the "piano stool" and ferrocene, which are constructed here in Scheme 9.4.

Pretending we have not seen it before, let us consider, as in Scheme 9.4a, constructing a complex of a Cr atom with one benzene (C_6H_6) molecule and as many carbon monoxide molecules as required. Since Cr has 6e, it is a six-fold connector atom, and since benzene is a three-fold connector molecule, we are going to also need three single-connector molecules of CO. And "click," we form the "piano stool" complex, in which Cr has acquired its requisite 18e.

In Scheme 9.4b, we are interested in constructing a complex of iron with the three-fold connector $C_5H_5^-$ anion, and we are wondering what should be the charge state of iron that should bring Nirvana to the complex. Since the ligand is a three-fold connector and the iron atom has 8e, the only way to make a complex in Nirvana is by usage of the Fe^{2+} cation. Since Fe^{2+} is a six-fold connector, it will require two three-fold connector ligands. And, "click," we generate ferrocene, $Fe(C_5H_5)_2$.

Let me proceed to Scheme 9.5, where we construct CO complexes of various transition metal atoms, from Cr to Ni. Part (a) shows the connectivity of CO and the

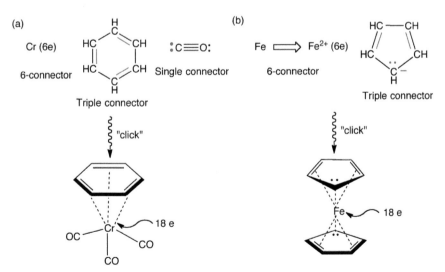

SCHEME 9.4 Construction of (a) the "piano stool" complex, $Cr(CO)_3(C_6H_6)$, and (b) ferrocene, $Fe(C_5H_5)_2$. On the top, the respective TM fragments and ligand molecules are shown along with their connectivities.

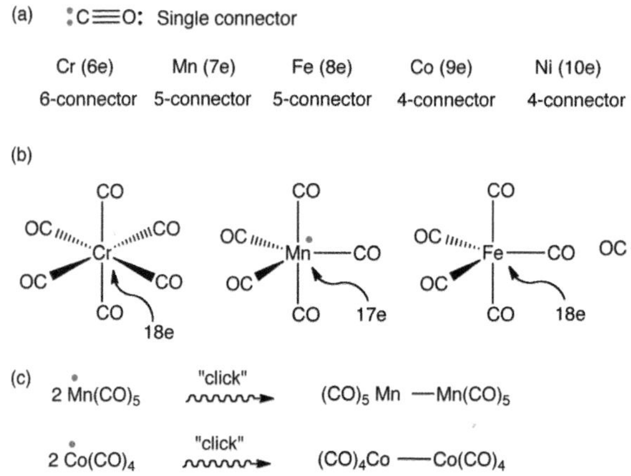

(a) $:C\equiv O:$ Single connector

Cr (6e)   Mn (7e)   Fe (8e)   Co (9e)   Ni (10e)

6-connector 5-connector 5-connector 4-connector 4-connector

(b)

(c)

SCHEME 9.5   Construction of transition metal carbonyl compounds, $TM(CO)_n$, where the TM is neutral, that approach the 18e requirement as much as possible: (a) The connectivities of CO and various TM atoms (TM atoms' connectivities are specified beneath the atoms). (b) Construction of $Cr(CO)_6$, $Mn(CO)_5$, $Fe(CO)_5$, $Co(CO)_4$, and $Ni(CO)_4$, and the corresponding electron counts on the TM centers. The different geometric structures will be further explored later. (c) Formation of complexes with TM–TM bonds from the radicals in (b).

maximum connectivities of the various TMs. Thus, based on the available number of electrons on the TM atoms and the 18e bar, Cr is a six-fold connector, and Mn and Fe are five-fold connectors, while Co and Ni are four-fold connectors (note that the terms 5-connector and five-fold connector are synonymous, and the terms 6-connector and six-fold connector are also synonymous). These $TM(CO)_n$ complexes are shown in Scheme 9.5b. Thus, only Cr, Fe, and Ni, which have even numbers of valence electrons, form complexes in which the TM achieves Nirvana. The other two TM complexes, $Mn(CO)_5$ and $Co(CO)_4$, have 17e and hence possess an unpaired electron residing on the TM. These are radical fragments, like those we saw in the lectures on organic molecules, such as $H_3C\bullet$. But unlike the organic radicals, the TM complex–based radicals are persistent and live sufficiently long to be easily probed. Nevertheless, like all radicals, they tend to reach Nirvana by coupling their odd electrons and forming new complexes with TM–TM bonds, as shown in Scheme 9.5c. Thus, $(CO)_5Mn–Mn(CO)_5$ is the transition metal analog of ethane ($H_3C–CH_3$), and so is $(CO)_4Co–Co(CO)_4$.

9.4   TRANSITION METAL COMPLEXES WITH 14–16e

As I mentioned already, there are quite a few complexes where the 18e rule is not obeyed. The most common electron count in these complexes is 16e,[6] but one can find complexes with even smaller numbers of electrons. These cases are found in particular among the early and late TMs, namely at the start and end of the period of the TM series (see Scheme 9.1).

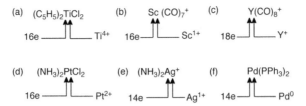

SCHEME 9.6   TM complexes with electron counts less than or equal to 18e: (a) a Ziegler–Natta-type catalyst, $(C_5H_5)_2TiCl_2$, (b) $Sc(CO)_7{}^+$, (c) $Y(CO)_8{}^+$, (d) cisplatin, $Pt(NH_3)_2Cl_2$, (e) $Ag(NH_3)_2{}^+$, and (f) the active form of a molecule catalyzing the Heck reaction, $Pd(PPh_3)_2$ (Ph is a shorthand notation for the fragment C_6H_5).

The number of valence electrons in the valence shells of early TMs (e.g., Sc, Ti, V) is small, and hence, the metals would need too many ligands to acquire 18e. But then the sphere of the TM would be too crowded and the electrons of the various TM–L bonds would repel one another too strongly to afford these many ligands. Therefore, these TMs take fewer ligands and form complexes with fewer than 18e in the valence shell of the TM. As we move toward the end of the TM period (e.g., Ni, Cu, Pd, Ag, Pt, Au), the number of electrons on the TM becomes quite large, and those that remain as lone pairs on the TM repel the bonding electrons and weaken thereby the propensity of the TM to acquire 18e.

Scheme 9.6 shows some of these deviant complexes and one with 18e. Let me start by finding out the charge on the TM, which can be deduced from the rules we discussed for the ligands in Scheme 9.3. Thus, take, for example, $(C_5H_5)_2TiCl_2$ in Scheme 9.6a. Since by convention, the ligands each have an octet, they must be both anions, $C_5H_5{}^-$ and Cl^-, and since the complex is neutral, then the titanium must be the cation Ti^{4+}. Applying the same reasoning can lead to all the charge states on the other TMs in the scheme, as Sc^{1+}, Y^{1+}, Pt^{2+}, Ag^{1+}, and $Pd^{0.}$

Let us now turn to discuss the total electron count on the TM. The complexes in parts (a) and (b) of Scheme 9.6 involve two early TM cations, Ti^{4+} and Sc^{1+}. The Ti^{4+} ion has no electrons in the valence shell, and therefore to achieve Nirvana, the ion must bind *nine* single-connector ligands! This is obviously too crowded, and the Ti^{4+} in Scheme 9.6a takes up two $C_5H_5{}^-$ ligands, which are three-fold connectors, and two Cl^- ligands, thus acquiring only 16e. This complex is by the way one of the Ziegler–Natta catalysts that performs polymerization of olefins (e.g., of $H_2C=CH_2$).

Similarly, in Scheme 9.6b, Sc^{1+} with only 2e in its valence shell requires eight single-connector ligands, and since a complex with coordination by eight ligands is crowded, what forms instead is actually the 16e complex $Sc(CO)_7{}^+$. Interestingly, the yttrium cation, Y^+, binds with eight CO ligands (in Scheme 9.6c) and forms the 18e complex $Y(CO)_8{}^+$.[7] As we shall see later (Section 9.8.3), as we move down the family column, the atom grows in size, and hence, the larger coordination sphere of Y^+ is more spacious and is likely to be better suited to bind eight ligands, while the smaller Sc^+ cannot accommodate as many ligands.

A quite famous complex is the *cisplatin* molecule shown in Scheme 9.6d, wherein Pt^{2+} with its 8e in the valence shell binds to four 2e-ligands and forms thereby a 16e complex. The complex is famous because of its usage in cancer chemotherapy. Its activity arises from the fact that it binds to two strands of DNA in the cancer cell and thereby causes cell "death." This binding involves the loss of the two Cl^- ligands and complexation of $(NH_3)_2Pt^{2+}$ to two DNA bases using the lone pairs of the nitrogen atoms of the bases.

Finally, in Schemes 9.6e and 9.6f, we show two 14e complexes of late TMs. One is $Ag(NH_3)_2{}^+$, and the other is a complex of Pd with two neutral phosphine ligands, PPh_3, where Ph is the way chemists denote the C_6H_5 fragment of benzene. The latter complex is famous too as it belongs to the family of catalysts made by Richard F. Heck, who received the Nobel Prize in Chemistry in 2010, together with Ei-ichi Negishi and Akira Suzuki, who together invented a group of catalysts that perform very useful reactions.

9.4.1   Comments on TM-Based Catalysts

Lack of Nirvana in complexes makes them potentially useful as catalysts of otherwise slow reactions, as they can readily pick up an additional molecule or two and activate them for further reactions. The principles of catalysis by the Heck and the Ziegler–Natta complexes in Schemes 9.6a and 9.6f are virtually identical. Since the TMs in these complexes are still electron-deficient, they tend to bind molecules that contain donors of electron pairs like double bonds. Once the molecule is bonded to the TM, it becomes more susceptible to attack by other molecules and undergoes a desired chemical change at a faster rate.

There are, however, complexes that satisfy the 18e rule, but are nevertheless powerful catalysts. Figure 9.5 shows a potent active species of CYP 450, in which the active part is a $Fe^{4+}=O$ unit, which is further bonded through the iron center to additional ligands. Thus, the iron is bonded to a sulfur of a cysteine residue, which is a thiolate group (RS^-) and hence a 2e-ligand, and to four nitrogen fragments of a porphyrin ligand (Scheme 9.3d), which is an 8e-donor. Note that the porphyrin is labeled as $Por^{\bullet+}$, where the sign ($\bullet+$) means that it is missing one electron from the double bonds of the ring, which becomes therefore a radical-cation moiety. Thus, the porphyrin lost one electron, which went to the iron to keep it as Fe^{4+} (instead of Fe^{5+}),

The $Por^{\bullet+}Fe^{IV}=O(S_{Cys})$ species of the enzyme CYP 450

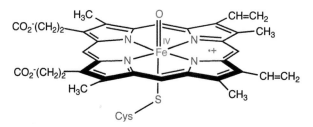

FIGURE 9.5   The active species of CYP 450 enzymes, $Por^{\bullet+}Fe^{IV}=O(S_{Cys})$.

such that the $Fe^{4+}=O$ unit possesses 8e around Fe^{4+}. Together then, the iron has 18e, and it formally attained Nirvana. Nevertheless, this complex is a very potent activator of molecules. It abstracts H atoms from a variety of alkanes, and then it transfers to the alkyl radical the OH group on the Fe to make an alcohol, which is secreted from the body. In this manner, the complex in Figure 9.5 neutralizes poisons that enter the body. Just imagine that you accidentally took in a little bit of octane (C_8H_{18}), which for a car is gasoline that makes it run, but for you is a potential poison. Here the active species of CYP 450 comes to the rescue and converts octane an alcohol that leaves the body. This activity and other reactions make the complex in Figure 9.5 one of the most versatile catalysts in Nature, being responsible for the biosynthesis of essential compounds such as the brain chemicals serotonin and dopamine, the sex hormones, and the essential fatty acids (ω-3, etc., see Scheme 5.9).

Thus, despite its 18e count, the activity of this complex is wide-ranging.[8] It arises because two of the electrons on the Fe=O moiety are not paired, and the O-center acts as a radical. Thus, it can abstract H atoms and perform other reactions that radicals initiate. As this is due to QM effects, we make some comments about it in the Retouches (see Section 9.R.3). Another aspect of catalysts is that they act like nanomachines using a cycle of reactions that turns over many times and performs the same final reaction such that the product of the reaction is produced with great efficiency. This aspect is also discussed in the Retouches (see Section 9.R.4).

9.5   3D SHAPES OF TRANSITION METAL COMPLEXES

You recall from Lecture 7 that the VSEPR (valence shell electron pair repulsion) rules predict the shapes of main-element molecules with nice compact reasoning, by simply counting the number of electron pairs, bond pairs and lone pairs around the central atom. In transition metal (TM) complexes, the lone pairs on the TM do not occupy space in the same manner as they do in the case of the main group elements, so what matters for the 3D shape is only the number of bond pairs the TM maintains with the ligands. This number is also known as the *coordination number* (*cn*) by the historical referral of Werner to these complexes as *coordination complexes*.

Figure 9.6 shows the most common 3D shapes of TM complexes, illustrating general concepts using specific examples. Thus, complexes with two bond pairs or $cn = 2$ have a linear geometry with a bond angle of 180°, as shown in Figure 9.6a for $Ag(NH_3)_2^+$. When $cn = 3$, as in $Cu(CN)_3^{2-}$ in Figure 9.6b, the most common geometry is a planar structure with bond angles of 120°.

Figure 9.6c shows the most common shape for $cn = 4$, which is the tetrahedral $Ni(CO)_4$ molecule with bond angles of approximately 109°. Nearby we show the less common geometry for $Pt(NH_3)_4^{2+}$, which has a planar structure with bond angles of 90°. This deviation from the tetrahedral shape is sometimes found in complexes where the TM has 8e, like the Pt^{2+} complex.

Of course it may be very bothersome for some of you that there are exceptions to the rule, but recall that this is the nature of all rules … Rules are needed because they paint

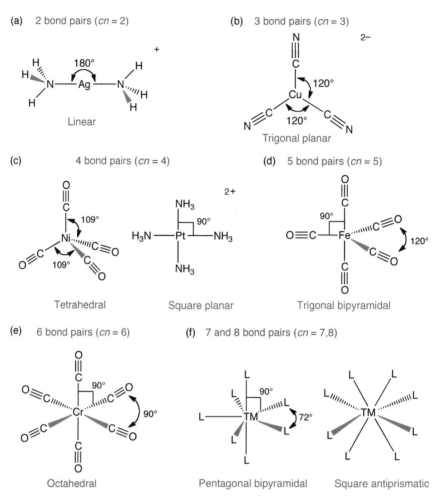

FIGURE 9.6 Common geometries and 3D shapes of transition metal (TM) complexes as a function of the number of bond pairs around the TM. This number is also called the coordination number (*cn*), which is shown within parentheses near each complex.

for us a focused worldview of patterns that help us find our way. We are not concerned when a rule is broken occasionally. On the contrary, we learn something new and further develop our worldview ... Thus, in the preceding example, the chemist can easily determine if a 3D shape is tetrahedral or planar. Imagine that you started with $Ni(CO)_4$ and you replaced two of the CO ligands by amine or phosphine ligands. If you did that, you would have found a single isomer. If, however, you started with $Pt(NH_3)_4^{2+}$ and you replaced two of the ammonia molecules by two Cl^- ligands, because the complex is planar, you would have ended up with two isomers, one where the two new ligands are in *cis* positions relative to each other, and the other when they are in *trans* positions relative to each other. When this is found, questioning what might be the difference between Pt and Ni will broaden our outlook on the 3D rules ...

Figure 9.6d shows that the 3D structure for a complex with $cn = 5$ is trigonal bipyramidal, as in $Fe(CO)_5$. Similarly, with $cn = 6$ in Figure 9.6e, the structure is octahedral. Try to think about isomers that arise by replacing ligands in these two 3D complexes.

Finally, in Figure 9.6f, we show complexes with $cn = 7$ and 8 using generic labels TM and L. The most common geometries are the pentagonal bipyramid for $cn = 7$ and the square antiprism for $cn = 8$. However, with these high coordination numbers, the 3D structure is very sensitive to the ligand size, and there are other variations. For example, $Y(CO)_8^+$ in Scheme 9.6c has an antiprismatic structure, but in $Sc(CO)_7^+$, the seventh ligand caps the octahedral face of three ligands.

9.6 BRIDGING TRANSITION METAL AND ORGANIC MOLECULES: BONDING CAPABILITIES OF FRAGMENTS OF TRANSITION METAL COMPLEXES

We already saw in Scheme 9.5b two complexes which behaved like $CH_3\bullet$ fragments and formed in Scheme 9.5c molecules with TM–TM bonds. One can therefore make an analogy between organic and TM complex fragments and create a bridge between these two branches of chemistry. This bridge was created by Roald Hoffmann (Figure 9.7), who talked about it in his Nobel Lecture.[9] Figure 9.7 shows the analogy

Roald Hoffmann

FIGURE 9.7 Roald Hoffmann near his analogy between CH_n ($n = 1–3$) fragments and TM complex fragments. The analogy is represented using a double-headed arrow with a teardrop attached on the bottom. The connectivity of the fragments is indicated. The photograph is used courtesy of Roald Hoffmann.

between various TM complex fragments and our well-known modular fragments of CH_4.

Originally,[9] this analogy was cast in QM terms (and was called the *isolobal analogy*), where the little teardrop in the middle of the double-headed arrow symbolized a lobe of these little density clouds that are called orbitals (see Retouches sections 7.R.2 and 7.R.3 in Lecture 7). Let me cast this analogy here in terms of the electron count on the respective fragments. Thus, H_3C has 7e around carbon, which misses a single electron to reach Nirvana, and hence it is a single connector, $H_3C\bullet$. The analogous TM complex fragment $(CO)_5Mn$ has 7e in the Mn valence shell and five CO ligands, which contribute each 2e to make five Mn—CO bonds. In sum, therefore, $(CO)_5Mn\bullet$ possesses 17e around manganese and is missing one more electron to reach Nirvana, and hence by analogy to the organic $H_3C\bullet$ fragment, the $(CO)_5Mn\bullet$ fragment will behave as a single connector. Similarly, $(CO)_4Fe$ possesses 16e around iron and is missing 2e to reach Nirvana, much as its analogous $H_2C\colon$ fragment, and since the latter is a double-connector fragment, so will $(CO)_4Fe\colon$ be. Finally, $(CO)_3Co$ possesses 15e around cobalt, and much like its analogous HC fragment, it will behave as a three-fold connector. There are many analogies one could make, but the ones in Figure 9.7 give a glimpse of the huge importance of the isolobal analogy.[9]

Scheme 9.7 shows some of the applications of this concept by constructing new molecules made from the organic and TM complex fragments. Thus, taking the two single-connector fragments $(CO)_5Mn\bullet$ and $\bullet CH_3$, we can form a new Mn—C bond and the new complex, $(CO)_5Mn—CH_3$, shown in Scheme 9.7a. Similarly, using the double connectors $(CO)_4Fe\colon$ and $\colon CH_2$, we can form a new complex with a double

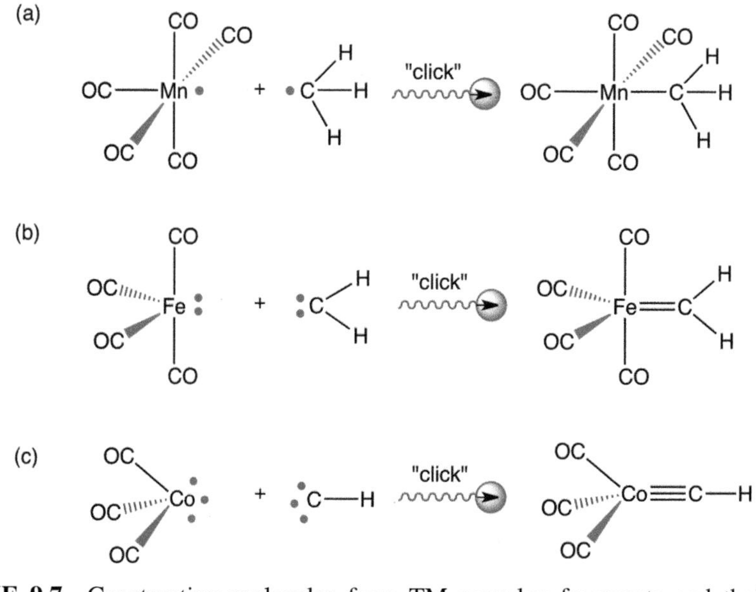

SCHEME 9.7   Constructing molecules from TM complex fragments and their organic analogs: (a) formation of $(CO)_5Mn—CH_3$ from the two single-connector fragments, (b) formation of $(CO)_4Fe=CH_2$ from the two double-connector fragments, and (c) formation of $(CO)_3Co\equiv CH$ from the two three-fold connector fragments.

bond $(CO)_4Fe=CH_2$, as shown in Scheme 9.7b. These are very useful complexes in chemical synthesis. They are called carbene complexes since the CH_2 species is called a carbene. These complexes are used, for example, to transfer the carbene fragment to molecules with double bonds to make three-membered rings. Finally, in Scheme 9.7c, we use the two three-fold connector fragments $(CO)_3Co$ and HC and "click," we make the $(CO)_3Co\equiv CH$ complex, which is an analog of acetylene, $HC\equiv CH$ and is called a carbyne complex. Again, this complex is also useful in chemical synthesis since it can transfer the carbyne HC to a variety of molecules. As free CH_2 and CH fragments are immensely reactive fragments, they cannot be handled as such, and the complexes in Schemes 9.7b and 9.7c constitute a means of taming these fragments and transferring them to other molecules in a facile manner.

To further demonstrate the potential of the isolobal analogy, let us generate TM cluster complexes from two of the fragments we generated in Figure 9.7. These are shown in Scheme 9.8. Part (a) shows how we can generate a four-membered ring from two double-connector $(CO)_4Fe$: fragments and two double-connector O: fragments. Click, and they form the four-membered ring made of two Fe and two O atoms. This molecule mimics enzymes which use the Fe_2O_2 "diamond core" to perform C–H bond activation reactions of fatty acids. Scheme 9.8b shows how four three-fold connector $(CO)_3Co$ fragments click to form a beautiful cage cluster of

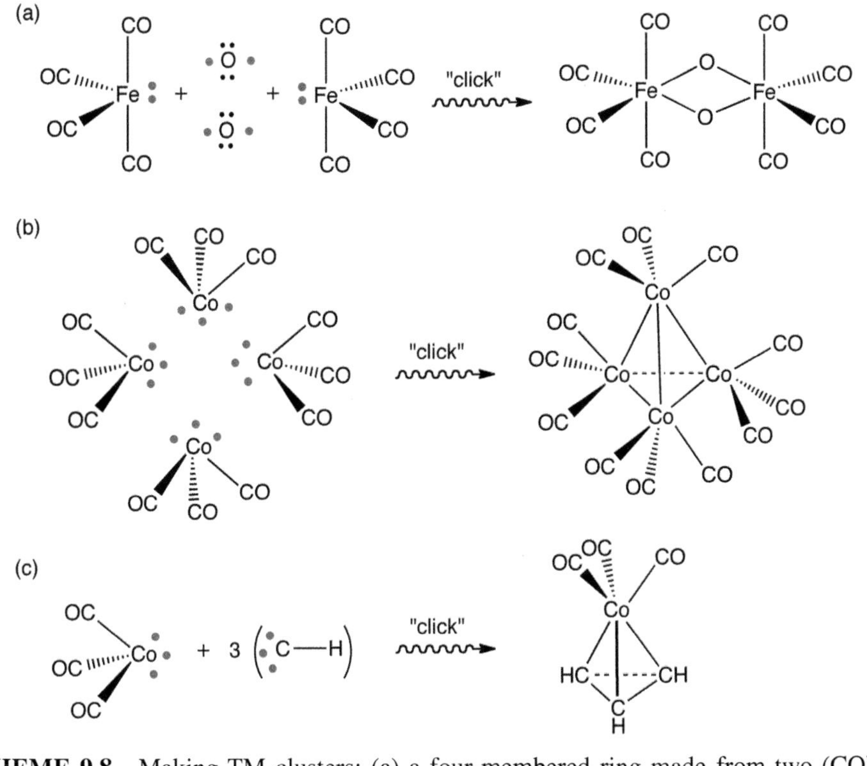

SCHEME 9.8   Making TM clusters: (a) a four-membered ring made from two $(CO)_4Fe$ fragments and two O fragments, (b) a cage made from four $(CO)_3Co$ fragments, and (c) a mixed cage made from a $(CO)_3Co$ fragment and three CH fragments.

four cobalt atoms. In Scheme 9.8c, we show how rich this game can be by taking a mixture of three-fold connector fragments, one $(CO)_3'Co$ and three HC fragments, which form a mixed carbon–cobalt cage.

9.7   SUMMARY OF TRANSITION METAL COMPLEXES

In this part of Lecture 9, we learned about some features of transition metal complexes, which form two important sub-branches of chemistry called inorganic and organometallic chemistry. These sub-branches include in fact also all the bioinorganic chemistry of metalloenzymes. We learned a few main principles:

☞ The state of Nirvana of transition metals (TMs) involves 18 electrons (18e). A TM with a given number of electrons in its valence shell (indicated by the respective family number in the periodic table) will form electron pair bonds with ligands (L), which are molecules or fragments that can donate electron pairs to the electron-deficient TM. The 18e rule limits the number of ligands that a given TM can bind.

☞ Complexes with fewer than 18e exist and generally serve as catalysts of chemical reactions, like the Ziegler–Natta-type and Heck catalysts.

☞ The number of ligands surrounding a given TM is the coordination number (cn). The 3D shape of the complex is generally determined only by its cn.

☞ The isolobal analogy creates a bridge between organic chemistry and transition metal chemistry. Using the analogy demonstrates once again that chemistry is the poetry of Mankind in matter!

9.8   SPECTROSCOPY OR HOW DO CHEMISTS "LISTEN TO MOLECULES"?

We discussed amply the notion of bonds in a molecule. How do we know that a molecule has indeed specific bonds like C–H, C=C, C=O, or Fe–C? We also argued with certitude that molecules have specific geometries or 3D shapes with specific bond angles. Can we actually "see" these 3D shapes? How can we tell the actual values of bond angles and bond lengths in a molecule? What about the total mass of the molecule? Can we actually "weigh" molecules? The answer to all these questions and many others is, "yes, we can" by means of spectroscopic techniques. Spectroscopy in a broad sense encompasses the interaction of matter (molecules) with radiation and other energy sources (e.g., bombardment by electrons, neutrons, etc.).

9.8.1   The Electromagnetic Radiation Spectrum

The space of our universe is not empty. It is laden with radiation that comes from the sun and from other luminous stars, as well as from general sources, called cosmic

THE ELECTROMAGNETIC SPECTRUM

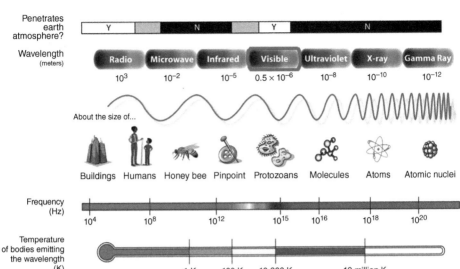

FIGURE 9.8   The spectrum of electromagnetic radiation: the various scales classify the waves in terms of wavelength (meters), frequency (Hz = oscillations per second), and temperature (in units of Kelvin, K) of the radiating bodies. Sizes of various objects corresponding to different wavelengths are noted for reference. This image was created by NASA (it is in the public domain and was obtained from http://commons.wikimedia.org/wiki/File:EM_Spectrum3-new.jpg).

rays. This radiation is composed of electromagnetic waves. This means that the waves involve both electric and magnetic fields, which oscillate in space and move at the speed of light.

As shown in Figure 9.8, these waves span a wide range (called spectrum) of radiation, going all the way from radio waves at one end to gamma (γ) rays at the other end, and through infrared (IR) rays, visible (VIS) rays, and X-ray waves in between. Near the spectrum, there are scales of units that measure the wavelength and frequency of the wave types. The wavelength is the distance between two oscillations, which is indicated here in units of meters, where each meter is 10 billion (10^{10}) Å. The other scales are the frequency of the wave, where the frequency is the number of oscillations per second the wave performs, and the temperature of the radiating bodies. *The energy of a wave is proportional to its frequency.* As we mentioned in the conversation on contents above, light is made of discrete bundles of energy called photons, whose energy is proportional to the frequency of the light (see Retouches section 9.R.5). Photons of long wavelength waves, like radio waves, have low frequencies and small energies, while X-ray photons have very short wavelengths and are hugely energetic.

As can be seen from the figure, the wavelengths and frequencies span more than 15 orders of magnitude. In the middle of the spectrum, we find VIS light, which is called visible because this is the range wherein our eyes "see" the world. Looking at the scale, one can see that VIS light spans a very narrow range in terms of wavelengths or frequencies. In fact, the VIS range constitutes a tiny fraction of the entire spectrum, 0.00000000000001 or even less. To appreciate how small is this fraction, imagine yourself standing in a room, which is 100 m long, 100 m wide, and 100 m high. Feel the immense space of the room for a few seconds. Then further imagine that at the very top of the room there is a tiny little window, even less than 0.001 m (1 mm) in size, through which you can get a glimpse of the outside world. This is approximately the slit through which we see our world. As much as this may be amazing, it is also a lesson for us to accept how limited our vision of "reality" is.

Why do we see through this narrow slit? Let me recall that the energy levels of molecules are discrete, and they can absorb light energy only in those frequencies where *the energy of the light photons matches the differences between the molecular energy levels.* Thus, "seeing" is achieved by absorption of visible light that precisely matches the energy levels *of our vision pigment,* the retinal molecule (which is connected to the receptor protein called opsin, a connection which is drawn in full in Scheme 1.6). And this is how the mechanism of "seeing" operates:

As Figure 9.9 shows, in the resting state of the retinal, which is the "off" position, the double bond of interest in the molecule has a *cis* geometry. The absorption of visible light provides precisely the energy needed to break one of the components of the *cis* double bond. This bond breakage allows the molecule to twist and be converted to the *trans* form. The *trans* isomer is the "on" signal; this signal is transmitted to the

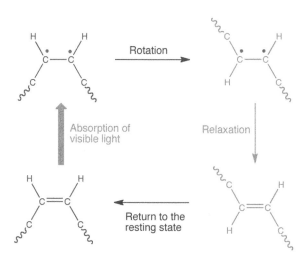

FIGURE 9.9   A schematic description of the mechanism of the conversion of *cis-* to *trans-* retinal caused by absorption of visible light. The wiggly lines pointing away from the C atoms indicate the rest of the molecule. The absorption of light disrupts the electron pairing in the *cis* double bond, and the resulting diradical species undergoes free rotation to the *trans* geometry and then relaxes electronically to create *trans*-retinal. This switching causes complex neuronal signaling, which is translated into seeing. The *trans* isomer then returns slowly back to the *cis* resting state in a multi-step process.

opsin protein, which then triggers neuronal signaling (ion flow), which is eventually translated to "seeing." Retinal certainly absorbs radiation from other regions, such as the IR region, but it is the VIS region that causes this *cis* → *trans* isomerization that leads to "seeing." Thus, Nature has cleverly used retinal as a *photoswitch* that allows us to get a glimpse of the "visible" world. This choice may be associated with the fact that, out of the entire spectrum of radiation, what reaches the Earth is primarily visible light. As such, natural selection found a molecule that is sensitive to this range of radiation and can be used a molecular switch for vision.

9.8.2 Energy Levels of Molecules as the Basis of Spectroscopy

In a nutshell, spectroscopy probes the molecular energy levels that are the typical and characteristic fingerprint of any molecule. As such, the molecular spectroscopist "listens" to the molecule and is able to translate the slits in which it absorbs light into structural and bonding features. Thus, molecules are made of a collection of electrons and nuclei, and all these particles can change their states; the electrons can change their modes of pairing, the bonds can vibrate (stretching and shrinking; angles can vibrate by opening and closing), the spins of the nuclei can flip directions, and so on and so forth. All these changes require some energy to be absorbed by the molecule.

As we mentioned already, *all the molecular energy levels* are discrete and have gaps of magnitudes that are characteristic of the molecule. As depicted in Scheme 9.9a,

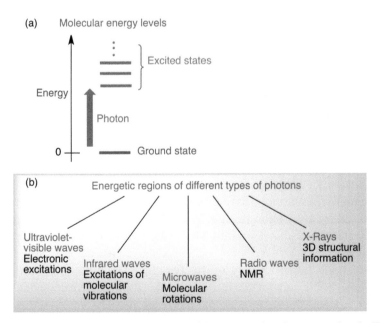

SCHEME 9.9 (a) Schematic representation of discrete molecular energy levels. The lowest level is called the ground state, and all the higher ones are called excited states. The transition from the ground state to an excited state is induced by absorption of a photon whose energy equals the energy gap between the two states. (b) The energy regions of various photons and the types of spectroscopy and molecular events to which they correspond.

the state of the lowest energy is called the *ground state*, whereas all the states of higher energy are called *excited states*. These excited states are accessed when the molecule absorbs a photon of light that possess the same energy as the gap between the ground state and one of the excited states, and this idea applies to all the types of changes we mentioned above, including electronic changes, vibrations, nuclear spin flips, etc.

Scheme 9.9b specifies spectroscopic methods that are associated with the various regions of the electromagnetic spectrum. Thus, photons in the VIS or UV regions send the molecule to excited states in which bond pairs are broken, and the electron pairs undergo rearrangement to form arrangements that are less energetically favorable than that of the ground state (e.g., the two unpaired electrons in the excited state of retinal in Figure 9.9).

Different molecules absorb in different regions of UV-VIS light, and the frequencies at which a molecule absorbs are its fingerprint. When a chemist makes a new molecule, she/he will run UV-VIS spectroscopy, and the chemist can immediately tell whether, for example, the molecule has isolated double bonds, or ones that are separated by only one single bond between them, or others that are arranged in a cycle like in benzene, etc. There are molecules that absorb in the VIS region, and they are colored, and their different colors mean that they absorb in different regions of visible light. TM complexes are usually richly colored, but there are other colored molecules, such as various organic dyes. Look around you and you will see colors; some are natural colors, and others are man-made. Color makes our world beautiful, and molecules that absorb the visible light create this beauty.

So far, we constructed molecules and reasoned their 3D structures in such a fashion that some of you may get the impression that these objects are motionless. The fact is that molecules are very much restless and perform many motions all at the same time. Scheme 9.10 shows the motions that a small molecule like water performs all at once. The molecule vibrates in three modes (Scheme 9.10a); one is a symmetric stretch where the two bonds stretch and shrink in the same direction, and the other is an antisymmetric stretch wherein one bond stretches while the other is shrinking, and finally the two O—H bonds perform a scissoring motion, closing and opening the HOH angle. Additionally, the molecule as a whole also performs rotations around three different axes.

All these molecular motions have discrete energy levels (like in Scheme 9.9a) that can be detected by absorption of light in different regions. Thus, absorption in the IR region excites molecular vibrations, while microwaves induce transitions between rotational modes.

IR spectroscopy is particularly useful for chemists, since it is sensitive to the strength of the bond and to the masses of the atoms performing the vibrations. As shown in Scheme 9.10c, in a simple molecule like $H_2C=CH-CH_2-HC=O$, the vibrations of the C—H, C=O, C=C, and C—C bonds absorb at distinctly different IR frequencies. Through years of practice, chemists can look at such an IR spectrum *and immediately identify the type of bonds the molecule has*.

Since the IR frequency is sensitive to the masses of the vibrating atoms, the chemist can further ascertain her/his assignment of, say, a C—H bond by substituting the H

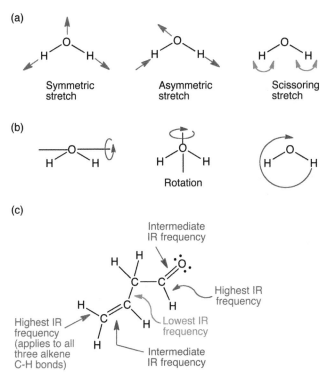

SCHEME 9.10 Molecular motions of H_2O and an aldehyde: (a) vibrational motions of H_2O, where arrows indicate the directions in which the atoms move, and (b) rotational motions of the whole molecule around three axes (the third axis goes through the plane of the page, pointing out of the plane). (c) Ranking of the expected infrared frequencies of the various bonds in the $H_2C=CH-H_2C-HC=O$ molecule.

with the heavier isotope D. This will immediately lower the absorbed IR frequency by a significant amount, which can be approximated from the square root of the mass ratio D/H (a factor of the square root of 2). This is a powerful form of spectroscopy for routinely detecting the active species in metalloproteins, where one has, for example, an iron ion bound to different ligands. Spectroscopy can even detect the presence of weak H-bonds.

Turning back to Scheme 9.9b, one can see another form of spectroscopy called NMR, namely nuclear magnetic resonance. In this form of spectroscopy, a molecule in a magnetic field experiences energy gaps between the different directions of the spins of its nuclei. The method is applicable to other nuclei as well, but it is most effective in H NMR (H has a single proton). The size of the energy gap in the magnetic field depends on the chemical environment of the nucleus, and hence different nuclei flip their spins at different radio wave frequencies. Again, after many years of practice, most chemists can look at an NMR spectrum and assign the proton as belonging to a CH_2 or CH_3 group, or to a CH in a benzene ring, and so on. The method today is so powerful that it can even reveal distances among various atoms in a molecule.

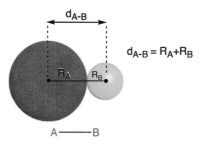

FIGURE 9.10   Viewing the bond distance (d) between fragments (atoms) as the sum of the corresponding radii (R) of the touching atomic spheres.

9.8.3   X-Ray Crystallography and 3D Molecular Information

There are many other spectroscopic techniques, but let me mention just one more, the X-ray crystallography method, which yield complete 3D information on a molecule (Scheme 9.9b). The technique is based on the phenomenon that when X-rays hit a *crystal*, which is an ordered solid form made of many molecules, the material diffracts the X-rays in patterns that reflect the distances between the atoms in the molecule. As such, the X-ray diffraction pattern can be analyzed and can provide us with the bond distances and the angles in a molecule. The rather complex statistical mathematical analysis was developed by Herbert Aaron Hauptman and Jerome Karle, who were awarded the Nobel Prize for their work in 1985.[10] In this manner, chemists, called X-ray crystallographers, are able today to determine the structure of any molecule that can be crystallized as an ordered solid (or half-ordered, as in the case of Dan Shechtman, who won the Nobel Prize in 2011 using X-ray crystallography and a related technique called transmission electron microscopy). This powerful method resulted in close to 30 Nobel Prizes.

The many solved structures provided a lot of information that eventually enabled chemists to define consistent sets of atomic and ionic radii, which are useful for an approximate estimate of bond distances. The idea behind this is illustrated in Figure 9.10, where the bond distance A–B is considered to be the sum of the radii of the two constituent fragments A and B, which are represented as spheres that touch one another. So, if we know the radii of bonded atoms or ions, we can approximately figure out before doing any experiment the bond length in any molecule.

This is a straightforward LEGO principle of combining atomic/ionic radii to construct the dimensions of the bonds in a molecule. Tables 9.1 and 9.2 list covalent and ionic radii of main group elements, and they allow us to highlight the power of the idea in Figure 9.10 with two applications.

Inspection of the tables shows the following trends:

☎   Atoms and ions (of a given charge) shrink as we move from left to right in a period (e.g., compare R_C to R_F in Table 9.1), and expand going down a given family (e.g., compare R_F^- to R_{Cl}^- in Table 9.2).

☎ The covalent radius depends on the multiplicity of the bond, such that the radius for a double bond is smaller than that for a single bond, etc.

☎ Anions are larger than cations, and the higher the positive charge on a cation, the smaller it is.

All these trends are extremely useful for chemists, but we cannot discuss here all this information. What we can do, however, is to demonstrate that the radii are useful for us to derive the complete 3D information for a given molecule, using the examples in Figures 9.11 and 9.12.

Figure 9.11 shows three molecules with 3D and bond length data. Thus, for CH_4 in Figure 9.11a, based on VSEPR rules, we can predict that the molecule will have a tetrahedral structure, with bond angles of about 109°. Using the sum of the covalent

TABLE 9.1 Covalent Radii^a of Some Main Group Elements in Å Units.

Atom	Single bonds	Double bonds	Triple bonds
H	0.30		
C	0.77	0.67	0.62
N	0.70	0.62	0.55
O	0.67	0.55	
P	1.10	1.00	
S	1.00	0.95	
F	0.65		
Cl	0.99		
Br	1.15		
I	1.35		

^{a}Note these are averaged values, and as such, there is some variation in these radii in different sources. Despite the lack of canonical values, these radii serve as useful general guidelines for gaining insight into molecular dimensions.

TABLE 9.2 Ionic Radii^a of Some Ions of Main Group Elements in Å Units.

		Li^+	Be^{2+}	
		0.68	0.30	
O^{2-}	F^-	Na^+	Mg^{2+}	Al^{3+}
1.40	1.33	0.98	0.65	0.45
S^{2-}	Cl^-	K^+	Ca^{2+}	Ga^{3+}
1.84	1.81	1.33	0.94	0.60
Se^{2-}	Br^-	Rb^+	Sr^{2+}	In^{3+}
2.02	1.96	1.48	1.10	0.81
Te^{2-}	I^-	Cs^+	Ba^{2+}	Tl^{3+}
2.22	2.19	1.67	1.35	0.91

^{a}The radius is indicated underneath each corresponding ion. Note these are averaged parameters, and as such, there is some variation in these radii in different sources. Despite the lack of canonical values, these radii serve as useful general guidelines for gaining insight into molecular dimensions.

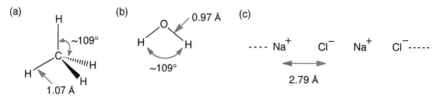

FIGURE 9.11 Complete structural predictions for three molecular systems: (a) CH$_4$, (b) H$_2$O, and (c) NaCl depicted as a periodic structure with a repeated Na—Cl distance common to all dimensions (only one dimension shown).

radii for C and H, we can also now predict the C—H bond length as being 1.07 Å. Similarly, for water, we predict a planar molecule with a bond angle of 109° and O—H bond lengths of 0.97 Å. Finally for NaCl, we know it is ionic and it has a periodic 3D structure with close packing of Na$^+$ and Cl$^-$ ions. Now, summing the respective ionic radii in Table 9.2, we can add that the distances between the oppositely charged ions are 2.79 Å.

Figure 9.12 shows three molecules made from C and H only. In part (a), we see ethane, and we know from Lecture 7 how to determine its bond angles and even its preferred conformation, but now we can add the C—C and C—H bond lengths, by summing the respective covalent radii. In parts (b) and (c), we show ethylene and acetylene. By comparison to ethane, we can see that as the bond multiplicity increases from a single to a triple bond, the C—C bond length decreases from 1.54 Å to 1.24 Å. The bond shortening is accompanied by bond strengthening, and the triple C—C bond of acetylene requires much more energy to split compared with the double bond of ethylene and the single bond of ethane.

The covalent and ionic radii of transition metals (TM) have also been determined, and we show some of them in Appendix 9.A.1. These data allow you to construct the complete 3D structure of TM complexes. It is important to remember that all the radii are averaged over many bonds, and therefore, we cannot expect that our predictions will be perfectly accurate for every molecule, ionic material, or TM complex we try to construct with Tables 9.1, 9.2 and 9.A.1. Such usage will provide, however, a very helpful way to imagine the structure of matter and acquire a more thorough

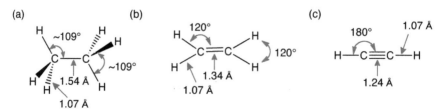

FIGURE 9.12 Complete predicted 3D information for: (a) ethane, (b) ethylene, and (c) acetylene. The angles in (b) are idealized, whereas, in practice, the H—C—H angle is 116.6°, and the H—C—C angle is 121.7°. All C—H bonds are single bonds, and hence, we use the corresponding radii from Table 9.1.

understanding. Other resources of covalent and ionic radii values are available upon request from the author.

Clearly, our game of pair bonding has become now quite refined. What is missing is a demonstration that the total bonding energy of a molecule is also given as a combination of individual bond energies. This material is placed in Appendix 9.A.2, along with a few examples which demonstrate that an entire molecular outlook can be based on the LEGO principle.

9.9   SUMMARY OF SPECTROSCOPIC METHODS

In summary, spectroscopy uses light and other radiation sources to probe the molecular energy levels that are typical and characteristic of the examined molecule.

☞ As such, spectroscopy enables the chemist to translate information from molecular light absorption and/or emission into structural and bonding features.

☞ X-ray crystallography allows the chemist to get complete 3D information about any desired molecule.

☞ There are plenty of spectroscopic methods that we did not discuss. Mass spectrometry bombards a molecule with electrons or photons that create a molecular cation (by ejecting an electron from the molecule), and by rushing this cation through combined electric and magnetic fields, a mass spectrometrist measures the molecular mass.

☞ Spectroscopy is constantly becoming more and more refined, and some techniques like atomic force microscopy enable the chemist to "see bonds" and to follow their rupturing and forming during a chemical reaction.[11]

In a nutshell, spectroscopy enables the chemist to "listen" to the molecule telling its innermost secrets.

9.10   REFERENCES AND NOTES

[1] M. McDonald and L. B. Hunt, *A History of Platinum and Its Allied Metals*, Johnson Matthey, London, 1982.

[2] E. C. Constable and C. E. Housecroft, *Chem. Soc. Rev.* **2013**, *42*, 1429.

[3] Read the fascinating story of Döbereiner's Lighter and the term catalysis in: R. Hoffmann, *American Scientist*, **1998**, *86*, 326.

[4] N. C. Thomas, *J. Chem. Ed.* **1991**, *68*, 631.

[5] V. Thomsen, *Spectroscopy*, **2006**, *21*, 32.

[6] C. A. Tolman, *Chem. Soc. Rev.* **1972**, *1*, 337.

[7] A. D. Brathwaite, J. A. Maner, and M. A. Duncan, *Inorg. Chem*, **2014**, *53*, 1166. For further discussion of another view concerning the possibility of $Sc(CO)^{8+}$, see: X. Xing,

J. Wang, H. Xie, Z. Liu, Z. Qin, L. Zhao, and Z. Tang, *Rapid Commun. Mass Spectrom.*, **2013**, *27*, 1403–9.

[8]   S. Shaik, S. Cohen, Y. Wang, H. Chen, D. Kumar, and W. Thiel, *Chem. Rev.* **2010**, *110*, 949.

[9]   R. Hoffmann, *Angew. Chem. Int. Ed. Engl.* **1982**, *21*, 711.

[10] E. Garfield, *Current Contents*, **1986**, *44*, 3.

[11] D. G. de Oteyza, P. Gorman, Y.-C. Chen, S. Wickenburg, A. Riss, D. J. Mowbray, G. Etkin, Z. Pendramrazi, H.-Z. Tsai, A. Rubio, M. F. Crommie, and F. R. Fischer, *Science*, **2013**, *340*, 1434.

9.A   APPENDIX

9.A.1   Radii Values for Transition Metals in Covalent and Ionic Bonds

Our click TM-L bonding model basically regards most of these bonds as covalent. There are, however, cases where L is highly electronegative, such as, L = Cl, and where we might suspect that the TM-Cl bond is more ionic, given the large electronegativity difference between the metal and the ligand. This is the reason we provide both types of radii in Table 9.A.1. Note that the ionic radii can sometimes be under 50% of the covalent radii: in an ionic bond, the greedy anion hogs the electrons, leaving the cation quite deprived and making it quite small. The covalent metal radii should only be used with the covalent ligand radii, and the ionic metal radii should only be used with the ionic ligand radii. When estimating TM-L bond lengths, one should try to use the two types of radii pairs and then compare the results to experimental bond lengths to better understand where a bond falls on the covalent–ionic spectrum. This can be a great exercise.

9.A.2   Bond Dissociation Energies and Their Usage as Building Blocks

We explained in the Retouches section of Lecture 8 the meaning of the energy unit 1 kcal/mol that chemists use for gauging bond dissociation energies (BDEs). The BDE is the energy required to break the bond to form its constituent fragments. For example, the BDE for the C–H bond is the energy invested to perform a reaction such as breaking the C–H bond in CH_4 to form $H_3C^\bullet + H^\bullet$, or breaking a similar bond in ethane, etc.

Table 9.A.2 provides the average BDE values for generic bonds. But since the spread of the specific values (e.g., C–H in methane vs. ethane, and so on) around this average is not too large, we may use these generic values as fixed estimates for a given bond type. Again, do not expect perfectly accurate results. Look for trends.

To demonstrate the usage of the table, let us consider the total bonding energies of two isomers, $H_3C–CH_2–O–H$ and $H_3C–O–CH_3$. We can immediately write this total bonding energy as a sum of BDE values for the respective isomers:

$$5BDE_{C-H} + BDE_{C-O} + BDE_{O-H} + BDE_{C-C} \text{ for } H_3C-CH_2-O-H$$

TABLE 9.A.1   Radii Values[a] **(in Å Units) for Transition Metals in Covalent Bonds (Part I)**[b] **and Ionic Bonds (Part II).**[c]

Part I

Sc	Ti	V	Cr	Mn	Fe	Co	Ni	Cu	Zn
1.70	1.60	1.53	1.39	1.50	1.42	1.38	1.24	1.32	1.22
Y	Zr	Nb	Mo	Tc	Ru	Rh	Pd	Ag	Cd
1.90	1.75	1.64	1.54	1.47	1.46	1.42	1.39	1.45	1.44
–	Hf	Ta	W	Re	Os	Ir	Pt	Au	Hg
	1.75	1.70	1.62	1.51	1.44	1.41	1.36	1.36	1.32

Part II

Sc^{3+}	Ti^{3+}	V^{2+}	Cr^{2+}	Mn^{2+}	Fe^{2+}	Co^{2+}	Ni^{2+}	Cu^{2+}	Zn^{2+}
0.745	0.67	0.79	0.80	0.83	0.78	0.745	0.69	0.73	0.74
Y^{3+}	Zr^{4+}	Nb^{3+}	Mo^{4+}	Tc^{4+}	Ru^{3+}	Rh^{3+}	Pd^{2+}	Ag^{+}	Cd^{2+}
0.90	0.72	0.72	0.65	0.645	0.68	0.665	0.86	1.15	0.95
–	Hf^{4+}	Ta^{3+}	W^{4+}	Re^{4+}	Os^{4+}	Ir^{3+}	Pt^{2+}	Au^{+}	Hg^{2+}
	0.71	0.72	0.66	0.63	0.63	0.68	0.80	1.37	1.02

[a]Note these are averaged values, and as such, there is some variation in these radii in different sources. Despite the lack of canonical values, these radii serve as useful general guidelines for gaining insight into molecular dimensions.

[b]Radius data from B. Cordero, V. Gómez, A. Platero-Prats, M. Revés, J. Echeverría, E. Cremades, F. Barragán, and S. Alvarez, *Dalton Trans.* **2008**, 2832. The values represent average covalent radii for elements, averaged across different effective charges for each atom (when applicable, we also averaged values for different spin states: see Section 9.R.2). These radii should be combined with covalent radii for main group elements to predict bond distances.

[c]Ionic radius data from R. D. Shannon, *Acta Crystallogr.* **1976**, *A32*, 751. Note how the ionic radii are smaller than the covalent radii: the ionic radii are most suitable for bonds where the electronegativity difference between the constituent atoms is large (see Scheme 8.1). These radii should be combined with ionic radii for main group elements to predict bond distances.

TABLE 9.A.2   Average Bond Dissociation Energies (in kcal/mol Units) for Selected Bonds.[a,b]

Type of bond	BDE (kcal/mol)	Type of bond	BDE (kcal/mol)
H—H	104.2	C=C	147
H—C	98.7	O—O	33
H—N	93.5	O=O	116
H—O	111	C—O	84
C—C	83	C=O (CO_2)	191

[a]Some values were taken from R. T. Sanderson, *Polar Covalence*, Academic Press, New York, 1983, and R. T. Sanderson, *Chemical Bonds and Bond Energy*, 2nd Ed., Academic Press, New York, 1976, and others represent the author's consensus values.

[b]Note these are averaged values, and, as such, there is some variation in these bond energies in different sources. Despite the lack of canonical parameters, these values serve as useful general guidelines for gaining insight into the energies of chemical reactions.

and

$$6BDE_{C-H} + 2BDE_{C-O} \text{ for } H_3C-O-CH_3.$$

This application demonstrates the principle: we are treating the bonding energy of the molecule as a LEGO of the individual BDEs of its respective bonds. Furthermore, inspection of the table shows that:

$$E_{bonding}(H_3C-CH_2-O-H) = 771.5 \text{ kcal/mol}$$

versus

$$E_{bonding}(H_3C-O-CH_3) = 760.2 \text{ kcal/mol}$$

Thus, the table predicts that the alcohol is more strongly bonded than its ether isomer. If we establish equilibrium between the two isomers, we shall end up finding that the alcohol is the major component of the mixture. We can also use the table to classify reactions as being exothermic or endothermic using the very same LEGO principle of BDEs.

9.A.3 Proposed Demonstrations for Lecture 9

• Colorful complexes of nickel

This demonstration shows the formation of colorful complexes of nickel by adding solutions of different ligands to solutions of nickel sulfate. The ligands bind to the Ni^{2+} ion and form the corresponding complexes. The formation of the complexes is accompanied by color changes. The observed colors are discussed in terms of the strengths of the Ni^{2+}–ligand bonds, which also affect the relative stabilities of the complexes.

The demonstration[1A] begins by dissolving nickel sulfate ($NiSO_4$) in water, which results in a green solution. Then, an ammonium hydroxide solution is added, and the color becomes blue. Adding ethylenediamine (en) changes the color to purple, which in turn changes to red when a solution of dimethylglyoxime (dmg) in ethanol is added. Finally, after adding a solution of potassium cyanide, the solution turns yellow.

The observed colors and complexes are written on the board, and the students are introduced to the concept of complementary colors, where the color seen by the human eye has a complementary relationship with the color absorbed by the complex. The color absorbed by the complex reflects the bonding strength of the ligands. The stronger the bonding, the shorter the wavelength of the color absorbed (and the higher the bonding energy). Thus, by using the color wheel of complementary

TABLE 9.A.3 Colors of Flames Produced by Various Solutions of Ionic Solids.

Metal salt	Color
LiCl	Pink
NaCl	Yellow/Orange
KCl	Purple
$CaCl_2$	Red-orange
$SrCl_2$	Red
$CuCl_2$	Green
$BaCl_2$	Green-yellow

colors, the relative strengths of the ligand–Ni bonding can be gauged to change in the following order:

$$H_2O < NH_3 < en < dmg < CN^-.$$

It is seen that CN^- forms the strongest bond, and H_2O forms the weakest one. This order also reflects the relative stability of the complexes these ligands form with the nickel ion.

- **Flame colors induced by different metals—spectroscopy with the naked eye**

This is a spectacular experiment that demonstrates the color of flames emitted from various metal atoms, which are the same colors made by sodium street lamps and many fireworks. The colors of flames are characteristic of the metal (or metal ion) and can be used to identify the metal and/or detect its presence in the solution. This constitutes spectroscopy with the naked eye.

In the experiment, solutions of various metal salts in methanol (15–20% by mass; Table 9.A.3 lists the salts[2A]) are placed within aluminum tea-light candle holders,[3A] the classroom lights are turned down, and the demonstrator ignites the solutions. A set of colorful flames is observed (Table 9.A.3), where each metal solution produces a distinctive color. While the students are busy observing the beautiful colors of the flames, the origin of the phenomenon is discussed. When atoms/ions are placed within the flame, they absorb energy that excites their electrons to higher energy levels (refer back to the text of the lecture on spectroscopy). As these electrons return to the lower energy levels, the atom/ion emits energy in the form of electromagnetic waves (light). Each atom emits a characteristic wavelength of light, and thus the light emitted serves as a "fingerprint" of the element. For some atoms, such as certain metals, these wavelengths are in the visible spectrum and are therefore observed as characteristic colors. Methanol can be flammable, so please take appropriate precautions.

- **The glowing pickle experiment**

Following the previous experiment, this is a fun, eye-catching demonstration where a home-pickled cucumber glows different colors by passing an electrical current through it.[4A] By using a specific electrical circuit apparatus described in Rizzo *et al.* (2005),[4A] both ends of a pickle are inserted into electrodes which are connected to a power supply. The power is turned on, and after a short induction period, the pickle starts to glow various colors for about a minute. The various colors reflect the way the pickled cucumber was prepared. If it was bought in the grocery store, it would glow a yellow-orange color. The demonstrator then asks the students: following the previous demo, what is it that makes the pickle glow this color? The answer is that the grocery store pickle was prepared in a concentrated sodium chloride solution, and Na^+ emits yellow light after absorbing a sufficient amount of energy. The literature source gives a recipe for pickling cucumbers in solutions of different salts (LiCl, KCl, $SrCl_2$, $BaCl_2$), which will produce different colors. The glow is a result of an electrical discharge from the electrode to the pickle, which is similar to the glow produced in a gas discharge lamp.

9.A.4 References for Appendix 9.A

[1A] B. Z. Shakhashiri, G. E. Dirreen, and F. Juergens, *J. Chem. Educ.* **1980**, *57*, 900.

[2A] G. M. McKelvy, *J. Chem. Educ.* **1998**, *75*, 55.

[3A] T. Mortier, A. Wellens, and M. J. Janssens, *J. Chem. Educ.* **2008**, *85*, 522.

[4A] M. M. Rizzo, T. A. Halmi, A. J. Jircitano, M. G. Kociolek, and J. A. Magraw, *J. Chem. Educ.* **2005**, *82*, 545.

9.R RETOUCHES

9.R.1 Why the 18e Rule, and Why Are Many TM Complexes Colored?

As we already mentioned in Retouches sections 6.R.1 and 7.R.1–7.R.3, electrons in atoms or molecules occupy orbitals. An orbital is a QM object that has *a specific energy level* and *a spatial shape* (for shapes, see: en.wikipedia.org/wiki/**Atomic_orbital**) that accounts for the probability of finding the electron in the space around the nucleus.[1R] As such, the electron will have the energy level of its orbital and will be distributed in space as a little cloud that has the orbital's shape. An orbital can house one or two electrons, but never more than two! Two electrons in an orbital must have opposite spins and are paired. A single electron can have either a spin pointing up or a spin pointing down.

Transition metals (TMs) possess nine orbitals in the valence shell, which are depicted in Scheme 9.R.1. Letters that describe the orbitals' shapes also coincidentally correspond to their spatial properties: s, spherical; p, polar (namely, pointed in a given direction); and d, dipolar (pointed in two different directions). The directed orbitals have also labels describing their directions in space: for example, p_x refers to a p orbital oriented along the *x* direction of the *x, y, z* coordinate system placed around

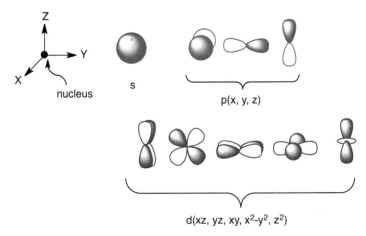

SCHEME 9.R.1   Atomic orbitals in the valence shell of a transition metal: the labels s, p, and d happen to denote the shape of the orbital (spherical, polar, and dipolar, respectively), while the subscripts (such as *x*, *y*, and *z*) indicate the orbital's directionality in the space around the nucleus.

the nucleus. The d orbitals have two directions: for example, d_{xy} refers to the orbital that lies in between the *x* and *y* directions, while d_{z^2} refers to the orbital that is mainly directed along the *z* axis.

With nine valence orbitals, the TM can accommodate up to 18e in the valence shell. After the TM makes bonds with ligands, new bond orbitals are created, while the five d-orbitals remain as the highest set of orbitals in the valence shell of the TM. This is where the electrons that do not participate in bonding end up residing. For example, in $Fe(CN)_6^{4-}$, there are 18 electrons: 12 are the Fe–CN bond pairs, and 6 remain on the Fe^{2+} in the d-orbitals.

Scheme 9.R.2 depicts the energy diagram of these orbitals for an octahedral complex like $Fe(CN)_6^{4-}$. The scheme shows that the ligands affect the d-orbitals and split them into two groups that differ in energy (actually, the ligand orbitals mix to

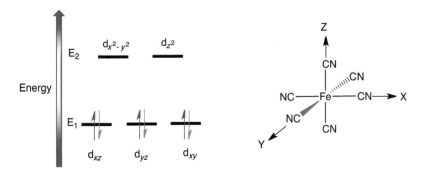

SCHEME 9.R.2   The d-orbital energy diagram for an octahedral complex, such as $Fe(CN)_6^{4-}$. The d-orbitals split into two energy levels, E_1 and E_2. In this particular complex, the Fe^{2+} ion has 6e, which occupy the three orbitals that are lowest in energy.

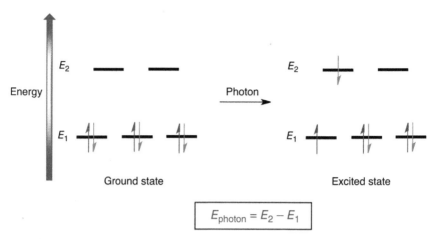

$$E_{photon} = E_2 - E_1$$

SCHEME 9.R.3   Changes in the d-orbital occupancy of a complex as a result of absorbing a photon whose energy E_{photon} equals the energy difference between the orbital energy levels, $E_2 - E_1$. Absorption of the photon causes one electron to jump to a higher d-orbital and gives rise to an excited state.

differing extents into the d-orbitals and thereby affect their energies). The orbitals d_{z^2} and $d_{x^2-y^2}$ that lie on the axes of the iron–ligand bonds are affected more, and their energy (E_2) gets higher than that (E_1) of the three orbitals that reside in between the axes. The scheme further shows that the 6e that remain on Fe^{2+} occupy in pairs the three lowest d-orbitals. As is commonly done, a single-headed arrow represents each electron.

As Scheme 9.R.3 shows, when our TM complex absorbs a photon of light that has precisely the energy difference of the two levels, E_1-E_2, an electron jumps (in the professional language, it is excited) from the lower level with E_1 to the higher one with E_2. In the case of $Fe(CN)_6^{4-}$, the photons with the right energy are photons of visible light, and therefore the complex is nicely colored (yellow).

The splitting of the d-energy levels depends on the bond strength of the ligand with the TM. When it is weak or only moderately strong, the complex will absorb in the VIS region and be colored. As the bond gets stronger, the splitting gets larger, and eventually only photons of UV light can be absorbed and the molecule will be colorless; for example, $Co(NH_3)_6^{3+}$ is yellow, while $Cr(CO)_6$ is colorless. The strength of the various TM–ligand bonds and the different modes of splitting which depend on the 3D structure of the complex are discussed in a theory called by chemists *ligand field theory*. A brief explanation of the theory can be found at http://en.wikipedia.org/wiki/Ligand_field_theory.

9.R.2 High-Spin Complexes

When the energy gap splitting of the orbitals becomes small, TM complexes exhibit an interesting phenomenon. As shown in Scheme 9.R.4, for small orbital splitting gaps, it is not anymore advantageous to pair the electrons in the lower orbitals, and

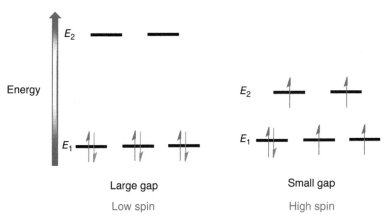

SCHEME 9.R.4   Different electron occupancy patterns when there are six electrons in the d-orbitals in the case of a large splitting gap (left) versus a small splitting gap (right). When the gap is small, four of the electrons occupy solo four of the d-orbitals and form a high-spin magnetic complex with four identical (parallel) unpaired spins.

instead the electrons prefer to occupy the orbitals solo, and to have the same spins (this is called Hund's rule after the name of its originator). In the case of the octahedral complex with 6e in the d-orbitals, as in the scheme, there are now four unpaired electrons with identical spins. Generally speaking, these complexes with unpaired electrons, which occupy orbitals solo and have parallel spins, are called high-spin complexes. They are *magnetic objects*. The number of unpaired and parallel spins can be measured by weighing the complex using balances where, underneath the arm that contains the complex, there is a strong magnet. How ingenious! Right?

9.R.3   The Active Species of CYP 450

The active species of CYP 450 obeys the 18e rule, but it is extremely reactive, and it will most happily abstract a hydrogen atom from most molecules having, for example, C–H bonds. The reason for this high reactivity is the diradical nature of the Fe–O bond in the complex, as shown in Scheme 9.R.5. Thus, we recall that the complex involves Fe^{4+}, and hence it has 4e in the d-orbitals. The lowest-energy occupancy in the three lowest orbitals involves pairing of two electrons in one orbital while leaving the two other electrons solo and with parallel spins in the other two d-orbitals. Since these orbitals have mixed d- and oxygen characters, the CYP 450 complex has in fact with one unpaired electron on the oxygen ligand. The $Fe^{IV}=O$ unit resembles the O_2 molecule we discussed in Lecture 3 (Section 3.R.4).

9.R.4   The "Life" of a Catalyst: The Catalytic Cycle

As we mentioned in the main text, a catalyst operates in a cycle, called a catalytic cycle. The cycle involves a sequence of steps whereby the catalyst helps reactions to occur faster, culminating in the formation of the desired product. During each step,

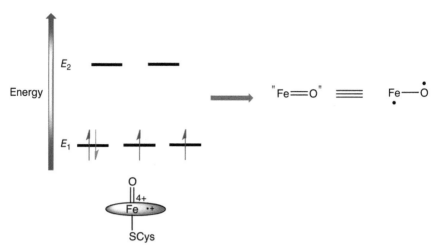

SCHEME 9.R.5   The d-orbitals and their electron occupancy in the CYP 450 active species, $Por^{\bullet+}Fe^{IV}{=}O(S_{Cys})$. The porphyrin ring is represented using the oval as a shorthand.

the catalyst transforms its molecular identity and structure, but, at the end of the cycle, it will be regenerated and ready to perform the reaction anew. The number of times a catalyst can achieve the same reaction within a certain unit of time is called the turnover frequency, which is a measure of the efficiency of the catalyst.

Figure 9.R.1 shows an example of a catalytic cycle that was deciphered by chemists using spectroscopic methods for the enzyme CYP 450. The catalyst involves an iron porphyrin complex. The naturally occurring porphyrin ring has different groups hanging on it. We skipped these groups at some points in the text since they were not necessary for discussing the 18e rule or the geometry of the complex. The $(CH_2)_2CO_2^-$ groups may actually have an ingenious role in coordinating the events of the cycle by acting as gates for water's entrance into and exit from the pocket where the complex is nestled. But this is not yet fully established.

The resting state in the cycle involves the Fe^{3+} ion coordinated to a water molecule (**1**). When an alkane (R—H) enters the pocket of the protein where the active species resides, the water molecule is displaced by the substrate (**2**). This loss of water triggers a transfer of an electron from another protein called a reductase, generating thereby complex **3** in which the iron is now in the Fe^{2+} state (since it accepted an electron, which has a negative charge). Unlike Fe^{3+}, Fe^{2+} binds O_2 quite strongly, and therefore it takes up an O_2 molecule (**4**) and activates it, so it can accept one more electron from the reductase to yield **5**. Because of its double negative charge, complex **5** is a powerful base and therefore sequentially abstracts two protons from nearby water molecules, and this leads to breakage of the O—O bond and to the formation of the active species **7**. This species is very reactive and can abstract the radical H$^\bullet$ (remember it is a diradical …) from the R—H molecule, and then the so-formed radical R$^\bullet$ grabs the OH from the iron and forms an alcohol molecule R—OH, which remains coordinated to iron (**8**). Finally, a water molecule displaces the alcohol and restores the resting state **1**. The enzyme is now ready for a second cycle …

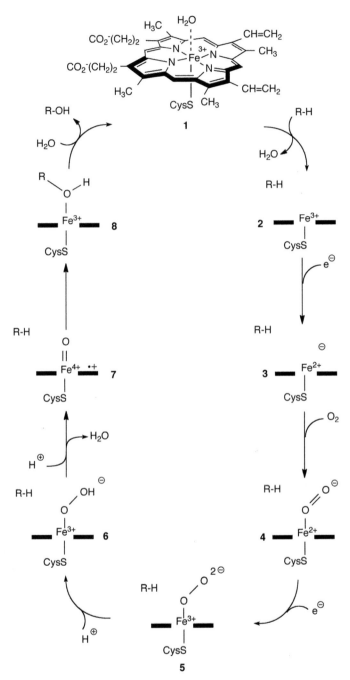

FIGURE 9.R.1 The catalytic cycle of CYP 450 during the hydroxylation of an alkane (R—H), in an overall process which replaces the C—H bond with a C—OH bond. Except for the resting state, **1**, where the porphyrin ring is drawn explicitly, for all the subsequent species, the ring is represented by two heavy bars flanking Fe. All the intermediate species are labeled with bold numerals. The active species, which breaks the C—H bond, is indicated as **7**.

This is an ingenious mechanistic cycle that nature and evolution concocted. I like its beauty and complexity. But not all cycles are so complex … It is very risky to have cycles with so many steps, because the catalyst can be destroyed in every such step when it is converted to some reactive intermediate. It is much more advantageous to have less eventful cycles. This is a challenge for chemists who design catalysts …

9.R.5   The Relation Between the Energy of the Photon and the Frequency of the Light

Niels Bohr postulated that light absorption occurs when an atom or a molecule absorbs a photon and one of its electrons jumps from one energy level to another. According to Einstein's theory of the nature of light, the light energy is distributed into little bundles or packets of energy called photons, whose energies are proportional to the frequency (ν) of the light:

$$E_{photon} = h\nu$$

where h is a proportionality constant called Planck's constant (it has units of energy•seconds), after the name of Max Planck, who postulated the idea that energy can be viewed as being subdivided into these bundles (later called photons).

It turns out that this photon-induced transition from one energy level to a higher one is the general mechanism of spectroscopy. In UV-VIS spectroscopy, the jump is between energy levels of electrons; in IR spectroscopy, it is between energy levels of molecular vibrations, and so on and so forth. What a powerful postulate!

9.R.6   References for Retouches

[1R] S. S. Zumdahl and S. A. Zumdahl, *Chemistry*, 9th Ed., Brooks College, Cengage Learning, 2010, Chapter 7; Chapter 8; Chapter 9. Available at: www.cengage.com/global

9.P   PROBLEM SET

9.1 Given the total charges on the following complexes, determine the electron counts and effective charges of the corresponding TMs: (a) $Fe(CN)_6^{4-}$, (b) $(OC)_5Mn(CH_3)$, (c) $Cr(C_6H_6)_2$, (d) $Fe(CN)_6^{3-}$, (e) $(C_5H_5)_2ZrI_2$, (f) $Pd(NH_3)_2Cl_2$.

9.2 Draw all the isomers of $Pt(NH_3)_2Cl_2$.

9.3 Draw a complex that obeys the 18e rule and contains an iron cation ligated to oxygen, O, to $H_2N(CH_2)_2NH_2$ and an additional CH_3CN ligand. Specify the effective charge of the cation and the total charge of the complex.

9.4 Draw the isomers that are produced from $Fe(CO)_5$ upon replacement of (a) one CO by a phosphine, $P(CH_3)_3$, and (b) two CO ligands by two phosphines.

9.5 Draw the isomers of $Co(Cl_2)[H_2N(CH_2)_2NH_2]_2{}^+$.

9.6 How many isomers can be anticipated if one CO ligand in $Ni(CO)_4$ is replaced by a phosphine?

9.7 Draw the 18e complex of Fe (with an effective charge of 0) with as many molecules of CO as required and one cyclic C_4H_4 molecule as ligands.

9.8 What kind of clusters can be imagined to form only from $Fe(CO)_4$ fragments? Draw two.

9.9 Try to describe the complete 3D structure for the cyclic C_3H_6 and C_6H_6 molecules.

9.10 Describe the 3D structure of the CH_3X (X = F, Cl, Br, and I) molecules. What is the major trend you can see in these structures?

9.11 Try to generate complete 3D information for $Ni(CO)_4$. Radii values for TMs are given in Appendix 9.A.1.

9.12 Calculate the energy of the reaction of CH_4 with O_2 (consult Table 9.A.2).

LECTURE 10

CHEMISTRY, THE TWO-FACED JANUS—THE DAMAGE IT CAUSES VERSUS ITS IMMENSE CONTRIBUTION TO MANKIND

10.1 CONVERSATION ON CONTENTS OF LECTURE 10

Racheli: Sason, this is the 10th lecture. Are you going to finish the book with this lecture like the Ten Commandments?

Sason: No, there is one more to go, Racheli.

But, anyway, this one is going to be a sober lecture, in which I cannot bask anymore in my glorification of chemistry. I have to confess that chemistry can potentially "upset the fine balances of life."[1]

Usha: Sir, you mean accidents like Bhopal?

Sason: This too. But not only tragic accidents like Bhopal. It is the ability of the chemists to reshuffle the atoms of Nature and make from them new molecules, like artificial DNA, artificial proteins, and all kinds of new "creatures" and excessive amounts of chemicals like CO_2, heavy metals, and so on, which may threaten the fine equilibrium in Nature.

Racheli: Sason, the readers may be quite confused how the love of chemistry turns into this sober assessment …

Sason: I hope not, Racheli … Chemistry is a magically great science, but we have to be very careful not to end up like the Sorcerer's Apprentice, in the symphonic poem by Paul Dukas.[2] He enchanted the broom to do work for him but ended up causing a great flood, because he could not control the broom anymore … I would want to argue here for stricter controls.

Chemistry as a Game of Molecular Construction: The Bond-Click Way, First Edition. Sason Shaik.
© 2016 John Wiley & Sons, Inc. Published 2016 by John Wiley & Sons, Inc.

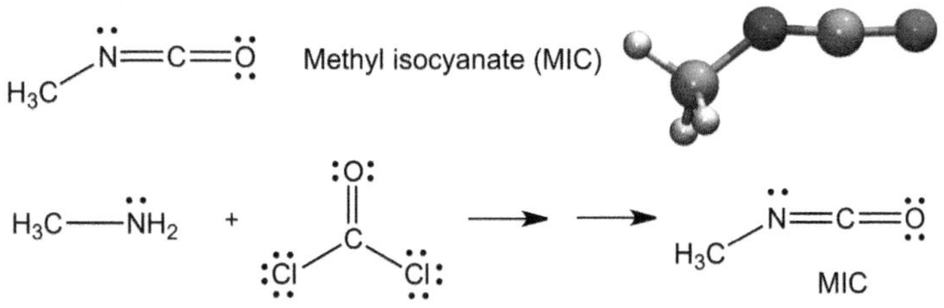

SCHEME 10.1   Methyl isocyanate (MIC) and its synthesis from methylamine (bottom left) and phosgene (bottom middle) (the two arrows indicate a synthesis involving multiple steps). MIC is shown also using a ball-and-stick model.

Usha: I want to tell you a bit about Bhopal. This was probably the worst chemical disaster, maybe until the Fukushima disaster in 2011.

Racheli: But the Fukushima disaster was not really a chemical failure.

Usha: Yes, you are right, Racheli. Bhopal was the worst chemical disaster ever. It occurred on the night of December 2, 1984, in the Indian state Madhya Pradesh in the Union Carbide India Ltd (UCIL) plant that produced the pesticide *Sevin* from methyl isocyanate (MIC). MIC is extremely toxic to living organisms, causing in humans anything from severe coughing to death by suffocation.

Sason: Let me interrupt you for a second and just show what MIC looks like in Scheme 10.1. Here MIC is synthesized in a few steps from the combination of phosgene, which is also very poisonous, and methylamine. Phosgene was used as a chemical weapon in World War I. So the UCIL plant was a Pandora's Box of lethal poisons.

Usha: Yes, lethal as they can come!

Racheli: You seem surprised, Usha. In fact, most molecules can be poisonous. Even oxygen at high concentrations …

Usha: As Sir always says, this is *the Janusian feature of chemical matter.* It is, at the same time, good where it is needed and bad where it is not.

Sason: Even MIC can be "good" when it is involved in producing rubbers and adhesives.

Usha: I cannot admit that MIC can also be good, Sir and Racheli. But this is another long story … Let me finish the one I started here.

It so happened that the demand for MIC-based pesticides dropped over the course of time, but the management of the plant continued its MIC production. The result was that the factory had amassed huge quantities that required storage.

Sason: This is always a problem when you need to store poisonous gases in huge containers. All this requires superb maintenance and tight safety measures.

Usha: If I may continue Sir … no one knows what precisely caused the accident. It is thought that the containers were cleaned with water, and the water spilled and reacted vigorously with the MIC, leading to an explosion and the subsequent leaking of many poisonous gases. Anyway, on the night of December 2, 1984, almost 40,000 kg of gas spread to nearby populous shantytowns and affected more than half a million people. The death toll was estimated at over 2,000 people, and it was rumored to be as much as 20,000–30,000.

Racheli: I read that the gas also contained HCN, which is deadly, and phosgene (Scheme 10.1), which is equally poisonous, as Sason was saying.

Sason: Another incident, which was defined as a chemical accident, happened on October 4th, 2010, in Ajka, Hungary, where a waste reservoir (which was an open pond) full of caustic material (such as Na_2O, CaO, Al_2O_3) broke open and spilled 1 million cubic meters of the waste material. The waste looked like an orange deluge since it contained also Fe_2O_3, which is orange-colored. Just imagine the sight: a calamity on a Biblical scale!

Racheli: It must have been scary to see all the land in sight covered with a blood-like mud … The main damage was caused by the high pH of the waste, which is harmful to humans and animals.

Sason: The cause of this accident and the Bhopal disaster was "human error," which is a whitewashed expression for negligence. The Web is laden with similar industrial accidents, and Figure 10.1 provides a list of some of the chemical accidents that have occurred between the years 1974 and 1993. This list is similar to one that appeared in *Chemical & Engineering News* in 1993.

Many of these accidents have been attributed to human errors. We cannot blame the chemicals …

Usha: Sir, the fact that such a report got published shows that the chemical community is well aware of the problem.

Racheli: Still, these accidents and disasters create a bad impression of, if not a terrible name, for chemistry.

Add to that all the air and water pollution, the climate change, the dissolution of glaciers, and so on, and you will understand the origins of the public image of our science.

Usha: And what about chemical weapons? Chemistry is to blame again …

Sason: Let me repeat, "We cannot blame the chemicals or the science of chemistry." This is just too easy.

Usha and Racheli: Spilling waste into rivers is premeditated. Using chemicals for warfare means that chemistry works in the service of evil. Can you please elaborate: what do you mean by "This is just too easy"?

Sason: OK, you got me there. I have to explain my statement better.

Behind all premeditated cases and accidents there are people who make decisions and policies.

Racheli: But they are chemists and should know better …

Sason: No! Very few of them are chemists, like Fritz Haber who designed and was in charge of the chemical warfare in World War I.[3] Others are politicians who accept such advice and press upon scientists to produce more powerful weapons, or allow sloppy practices in chemical firms in the name of caring for employment.

A partial list of chemical accidents

Year	Place	Nature of the accident
1974	Flixborough, England	Cyclohexane release from the pipes at a Nypro facility killed 28.
1976	Seveso, Italy	Dioxin release at an ICMESA plant (Hoffmann-La Roche subsidiary) affected thousands.
1984	Cubatao, Brazil	Leakage and explosion of 700 tons of high-octane gasoline killed 508 shantytown residents.
1984	Mexico City, Mexico	542 were killed and over 4,000 were injured when a liquefied petroleum gas facility in San Juanico suffered a propane and butane explosion.
1984	Bhopal, India	At a Union Carbide pesticide plant, a runaway reaction gave rise to a methyl isocyanate leak that killed over 2,000 and affected hundreds of thousands more.
1985	Institute, West Virginia, US	135 were harmed after a Union Carbide plant emitted a cloud of gaseous aldicarb oxime and methylene chloride after a storage tank malfunction.
1986	Basel, Switzerland	The Rhine River was contaminated by 30 tons of pesticides from a Sandoz plant's spill.
1987	Ras al-Ju'aymah, Saudi Arabia	Up to 22 were believed to have been killed and 15 others were injured after a natural gas plant experienced an explosion.
1987	Texas City, Texas, US	Over 1,000 needed medical attention after Marathon Oil Refinery released a vapor cloud of HF acid.
1988	Henderson, Nevada, US	An explosion at a PEPCON plant, due to ammonium perchlorate, killed two and harmed over 350, causing $74 million in damage.
1989	Pasadena, Texas, US	At a Phillips 66 plant, 23 were killed and 132 were injured after a reactor broke and a mixture of ethylene and isobutane ignited.
1990	Channelview, Texas, US	A city block-sized area was scorched and 17 were killed when an ARCO wastewater tank exploded.
1990	Cincinnati, Ohio, US	One died and 71 were injured as a result of a cleaning solvent explosion at a BASF resins plant.
1991	Charleston, South Carolina, US	A cooling system failure caused an explosion killing six and injuring 33 at an Albright & Wilson Americas phosphorus chemicals plant.
1991	Port Lavaca, Texas, US	At Union Carbide's Seadrift plant, one was killed, and 19 were injured, after an explosion involving the ethylene oxide division.
1991	Sterlington, Louisiana, US	An explosion and fire killed 8 and gave rise to 128 injuries at an IMC Fertilizer facility.
1991	Corpus Christi, Texas, US	Two employees at a Kerr-McGee plant died due to exposure to HF acid vapors that escaped after a pump leaked.
1993	Frankfurt, Germany	At a Hoechst poylvinyl alcohol plant, one died and one was injured after an explosion.

FIGURE 10.1 A list of chemical accidents that occurred during the period 1974–1993. Details were taken from *Chemical &Engineering News*, 1993, November 29, p. 14.

Still others, in management positions, make poor decisions to save money and earn more of it by cutting down on safety measures and by spilling waste into rivers, seas, and oceans. In other cases like the Bhopal tragedy, those involved exhibited carelessness and poor judgment to continue producing MIC despite the declining demand for the pesticide *Sevin* and in view of the risks involved.

Still in other cases, greed is behind the premeditated decision to use harmful chemicals. You may recall the movie *The Insider*, which tells the story of how tobacco companies added to the tobacco mixture some chemicals that cause addiction to smoking. The movie was based on events that actually occurred.

Racheli: This is a story about greed, power, and individuals who got caught among these mighty interests and were almost ruined by them. Fortunately, there are also brave individuals like Lowell Bergman, who was a journalist, and despite the great risks, he was tenacious enough to follow the story. Another brave individual was Dr Jeffrey Wigand, a PhD in biochemistry, who worked for a tobacco company as the corporate VP for research and development and was willing to give testimony about the practices of tobacco firms, even though the personal cost ended up being very high. At the end of the story, after lengthy legal procedures in various states in the US, four tobacco firms agreed to pay about 206 billion dollars to the states. Most of the money (about 183 billion dollars) is payable in annual installments from 2000 through 2025.

Usha: Racheli, you forgot to mention that after the ordeal was over, Dr Jeffrey Wigand became a chemistry teacher and was awarded the Sallie Mae First Class Teacher Award, and was one of 51 such recipients nationwide (in the US).

Sason: A good end for the story indeed!

Anyway, we can hope that organizations tightening the regulations, like the Chemical Safety and Hazard Investigation Board (CSB) established in the US in 1990 or the Chemical Reaction Hazards Forum (CRHF) (in the UK), will work out. Also, airing all practices, like the Emergency Planning and Community Right-to-Know Act (EPCRA), which was passed in the US in 1986 promotes, along with chemical education for the public and for elected politicians, may together eventually minimize "human and nonhuman errors." This is what I meant by saying that it is too easy to go on blaming chemicals or the science of chemistry.

Usha: Still what is the role of chemists and their chemicals? Are they only "victims" of decisions made by others?

Sason: Of course not! Chemistry has either to find the means to create molecules that are less hazardous than MIC or to replace such molecules (e.g., biological pesticides can replace the chemical ones). Chemists have to develop procedures that do not yield pollutants and safe means of handling waste. There are so many things chemists can do. And chemistry is in the process of pursuing these goals.

I am willing to go even farther and suggest that all chemists in the world should sign a protocol like the Hippocratic Oath taken by physicians to refuse to practice harmful chemistry. It is a far-fetched dream, and perhaps an unachievable one, but without dreams, there is no reality.

Racheli: So how do you see the place of chemical education, such as your lectures, within these goals?

Sason: For one thing, Racheli, the central message of this lecture and the entire book is that everything, including us, is made of molecules. *We are chemical matter that makes chemical matter,*[4] and by creating new forms of matter, we are enhancing and reshaping evolution on our planet, including our own evolution, and we are doing this in a nonnatural way. By its nature, chemical matter changes and mutates, not always for the better. Therefore, we must all ultimately come to terms with the Janusian effect of matter and learn to minimize its harmful aspects and benefit from its positive ones. *Only chemistry can do it for us!*

Usha: It is important for chemists to show that chemistry is a central human culture in which we can imagine new things, import our imagination to create new materials, and enjoy the beauty of a new idea, of a new creation …

Racheli: And at the same time, this culture created by chemists has advanced humans immensely,[1,4] as you showed, Sason, in all the previous lectures.

Sason: This is a matter of education. I sincerely hope is that this book will make a contribution toward the eventual appreciation of the wonderful science of chemistry and will at the same time alert young students to the other side of the chemical Janus.

10.2   TYPES OF POTENTIAL CHEMICAL DAMAGE

Chemistry has two faces: It is the constituent of life, on the one hand, and on the other, it is the root cause of the destruction of life. When Mankind learned to make new materials/molecules, this know-how caused enormous progress. But at the same time, it created copious amounts of material that can endanger or alter the very continuation of life as we know it on this planet. Here are some of the problems that Mankind created.

10.2.1   The Ozone Hole

In its most stable form, elemental oxygen exists as diatomic molecules (O_2). Ozone is another form of oxygen that contains three oxygen atoms (O_3) and is less stable than O_2 (see Retouches section 10.R.1). At room temperature, ozone is a pale blue gas with a sharp odor, characteristic of the air after a thunderstorm or near an old Xerox machine (copying machine). Ozone is a very reactive gas, and, even at low concentrations, it is irritating and toxic.

Ozone occurs naturally in small amounts where it should not be present, namely in the Earth's upper atmosphere and in the air of the lower atmosphere after a lightning storm. However, at much higher altitudes of 16–50 km, called the stratosphere, O_3 is formed from O_2, as shown in Scheme 10.2, and O_3 is present in larger amounts.

In the stratosphere, the presence of O_3 is absolutely essential since this molecule, which is, otherwise toxic to humans when is present in the atmosphere, *acts in*

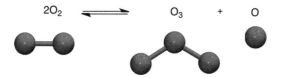

SCHEME 10.2   The reaction of two molecules of O_2 produces O_3 and an atom of O. The two arrows mean that the reaction is in equilibrium, going to and fro with a constant ratio of O_3:O_2 under the conditions existing in the stratosphere. Ball-and-stick structures of the molecules are shown underneath the equation.

the stratosphere as a shield that protects all life from destructive ultraviolet (UV) radiation. This radiation emanates from the sun and crosses the ionosphere (the layer of the atmosphere where the radiation is so strong that there exist only ions) *en route* to other regions of the atmosphere. Thus, when O_3 absorbs UV radiation, it decomposes back to O_2 and O, going in the backward direction of the chemical reaction in Scheme 10.2. (The O formed from the decomposition combines with another oxygen atom to become O_2, for two molecules of O_2 overall, as shown on the left-hand side of Scheme 10.2.) In this manner, most of the UV radiation is converted to heat and does not reach the surface of the planet, where it is extremely harmful to humans (causing diseases such as skin cancer), animals, vegetation, and to the oceanic plankton, which are the foundation of the ocean food web (e.g., algae and other drifting single cell species, which are used as foodstuff for fish and whales). After UV absorption, the equilibrium between O_2 and O_3 is restored and the shield remains intact.

One of the causes of the destruction of the ozone layer is the presence of free radicals. For example, chlorofluorocarbons (CFCs),[5] like CF_2Cl_2 or $CFCl_3$, which are very stable in the atmosphere, drift to the stratosphere, where they absorb UV light and break into free radicals.

The free radicals destroy ozone through a mechanism that is depicted in Scheme 10.3. We will be using CF_2Cl_2 for illustration. Thus, after UV absorption by the molecule, its C–Cl bond breaks (reaction (a)) and generates two radicals,

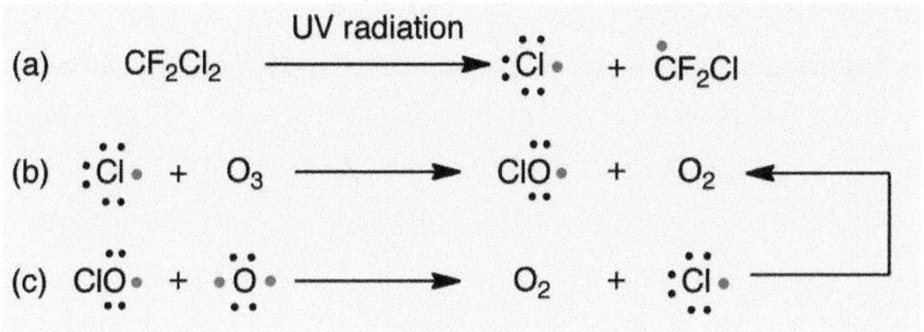

SCHEME 10.3   The three steps of the chain reaction by which CFCs, such as CF_2Cl_2, destroy ozone. Initially, the CFC molecule absorbs UV radiation and generates two radicals. In step (b), the Cl• radical destroys an ozone molecule, and this reaction forms a new radical, ClO•, which combines with O in step (c) and regenerates the Cl• radical. The newly generated Cl• radical repeats step (b). And the chain process continues, in principle indefinitely.

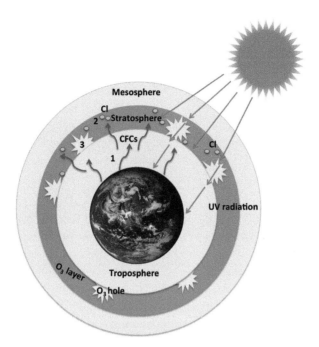

FIGURE 10.2 A schematic diagram showing the creation of holes in the ozone layer, through which destructive UV radiation from the sun reaches the Earth's surface. First, CFCs are emitted into the stratosphere, represented by the blue arrows (1), and then they produce chlorine atoms, represented by green circles (2). The reaction of ozone with chorine depletes ozone, forming holes (3) that enable UV radiation from the sun to arrive on the Earth.

one being $F_2ClC^\bullet$ and the other being $Cl^\bullet$. It is known that the $Cl^\bullet$ reacts with an ozone molecule (reaction (b)), abstracts from it an O atom, and thus generates $ClO^\bullet$ and O_2. Subsequently, in reaction (c), an O atom from the stratosphere (see this layer in Figure 10.2) attacks $ClO^\bullet$ and generates O_2 and a new $Cl^\bullet$ atom. This $Cl^\bullet$ radical then repeats the ozone destruction reaction (b), and it next proceeds to reaction (c) to regenerate the $Cl^\bullet$ radical. This *chain reaction* can continue for many iterations such that one $Cl^\bullet$ radical can destroy many O_3 molecules and thereby depletes the ozone layer. Since the mixing in the stratosphere is horizontal, a CFC-originating radical starting in any place on Earth eventually reaches many other places. CFCs are the main gases in deodorants and also widely used in refrigerants and solvents.

This chain process is so destructive that it has created a huge hole above Antarctica and the South Pole, estimated to have a radius of 1,000 km or more. Figure 10.2 shows schematically the ozone hole generation. The CFC (1) rises to the stratosphere, and its C—Cl bond is broken by sunlight, releasing atomic chlorine (2), which destroys the ozone layer and creates a hole (3). Through this hole, intense UV radiation reaches the Earth's surface and leads to dire consequences.

Three chemists, Paul J. Crutzen, Mario J. Molina, and F. Sherwood Rowland (Figure 10.3), formulated the theory of the ozone hole and shared the Nobel Prize

FIGURE 10.3   The 1995 Nobel Laureates in Chemistry for their work on the ozone hole. Shown from left to right are Paul J. Crutzen, Mario J. Molina, and F. Sherwood Rowland. The photographs were obtained from the following sources: http://commons.wikimedia.org/wiki/File:Paul_Crutzen.jpg?uselang = de, http://commons.wikimedia.org/wiki/File:Mario_Molina _1c389_8387.jpg, and http://commons.wikimedia.org/wiki/File:F._Sherwood_Rowland.jpg.

in 1995. This shows how instrumental chemistry is in understanding the source of molecular damage.

10.2.2   The Montreal Protocol

In 1987, at a convention in Montreal, that was organized by the United Nations Environment Programme,[5] it was agreed to reduce the amount of CFCs by 50% by removing them from deodorants and other products. It is said that this accord has been followed and it resulted in reducing the growth rate of the hole.

However, it turns out that hydrochlorofluorocarbons (HCFCs), like CF_2ClH, C_2F_4ClH, etc., which are used as refrigerating gases, have the same effect on the ozone hole through the generation of $F_2ClC^{\bullet}$ radicals or analogous species. On March 9th, 2014, the journal *Nature Geoscience* published the results of a study showing that in fact the concentrations of three CFCs and one HCFC have grown in the atmosphere. Once reaching the stratosphere, these substances will continue to consume the ozone layer. The danger is still looming, and it is clear that tighter protocol is going to be needed.

10.2.3   Climate Change

Our atmosphere is composed of 78.1% of N_2, 20.9% of O_2, 0.965% of a variety of other gases (Ar, He, H_2, ozone, Rn, etc.), and 0.035% of CO_2. CO_2 is important, as one of its functions is to absorb the heat radiated from the surface of the planet (in the IR region), and to emit it back to the Earth. In this manner, CO_2 acts as a warming insulator for the earth, thus contributing to making life possible on the planet during cold times (nights and winters). CO_2 is therefore good for life …

CO_2 is generated on Earth by burning organic material, breathing, and emissions from volcanoes and man-made machines, which includes emissions due to industrial processes and transportation. On the other hand, CO_2 is consumed by the process of photosynthesis and by dissolution in the oceans. On a balance, its quantity should more or less be constant and not deviate much from the 0.035% value. However, in between the years 1900–1960, the CO_2 percentage increased by 7.4%, which still does not look too impressive. However, it is calculated that an increase of 15% in the percentage of CO_2 will cause warming of the Earth by 4°C, which will result in major disasters on the planet.

Since there are correlations between the planet's average temperature and the percentage of CO_2 in the atmosphere, many scientists conclude that the climate change is man-made and is due to increased CO_2 emissions because of increased industrialization and the usage of coal. If this is indeed the case, then decreasing CO_2 emissions is an essential mission. The United Nations Framework Convention on Climate Change (UNFCCC) has convened several times to address the issue of reducing greenhouse gases (GHG) emissions. The UNFCCC recognized that "developed countries are principally responsible for the current high levels of GHG emissions … as a result of more than 150 years of industrial activity…" The last meeting, that took place in Kyoto, produced the Kyoto Protocol, in which many developed countries agreed to legally binding limitations in/reductions in their emissions of GHGs in two commitment periods, stretching from 2008 to 2020. However, for all practical purposes, the accord to lower carbon emissions has not been very successful, as major countries did not ratify the agreement and many countries renounced it for political and economic reasons.

Recently, in 2010, Molina suggested that the major effect of climate change is not necessarily CO_2, but rather associated with the ozone hole, that allows increased radiation to reach the surface of the Earth. He therefore proposed that tackling these non-CO_2 warming agents is an important complement of not substitute for reductions in CO_2 emissions. He suggested expanding the already existing Montreal Protocol for CFCs and adding to it other ozone hole-promoting gases like HCFCs (which are refrigerants in air conditioning), black carbon soot, methane, etc. But the plan may not be adopted very quickly. Large developing countries object to it on the grounds that the timetable is too rapid and that the financial cost of eliminating the refrigerants may be too high. Thus, the role of politics and financial interest groups is very crucial. One may hope that the most recent accord sing in Paris (in December 2015) will be effective.

It is fair to say that almost all agree that our planet is experiencing climate change. However, not all scientists subscribe to the theory that this is because of CO_2 and other greenhouse gases. For example, Nir Shaviv, a physicist from the Hebrew University of Jerusalem, has written amply about the topic.[6] His analysis led him to conclude *that the rise of CO_2 is the result of climate change and not its cause*: for example, warming causes the CO_2 dissolved in the ocean to be emitted into the atmosphere. This resultant rise in CO_2 has indeed contributed in turn to climate change but to a negligible extent of less than 1 °C. Shaviv maintains that the incrimination of CO_2 (and other GHGs) is primarily because "we expect it to warm and we do see warming," but not because we have proof of their effects. He argues that there are other much more powerful suspects, which are associated with increased solar activity in the twentieth century, that are responsible for climate change. A more active sun inhibits the formation of cloud condensation nuclei, and this thereby leads to the warming of the Earth. However, this argument ignores reports that the warming is localized in the troposphere (Figure 10.2), while the stratosphere, which is closer to the sun, is cooling down.

Proving causality is not simple, and the problem may never be absolutely settled. My own position on the problem is rather simple. It is a fact that we are polluting the environment with many hazardous compounds, so much so that there are major cities in the world which are becoming less and less suitable for living. Therefore, a protocol that reduces the formation of pollutants in the atmosphere is certainly a desired one, and this is so whether or not GHGs and other pollutants are the cause of global warming. Being able to breathe is also important …

10.2.4   Acid Rain

There are many other pollutants that go into the atmosphere and cause health issues and other damage. For example, SO_2 is a common product of many industries and a major pollutant in large cities. The largest sources of SO_2 emissions are fossil fuel combustion at power plants (73%) and other industrial facilities (20%). Smaller sources of SO_2 emissions include industrial processes such as extracting metal from ores and the burning of fuels with high sulfur contents by locomotives, large ships, and nonroad equipment. SO_2 is linked with a number of adverse effects on the respiratory system, including bronchoconstriction and increased asthma symptoms, especially in children and the elderly.

Another effect of SO_2 emission is acid rain. The emitted SO_2 rises to the atmosphere, where it is oxidized to SO_3, as shown in Scheme 10.4a. When it rains, SO_3 is converted to the acid H_2SO_4 and is dissolved, and it falls with the rainfall. Similarly, NO, which is a byproduct of airplane fuels and of industrial processes and is also produced through lightning, is oxidized to NO_2, and, as the rain starts, NO_2 becomes the strong acid HNO_3 and the weak acid HNO_2 (Scheme 10.4b), resulting in more acid rain.

Acid rain is a serious environmental problem that affects large parts of many countries in Europe and North America. As shown in the diagram in Figure 10.4,

(a)   (b)

$$SO_2 + [O] \longrightarrow SO_3 \qquad\qquad NO + [O] \longrightarrow NO_2$$

$$SO_3 + H_2O \longrightarrow H_2SO_4 \qquad 2NO_2 + H_2O \longrightarrow HNO_3 + HNO_2$$

Acid rain   Acid rain

SCHEME 10.4   Formation of acid rain from pollutants: (a) from SO_2 and (b) from NO. The acids formed in these reactions give rise to acid rain. The symbol [O] represents an oxygen-based species (such as O_2) without specifying its exact composition.

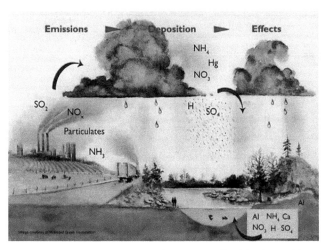

FIGURE 10.4 A drawing summarizing the formation of acid rain and its consequences for the environment. Adapted with kind permission from the Hubbard Brook Research Foundation, 32 Pleasant Street, Woodstock, VT 05091.

acid rain is particularly damaging to lakes, streams, and forests and to the plants and animals that live in these ecosystems. For example, acid depletes the O_2 in lakes and causes the deaths of their organisms. Acid dissolves essential ions (Al^{3+}, Mn^{2+}) which are required for growth of trees and other vegetation, and it thereby causes the destruction of forests. Acid rain can cause the peeling of paint, the corrosion of steel structures such as bridges, and the erosion of stone statues made from limestone and marble ($CaCO_3$).

10.2.5   More Evils and the Other Side of the Chemical Janus

Our environment is polluted in many other ways, such as pollution caused by heavy metals, including mercury (Hg), lead, plutonium, cadmium, and so on; biological material pollution; CO pollution; methane pollution; benzene and toluene pollution; and radon, sulfur, chlorine, and ash pollution. All these pollutants have dire health consequences.

Most of these pollutants are produced by industries and by machinery that maintain our infrastructure (such as electricity, etc.), and others are natural (such as radon). Necessarily, all of these pollutants have to do with chemistry because all matter is made of atoms and molecules, and dealing with matter means that usage of chemistry produces new molecules, including these pollutants.

These consequences and the chemical accidents (see Figure 10.1), may create the impression that chemistry causes damage to the planet and to mankind, and that chemists are a sloppy group of people who permit these consequences to happen. And all we have to do to end this misery is to stop or minimize chemistry! However, at this point in the course, we are a bit smarter and know that *there is a much more profound and loftier conclusion that we can draw about chemistry.*

Mankind is made of molecules, and He acts within a material world as the "ruler of matter." Chemistry is the know-how that mediates between the material essence of Mankind and His mastery over matter. As such, chemistry is two-faced, like the Roman God Janus depicted in Picture 10.1:

PICTURE 10.1   The Roman God Janus. (This public domain image was obtained from http://commons.wikimedia.org/wiki/File:Janus_coin.png.)

And the two faces of the chemical Janus coexist in a never-ending conflict:

☞   On the one hand, chemical know-how is responsible for our cognitive evolution, as it provides us with a glimpse into our chemical essence. Even knowledge is chemically encoded into our long-term memory.

☞☞   On the other hand, a chemical reaction can also be destructive, as is actually occurring continuously in living systems where there is a balance between the construction and destruction of matter.

☞   On the one hand, chemical know-how is immensely important for the welfare of mankind and the well-being of society. Try to imagine life without chemistry![1]

☞☞   On the other hand, despite all this acquired know-how, still mankind has not learned how to be a compassionate master of matter. He is still the Sorcerer's Apprentice.

The historians Bernadette Bensaude-Vincent and Isabel Stengers wrote the following beautiful words about the relationship of man and matter through chemistry:

"Chemistry defines very specific relationships between man and matter: neither domination nor submission, but a perpetual negotiation—through alliances or hand-to-hand struggles—among individual materials and human demands."[7]

Some of the most beautiful reflections on such negotiations were written by the Italian chemist and acclaimed writer Primo Levi in his book *The Periodic Table*.[8]

Chemists are still not there, but they are making an effort to become benevolent masters of matter by thinking up new ways to make chemicals, while minimizing the dire consequences for the environment. Green chemistry and renewable energy sources are just two of the new directions in which chemistry is heading. Some industries are trying to reuse all waste materials for new processes rather than spilling them (for example the company BASF uses the waste N_2O produced from one useful process to carry out another useful process that consumes this waste and creates only N_2, which for all we know now is benign). The Sorcerer's Apprentice may after all become a true master ...

10.3 SUMMARY

In this lecture, we discussed potential chemical damage. We told the tragic stories of Bhopal, the accident in Ajka, Hungary, the ozone hole, the greenhouse effect of CO_2 and other gases, air pollution, and the destabilization of ecosystems. We cannot blame chemicals for these effects since chemicals are like Janus, the Roman God; they are beneficial where they are needed and harmful where they are not. For example, the "evil" free radicals, which are formed by UV radiation, were just recently shown to be highly beneficial in some contexts. The formation of free radicals in the eyes of birds, in a process caused UV radiation from the sun, is what enables birds to navigate, and not only birds ...[9] So, every molecule is both good and bad, depending on its usage and where it is formed ...

As we showed, these negative aspects of chemistry are also not necessarily the fault of chemistry. It is primarily a political issue.[10] Be it as it may, branding chemistry as "the source of all evils" is not a sensible approach. We are made of molecules. We are chemical creatures. Everything around us is chemical, and the periodic table is the most compact census book of the planet. In fact, we are in a way material entities that learned to manipulate matter. In so doing, we created a central human culture, which we call chemistry, and in which we can import our imagination to create new useful materials, and enjoy the beauty of a new idea, of a new creation. This is what chemistry is all about!

Still we, the chemists, have to sharpen our skills and negotiate with matter differently than we do now. We are trying ...

10.4 REFERENCES AND NOTES

[1] P. Atkins, *What Is Chemistry?* Oxford University Press, Oxford, United Kingdom, 2013.

[2] Paul Dukas was a French composer who wrote the symphonic poem in 1896–97, based on a ballad written by Goethe in 1797. The symphony was animated in 1940 by Disney's film *Fantasia*.

[3] F. Stern, *Angew. Chem. Int. Ed.* **2012**, *51*, 50.

[4] S. Shaik, *Angew. Chem. Int. Ed.* **2003**, *42*, 3208.

[5] See United Nations Environment Program (UNEP): Available at: http://ozone.unep.org/new_site/en/index.php

[6] N. Shaviv. Available at: http://www.sciencebits.com/CO2orSolarHYPERLINK

[7] B. Bensaude-Vincent and I. Stengers, *A History of Chemistry*, Harvard University Press, Cambridge, MA, 1996, p. 10

[8] P. Levi, *The Periodic Table*. G. Einaudi Ed., Penguin Books, Torino, Italy (English translation), 1984.

[9] A. King, *Chemistry World*, 2013, December, 49.

[10] See, N. Oreskes and *E. M. Conway, Merchants of DOUBT*, Bloomsbury, New York, 2010.

10.R   RETOUCHES

10.R.1   The Electronic Structure of Ozone

Ozone is one of those odd molecules that "negotiate" with the octet rule. Its electronic structure can be written following the rule as two structures, which are depicted in the bottom part of Scheme 10.R.1, with a double-headed arrow between them. Recall that the double-headed arrow means that the two structures contribute equally to the description of the molecule, which has thereby four mobile electrons (recall benzene in Scheme 4.R.3). It turns out, however, that quantum mechanics (QM) calculations predict that these two descriptions are minor, and the major description of ozone is the structure that is seen in the scheme to possess two regular O—O bonds and one long bond between the two terminal oxygen atoms.[1R] The bond lengths of the molecule show that this is not a short bond like the other two. I have put the reference of a scientific book, not as an actual source of reading material, but rather just to show you that this electronic structure *is* discussed somewhere in these terms.

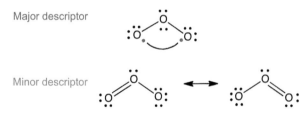

SCHEME 10.R.1   The major and minor electronic structure descriptions of ozone, as elucidated by quantum mechanical calculations.

10.R.2   Reference for Retouches

[1R] The ozone problem is worked out in S. Shaik and P. C. Hiberty, *A Chemist's Guide to Valence Bond Theory,* Wiley Interscience, Hoboken, NJ, 2008, pp. 92–93.

10.P PROBLEM SET

10.1 (Bonus question) The Cl• radical destroys the ozone molecule in a chain process. Other radicals, such as HO•, destroy the cell membrane (made in part from fatty acids) in a similar manner. Write the mechanism of destruction for the HO• radical's reaction.

10.2 Acid rain causes erosion of statues made of limestone. Write the corrosive chemical reaction.

10.3 Cl• can form acid rain when CH_4 pollutes the atmosphere. Write an equation that explains the origins of this acid rain.

10.4 Using Web sources, write an essay of your choice on chemical damage. Be critical!

LECTURE 11

CHEMISTRY IS EVERYTHING AND EVERYTHING IS CHEMISTRY

11.1 CONVERSATION ON CONTENTS OF LECTURE 11

Racheli and Usha: So that's it, this will be the last lecture of the book?

Sason: Yes. It is better to end the book while it is still short …

Usha: Sir, given the sober tone of Lecture 10, maybe an upbeat lecture should be the right way to end this book?

Racheli: After all, chemistry is a great science, and it is also a human culture …

Sason: Hey, who is giving the lecture, you or me?

The last lecture is going to be a very short slide presentation, with the goal of leaving the rest of the time for demos that will wow the students. I heartily recommend this to any teacher.

Racheli: Yes, I agree. Lively chemistry is the way to end the course.

Sason: Sure, this is a fun lecture. No problem set …

Usha: But what is your message *going to be*, Sir?

Sason: I am single-minded about that: that Mankind and Matter are not separate entities, but rather two related entities that affect each other.

Molecules in the air and inside Earth were once parts of living beings, and those that have not been yet may become so one day. And, vice versa, our own molecules were once part of the atmosphere or the rivers, the trees and the flowers, or the proteins

Chemistry as a Game of Molecular Construction: The Bond-Click Way, First Edition. Sason Shaik.
© 2016 John Wiley & Sons, Inc. Published 2016 by John Wiley & Sons, Inc.

of a chickpea or a chicken. Because of all these features, knowledge of chemistry is a must for everyone.

I now wish to cap it all in a relaxed manner and show that chemistry is everything and everything is chemistry.

11.2   THE BIRTH OF CHEMISTRY IS THE NASCENCE OF MANKIND

In a way, chemistry is 236 years old (at the time of completing this chapter). Its year of birth is 1778, when Lavoisier (in Figure 11.1) analyzed the components of air, determined their weights, and dubbed one of them "oxygen" (the generator of acids). Until that moment, chemists had a spiritual/metaphysical outlook on matter, and they used concepts like "phlogiston" to discuss the inflammability of materials (e.g., of coal, which is made of carbon mostly). Lavoisier showed that what happened during the combustion of materials was simply a reaction of these materials with the newly discovered and isolated gas he called oxygen. By carrying out reactions in sealed vessels and weighing the reactants and products, he demonstrated that the oxygen is "a material" that has weight/mass and that all "chemicals" had in fact weights/masses, which could be followed during a reaction. He also formulated the Law of Conservation of Mass in a chemical reaction. He further argued that if one assumed that there was a phlogiston that escaped the inflammable material during combustion, then *this phlogiston must have had a negative weight*! Which is logically impossible. In so doing, Lavoisier recreated "chemistry" as a materialistic science and freed it from metaphysics.

In another way, chemistry is as old as Mankind, and it dates back to the days when Mankind learned to utilize fire as a tool. What a portable tool was fire! It could

FIGURE 11.1   A portrait of Lavoisier by Louis Jean Desire Delaistre, using a design by Julien Leopold Boilly. Courtesy of Chemical Achievers. (This image is in the public domain.)

(a)   (b)

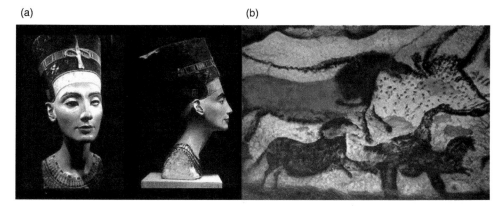

FIGURE 11.2   (a) The bust of Nefertiti, created in approximately 1340 BC. (b) A cave painting from Lascaux, thought to have been created up to 20,000 years ago. These images were obtained from http://commons.wikimedia.org/.

have been taken from one place to another by the wanderer, or set in any place, and importantly, it had the power to change matter. Mankind then learned to reshape matter and fabricate new materials.

Figure 11.2a shows the enchanting bust of Nefertiti, the royal wife of the Egyptian Pharaoh Akhenaten. She lived from approximately 1370 to 1330 BC. Look at the beautiful colors—they show knowledge of chemistry of pigments. The artwork pictured in Figure 11.2b is much older, estimated to be up to 20,000 years old. This is one of the many cave paintings found in the French village Lascaux. It shows knowledge of using pigments and stabilizing their colors: very early chemical know-how!

Another beautiful picture of an Egyptian girl, dated as being from approximately 2,000 BC, is the one shown in Figure 11.3a. It is often pointed out that the human faces drawn on mummies wear generally heavy black-green makeup for women as well as men. At the same time, Egyptian papyri often contained the phrase: "Horus and Ra will protect the eyes." The Roman historians Pliny the Elder and Dioscorides wrote in the first century AD that the procedure for preparing this Egyptian makeup uses a variety of lead-derived compounds, like Litharge and Green Galena (which are now known today to be PbO and PbS, respectively). Dioscorides added that this material is ideal for being applied on the eyes, suggesting perhaps it cures and/or prevents eye inflammation.[1] Much chemistry goes then into this procedure, which in ancient Egyptian times was used for making "antibiotics" to cure eye inflammation. The investigation of this makeup in 2010[1] followed Dioscorides' procedure and generated this ancient mixture of lead compounds. Injection of a lead compound into a single living cell (Figure 11.3b) caused the cell to spew species associated with the immune system's defensive response, namely O_2^- and NO. Apparently, practical know-how of medical chemistry was available thousands of years ago. It was not chemical knowledge as we have it now, but it was good and practical proto-chemistry.

(a)

(b) A Pb salt induced an immune system defensive response.

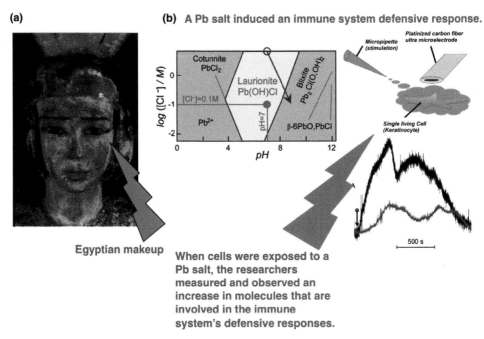

Egyptian makeup

When cells were exposed to a Pb salt, the researchers measured and observed an increase in molecules that are involved in the immune system's defensive responses.

FIGURE 11.3 (a) A face on a girl in an approximately 4,000-year-old piece of art showing heavy makeup around the eyes. (b) Preparation of the material used for the makeup shows that it is a mixture of lead salts. Injecting a lead salt into a living cell causes it to produce molecules involved in the immune system's defensive responses. Reproduced with permission from Figures 1–3 in Ref. 1 (*Anal. Chem*, **2010**, *82*, 457).

The chemistry of metallurgy has also been known for millennia. Here is written evidence from the Bible about a sophisticated process of separating gold ores from cheaper metals:

> "... They are brass and iron; they are all corrupters. The bellows are burned, the lead is consumed of the fire; the founder melteth in vain: for the wicked are not plucked away. Reprobate silver shall men call them ..." (Jeremiah, 6, 28–30).

Here the prophet Jeremiah likens the process of purifying gold from cheaper metals as a metaphor for the character of his people. What he describes is a process of purification of ores that contain, in addition to gold, also some silver, iron, copper, etc. The usage of fire intensified by the bellows (which provide a stream of oxygen, as we know now) causes the oxidation of lead (which becomes, as we know now, PbO/PbO_2). All the cheap metals dissolve into the oxidized lead, except for gold, which floats over the solution and can thus be separated and collected. But in Jeremiah's allegory, the cheap metals (the wicked) were not plucked away (into the PbO_2 solution). His people were adamant to stay bad. The fact that Jeremiah preaches to the masses and uses this metallurgical allegory surely means that many knew about this sophisticated purification process of gold. Hence, we have here a written testimony on the widespread know-how of a chemical process.

Chemistry has accompanied humans throughout their existence in their exercising imagined usages of matter. Matter has been everything for Mankind and has been the chief medium for Mankind's creative manipulations. Matter is chemistry, and chemistry is the science of Man, who is made of matter!

11.3   CHEMISTRY IS EVERYTHING

"Chemistry is everything" was the slogan of the International Year of Chemistry, in 2011, and it has been a motto in this book. It highlights that Mankind and all living organisms are made of molecules and atoms, much the same as any kind of matter on this planet. You have seen already the periodic table. According to the International Union of Pure and Applied Chemistry (IUPAC), the table in Figure 11.4 contains now 114 elements that chemists and other scientists found or produced since Mankind has become conscious of Matter. As I already mentioned, the table is the most compact "book of census of this universe."[2] When we die, our material body disintegrates into smaller molecules (amino acids, hydrocarbons, CO_2, minerals, etc.) and "our"

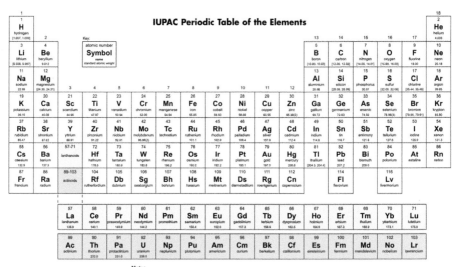

INTERNATIONAL UNION OF
PURE AND APPLIED CHEMISTRY

FIGURE 11.4   The periodic table published by the International Union of Pure and Applied Chemistry (IUPAC). Used with permission of IUPAC. This is the compact census of our planet.

atoms find their way back into the material stock of the planet. Some of them become part of the Earth, others become part of the atmosphere, and still others "incarnate" as part of other humans, living organisms, or plants. *Our atoms are conserved! This is a rule of nature.*

We have already seen quite a few examples that demonstrate this truism that chemistry is everything and everything is chemistry. For example, the chemical transport mechanism of the sperm to reach the egg appears to be dominated by weak interactions of molecules which are released in the female reproductive system with receptors (similar to odor receptors) in the sperm (Figure 1.5 in Lecture 1). The future generation is chemically selected ... Another example is the genetic code, which is regulated and dominated by specific H-bonding between pairs of DNA bases (Figure 7.16 in Lecture 7). In a nutshell, what we are, what we look like, and what our children may or look like, all this is regulated by a H-bond code. Chemistry is everything!

Let me add here a few more examples that underscore this assertion. Since we mentioned DNA, and its little sister RNA, we should mention here the new field of epigenetics (in Greek, *epi* = beyond), which is now one of the hottest fields in science. The mechanisms of gene expression (the ability of the gene to be copied to form RNA and then be used to make the appropriate protein), the mechanisms of gene silencing (a situation where a gene sequence cannot be read and the corresponding protein cannot be made), and the development of the embryo and subsequently of the individual are all parts of epigenetics.

When cell division starts en route to the formation of an embryo, the DNA molecules of all cells are virtually identical. Nevertheless, the cells develop to become specialized brain cells, liver cells, muscle cells, etc. The manner by which this happens is a reaction of DNA with enzymes that modifies the DNA bases and silences particular genes. A particular enzyme, called DNA methyltransferase, attaches a methyl (CH_3) group to some of the cytosine (C) bases of the DNA and generates methylated cytosines, as shown in Scheme 11.1. Those particular genes in which C-methylation occurred are silenced, and they do not produce anymore their specific proteins, most likely because of the disruption of the H-bonding capability of the methylated bases. At the same time, all other genes that did not undergo methylation

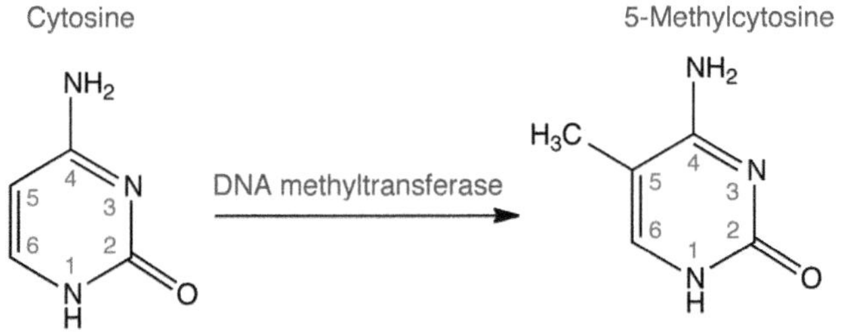

SCHEME 11.1 Cytosine and the form in which it is methylated on the carbon in position 5, a transformation which is accomplished by the enzyme DNA methyltransferase (NH is in position 1, and the numbers increase anticlockwise).

are active. This is one of the ways in which some cells can become brain cells, heart cells, liver cells, etc., which produce their specific proteins and exercise their specific functions.

Furthermore, since DNA is common to all organisms, it may well be that the silencing mechanism is special such that silencing leads to speciation of organisms. Humans have special sets of genes that are different from those of other species.

There are other chemical mechanisms which attach acetyl groups to the protein beads around which DNA double helices are wrapped like a cocoon. These proteins are called histones, and the wrapping serves to compact the DNA and defend it from inflicting self-damage. This is the form in which DNA resides in the nucleus of the cell. Therefore, to be copied, particular DNA genes have to get loose of the tight grip of the histone and become accessible in order to serve as a template to create DNA and RNA molecules. Enzymes that acetylate the histones enable this. Histone acetylation acts as a signal that activates specific parts of the DNA. Other histone methylation patterns serve as a means of deactivation of specific genes. It turns out that if the methylation mechanism is disrupted during the formation of the embryo, this causes the death of the embryo after 10 days! The mechanism of methylation/acetylation is the manner by which enzymes regulate and articulate the genetic plan by chemical means. This regulation accompanies us throughout our lives and is dependent on our diet and environmental conditions. This is an interesting example of Nature and Nurture that is all chemical!

A recent exciting study[3] looked at the gene called *CNTNAP2*, in the cortices of humans versus chimpanzees. *CNTNAP2* is one of the largest genes in the human genome. *It has been linked to human-specific language abilities.* Language ability is considered to be the key factor that differentiated Mankind from the great apes during evolution. Indeed, the *CNTNAP2* study revealed a striking difference in the degree of methylation of the gene, with 28% methylation in the human cortex and 59% in the chimpanzee cortex (the cortex is the part of the brain responsible for thinking abilities). As the authors write, the results suggest that widespread differences in cortex DNA methylation, which were found in comparing *CNTNAP2*, in the two species *were associated with the human–chimpanzee split* and the development of human-specific language and communication traits. If this is indeed the case, then a simple methylation is behind a major evolutionary development. Then, chemistry is indeed everything!

Another interesting example, which may be less awe-commanding than the human–chimpanzee split, but is still hugely thought-provoking, is the story of forest fires. Very recently, the Carmel Forest in the north of Israel was almost consumed by a fire that lasted many days. Not too long afterward, the forest seemed to recover and return back to life. This experience is universal, as Figure 11.5 tells us.

It turns out that smoke-derived molecules will jumpstart the forest's regrowth by promoting fast seed germination.[4] Minuscule concentrations of these molecules, which are called *karrikins* (*karrik* is the Australian Aboriginal word for smoke), can control seedling growth. One way karrikins are generated is by burning cellulose, which is a polymer of sugar and a main component of plants, which decomposes when burning, and its sugar components react with amino acids (coming off of the

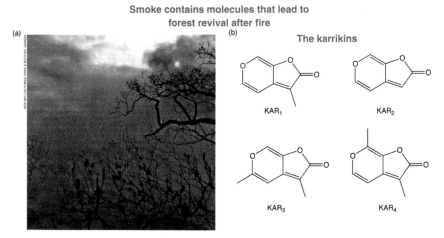

FIGURE 11.5   Small signaling molecules, called karrikins, are found in the smoke of forest fires. They awaken growth genes and are responsible for the fast recovery of a forest after a fire. The photograph is reproduced with the permission of Dr. Ben Miller, an ecologist at the Botanic Gardens and Parks Authority in Western Australia.

plants' proteins). One of the research groups in this field showed that these small molecules stimulate the awakening of growth genes. Most likely, those trees and plants that survived many fires over millennia were those in which this molecular mechanism of gene awakening was most effective. Evolution, life, and chemistry are all entangled. Everything is chemistry!

Again and again, we can see that in the basis of life, there exists chemical information of molecular architecture and interaction, and of molecular motion and chemical reactions. Chemistry is indeed everything, and with this recognition, we are standing at a new dawn: We have created the alphabet of the chemical origins of life and its attributes. *We now have to create the words, sentences, and the text that will bridge between our chemical matter and the less tangible spiritual aspect of living.*

11.4   THE MAGIC OF CHEMISTRY AND PATHOLOGICAL SCIENCE

The entanglement of life and chemistry and the ability of chemists to make discoveries with an immediate impact on humanity have always been strong motivating powers, which have occasionally drifted to pathological science. In the old days, the alchemist was a demiurge whose role was to emulate heavenly perfection and achieve it in her/his lab. In the course of time, this lofty goal has deteriorated into the alchemy of gold making (the "perfect" metal) from the base metals.

In many ways, the chemist has remained a demiurge who can make life-impacting transformations. The chemical magic (which is the essence of the transformation) constitutes a great temptation to seek the unusual. So much so that even well-intentioned scientists sometimes make drastic mistakes. Such was the case with "polywater," a seemingly remarkable substance discovered by Russian scientists in the mid-1960s.

The story was told very well in a book published in 1981,[5] which is reminiscent of the science fiction novel *Cat's Cradle*, written in 1963 by Kurt Vonnegut. In Vonnegut's story, a scientist discovers a new form of water, called ice-9, which freezes at room temperature. Moreover, when ice-9 comes in contact with ordinary water, that water turns into ice-9 and solidifies too. Thus, the dramatic tension comes from the risk that all the world's water could turn into ice-9 and solidify. Imagine life without flowing water!

The scientific story of polywater is extremely similar. It took place only a few years after Vonnegut's fictional story. In the mid-1960s, a Russian scientist named Fedyakin was performing experiments that involved putting water in sealed thin glass tubes (called capillaries). He noticed that in some of the capillaries, a second column of water formed above the water. This second column of water, which kept increasing at the expense of the regular water, was thick and viscous, and it boiled at a much higher temperature than regular water. Fedyakin believed he had discovered a new form of water. He must have been very excited …

After the publication of the paper, the prominent Moscow-based scientist Boris Deryagin took over Fedyakin's research and was also able to produce this "new water." He reported his findings, and being one of the few Russian scientists allowed to leave to the West, he began traveling abroad to present his work in scientific meetings. Some American and English scientists succeeded in reproducing this unusual "water," though not in large quantities and not on every attempt. But in 1969, the American Bureau of Standards examined a sample and determined that it was indeed "a new form of water." They believed that the water had polymerized, or formed long chains or rings of molecules. *They dubbed the substance polywater.*

Meanwhile, the notion of a new form of water sparked an enormous controversy. Those who trusted the experimental results came up with theories to explain what might have caused polywater to form. The most famous theory was based on early quantum mechanics (QM) calculations, which mistakenly (and unintentionally) showed that the water molecules generated "symmetric bonds," –H–O–H–O–H–O– which connect oxygen to hydrogen, and these interactions thereby turned the water molecules into a polymer rather than being the collection of discrete molecules that makeup water as we know it.

Scientific and popular furor followed. Apparently, someone recalled Vonnegut's novel, and the worry arose and quickly escalated into a panic. On September 22nd, 1969, *The New York Times* published an article in this respect, expressing the concern that if polywater were to come in contact with ordinary water, the plain water on our planet would all turn into a viscous polymer too! Scientists were urged to treat polywater as a deadly substance until it was shown definitively to be safe—even though no one could make more than a drop of polywater at a time, and even though no one had ever demonstrated that polywater had any effect on plain water.

In the end, "polywater" samples were subjected to much closer scrutiny and were all shown to contain some contamination with impurities of substances suspended in ordinary water. When the original experiments were repeated with extraordinary care given to cleaning the apparatus, "polywater" could no longer be produced! By 1972, most of the world's scientists considered the case closed, and by 1973, even Deryagin conceded that "polywater" did not exist.

Even as we know that a chemical transformation involves the reorganization of bonds such that the initial molecules are transformed into other ones, still the magic of the chemical transformation is so enchanting that we sometimes fall into believing in the unusual and grandiose, without checking the results with extreme care. Throughout the modern history of chemistry and now also of chemical biology, we find such enchanting stories on a grandiose scale. The 1989 finding of "cold fusion" was one of these stories in which chemists were led to believe that they had found a cheap way of causing the fusion of two hydrogen atoms to form an atom of helium.[6, 7] This is a nuclear fusion process that produces immense quantities of energy. In fact, it happens regularly on the sun, but it requires the surmounting of huge energy barriers. This is possible on the sun, which is a place of intense energy, but not by simple means, as found by the cold fusion researchers. Despite the majority judgment that there is nothing like cold fusion, the prospects of finding cold fusion are so tempting that the topic is still being kept alive by a group of believers.[7]

Another story is the "memory" of water, which was published by a very respected group of scientists in a leading international journal, *Nature*.[8] This paper showed that a water solution with a "remedy" (namely, an antibody) dissolved in it underwent a serial dilution until it contained *not even a single molecule of the remedy*. But nevertheless, this "pure water" still retained a memory of this substance and exhibited the corresponding therapeutic effects. This fantastic claim that "nothingness can produce an effect of something" has been proven wrong and defined by the journal's editor as a "delusion."[9] Nevertheless, more recently, another highly acclaimed scientist claimed that water solutions of bacterial and viral DNA, that were diluted by a factor of 10^{121} and hence could not contain even a single molecule of DNA, were nevertheless radiating electromagnetic waves.[10] This equally fantastic claim is still awaiting its stringent testing. Assuming that nothingness cannot cause something, the observation may be proven to result from some unchecked problems in the experiment. Still, testing such a claim is a must in science!

A fascinating story that was also published in 2010 in the leading journal *Science* is the story of the discovery of "arsenic bacteria."[11] Figure 11.6 shows the essence of the claim, namely that some bacteria, which grew in a pond that contains a high concentration of arsenic, could use arsenic instead of phosphorus in their DNA. The

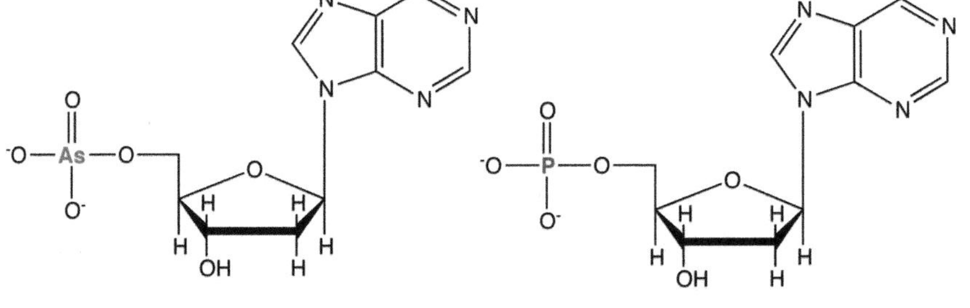

FIGURE 11.6 A hypothesized new DNA nucleotide that contains arsenic, shown side by side with the standard one containing phosphorus.

claim defies everything that we believe about the unity of the DNA world on the planet Earth. At the same time, the prospects of this being true are scary. No wonder that when the news about arsenic DNA broke, their echo and publicity were huge.

Just looking at the arsenic-based nucleotide brings to mind the fear of alien creatures and the risk of having even a single arsenic DNA molecule that will contaminate the natural DNA on the planet. Of course, such a claim/suggestion will undergo many, many stringent tests by chemists and biologists. So far, most of the investigations show that the "arsenic bacteria" have actually a strong preference for phosphorus over arsenic, and that the bond of the AsO_3 to the nucleotide is unstable and cannot sustain a DNA strand.

It seems that the moral of this story is that the magic of chemistry is playing its tricks again … The magic of chemistry creates fantastic ideas. Some turn out to be correct, while others turn out to be false. But, there is no worry; scientists will examine any unusual new idea in the most stringent ways until it fails or passes all the possible tests. Chemists, in particular, still remember in their collective consciousness the days of alchemy, where fantasy eventually took over and drove this science almost into oblivion. They will not let their science go astray.

11.5   THE LOVE OF CHEMISTRY

Most chemists are crazy about chemistry and express their love in writing. Let me give you two examples of this great love of chemistry and science. Carl Wilhelm Scheele (Figure 5.R.2 in Lecture 5), the great Swedish chemist who discovered oxygen but was almost forgotten, wrote to his scientist friend, Johan Gottlieb Gahn:

> "Oh, how happy I am! No care for eating or drinking or dwelling … But to watch new phenomena this is all my care, and how glad is the enquirer when discovery rewards his diligence; then his heart rejoices."

These touching phrases were written by Scheele in 1775, three years after the discovery of oxygen. (You may notice that this corresponds to a different year of oxygen's discovery than that given in section 11.2: Retouches section 5.R.2 gives more detail on the history of the different phases of the discovery of oxygen.)

Another famous statement is attributed to the alchemist Johann Joachim Becher in Figure 11.7, who said:

> "… Chemists are strange class of mortals, impelled by an almost insane impulse to seek their pleasure among smoke and vapor, soot and flame, poisons and poverty, yet among all these evils I seem to live so sweetly, that [I'd die before I'd] change place with the Persian King."

FIGURE 11.7   Johann Joachim Becher (1635–1682). (This image, which is in the public domain, was obtained from http://commons.wikimedia.org/wiki/File:Jjbecher.jpg.)

Let me end this lecture on a somewhat personal note. A few years ago, my spouse, Sara, and I got stranded for one night in a small village of mountain-dwelling minorities of Vietnam, en route from Dien Bien Phu to Hanoi. On the way to this village, we saw little Vietnamese children going to school and carrying their chairs along. In the morning, I went down to have breakfast in the hotel and saw a briefcase and chemical tools: a stative that holds a burette and a *kolba* (a little flask) beneath the burette, as shown in Figure 11.8. This was the equipment that a chemistry teacher was carrying to her/his class …

I was very moved, and as you can see from the blurred photo, I also moved physically. It suddenly dawned on me in this bizarre set of circumstances that chemistry is a universal human culture knowing no boundaries …

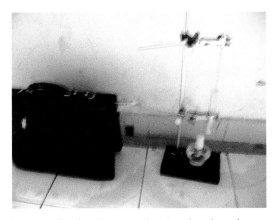

FIGURE 11.8   Equipment of a chemistry teacher in a hotel on the way to Hanoi from Dien Bien Phu (photograph taken by the author).

(a)   (b)

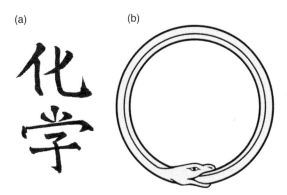

FIGURE 11.9   Symbols for chemistry: (a) the Chinese symbol and (b) the contemporary symbol. (Image (b), which is in the public domain, was obtained from http://commons .wikimedia.org/wiki/File:Ouroboros-simple.svg.)

11.6   SUMMARY

Chemistry is a great human culture. Figure 11.9 shows two symbols for chemistry. Part (a) shows the Chinese symbol for chemistry, which means "the science of change." Furthermore, if one interprets the symbol for chemistry with traditional Chinese philosophy (e.g., Daoism), the meaning of chemistry can be understood as being, "the change of the everlasting." What a beautiful description of the magic of chemistry and the conservation of the atom! Part (b) shows the contemporary symbol, the sign of the Ouroboros,[12] which means "my beginning is my end and my end is my beginning"—the everlasting periodicity of chemical matter in nature.

FIGURE 11.10   Sir Humphry Davy's portrait (left). Davy in a chemistry demonstration (he is shown on the right, behind the bench) presented to the Royal Institution (right). These images, which are in the public domain, were obtained from http://commons.wikimedia.org.

Indeed, everything is chemistry, from the air that we breathe and the food that we eat to our thoughts, joys, and sorrows, and all the way to the protein plaque that may one day cover our brains like a viscous wad. And chemistry is also everything: it is, on the one hand, the intimate knowledge of matter, and on the other hand, it is the constituent of living creatures and the matrix for their well-being. Chemistry is a call for the owner of this knowledge to handle matter with care. We (the chemical community) are still not there, but we are on the way …

Figure 11.10 shows a painting of the great English chemist Sir Humphry Davy. It also shows him participating in a demonstration showing an audience the magic of chemistry in the lecture hall of the Royal Institution, in the middle of London. In our days, when the public image of chemistry is low, it is perhaps most fitting to demonstrate to students that the magic of chemistry is still enchanting …

11.7   REFERENCES AND NOTES

[1] I. Tapsoba, S. Arbault, P. Walter, and C. Amatore, *Anal. Chem.* **2010**, *82*, 457.

[2] F. Tibika, *Molecular Consciousness: Why the Universe Is Aware of Our Presence*, Park Street Press, Rochester, VT, 2013, p. 17.

[3] E. Schneider, N. El Hajj, S. Richter, J. Roche-Santiago, I. Nanda, W. Schempp, P. Riederer, B. Navarro, R. E. Bontrop, I. Kondova, C. J. Scholz, and T. Haaf, *Epigenetics*, **2014**, *9*, 533.

[4] For a cover story, see: *Chemical & Engineering News*, April 12, 2010, p. 37.

[5] F. Franks, *Polywater*, MIT Press, 1981.

[6] J. R. Huizenga, *Cold Fusion: The Scientific Fiasco of the Century* (2nd Ed.), Oxford and New York: Oxford University Press, 1993.

[7] See also: http://en.wikipedia.org/wiki/Cold_fusion

[8] E. Dayenas, F. Beauvais, J. Amara, M. Oberbaum, A. Robinzon, A. Miadonna, A. Tedeschit, B. Pomeranz, P. Fortner, P. Belon, J. Sainte-Laudy, B. Poitevin and J. Benveniste, *Nature*, **1988**, *333*, 816.

[9] See also: http://en.wikipedia.org/wiki/Water_memory

[10] See interview with Luc Montagnier (the discoverer of HIV): *Science*, **2010**, *330*, 1732.

[11] See interview with F. Wolfe-Simon: *Science*, **2010**, *330*, 1734.

[12] W. H. Brock, *The Norton History of Chemistry*, W.W. Norton & Co., New York, 1992, Chapter 16, p. 662.

11.A   APPENDIX

11.A.1   Proposed Demonstrations for Lecture 11

What is more appropriate than ending a course in chemistry with fun demonstrations that make the students feel how enchanting chemistry can be? Each of the following demos is briefly described, and a reference is provided for experimental details and explanations.

- ## **Traffic lights in a flask**

In this demo, one creates a solution that changes its colors like the three colors of a traffic light, yellow, green, and red. A solution of sugar (sucrose or dextrose) is mixed with the base NaOH and the indicator indigo carmine, and the resulting solution is allowed to stand for a few minutes.[1A] Indigo carmine changes its color according to the levels of dissolved oxygen in the solution. The original color of the solution is yellow, and when O_2 oxidizes the indicator (oxygen abstracts H atoms from the indicator), the solution changes color to red and then to green, while when it is reduced by the sugar (the opposite of oxidation; a process called reduction, in this case involving H addition), it turns back to yellow.

To cause these changes, take the yellow solution and slowly pour it from one beaker into another (thus allowing oxygen from the air to dissolve in). The solution will turn red, and then green and then back to yellow. These changes in color are the result of reversible oxidation and reduction reactions that indigo carmine undergoes under the experimental conditions.

- ## **The oscillating iodine clock**

This vivid demonstration belongs to the class of *oscillating reactions*, namely ones that change periodically with time. In this demo, three colorless solutions (A: potassium iodate dissolved in dilute sulfuric acid; B: a malonic acid solution mixed with manganese sulfate and starch; and C: concentrated hydrogen peroxide) are mixed together, producing a mixture that changes among clear, amber, and deep blue colors for about 10 minutes.[2A]

The mechanism of this reaction is not precisely known. Most likely, it involves a series of steps of oxidations and reductions, which proceed in a periodic manner in time. As the species that are formed (e.g., I_2, I^-, and Mn cations of different charges) absorb lights of different wavelengths in the visible region, the color of the solution also changes periodically.

- ## **The silver mirror**

In this experiment, a glass vessel is silver-plated from the inside by mixing together a solution of a silver complex (Tollens' reagent, which contains Ag^+ ions) with a basic reducing sugar (glucose) solution in it.[3A] The sugar converts Ag^+ to metallic Ag, and the latter is deposited on the glass walls. It is advised to rehearse this experiment at least once before demonstrating it …

- ## **The orange tornado**

In this demonstration, potassium iodide (K^+I^-) is slowly added to a vigorously stirred mercuric nitrate ($Hg^{2+}(NO_3^-)_2$) solution. This results in the formation of an orange precipitate of HgI_2 in the form of an "orange tornado," which disappears and re-forms alternatingly for a few minutes. Eventually, when a sufficient amount of potassium iodide solution is added, the orange tornado ceases to re-form, as a soluble, colorless mercuric iodide complex ($[HgI_4]^{2-}$) is formed instead.[4A]

- **The vanishing glassware objects**

In this demo, small glassware objects are made invisible when they are immersed within a beaker filled with vegetable oil.[5A] Borosilicate glass (also known as Pyrex) and vegetable oil have similar refraction indices. Thus, when a small beaker or a stirring rod is immersed within a larger beaker filled with vegetable oil, light does not reflect or refract from the surface between the oil and the object. This makes the object "vanish."

- **A candle in a jar**

In this demo, a lit candle placed within a water container is extinguished by lowering a cylinder over it. As the candle is extinguished, lo and behold, the water level in the cylinder starts rising.

Over the years, there were misconceptions regarding the explanation for this demonstration. The most plausible explanation is given by Birk and Lawson (1999).[6A] In brief, the flame is extinguished partially by the local depletion of oxygen and the increase in accumulated gases (mainly CO_2 and water) given off by the combustion. As the candle burns, the gas inside the beaker expands and escapes from the beaker. When the candle is extinguished, the remaining gas in the beaker cools off, and its volume contracts, thus creating a partial vacuum, which thereby causes a rise in the level of the water.

11.A.2   References for Appendix 11.A

[1A] http://www.elmhurst.edu/~chm/demos/trafficlight2.htm;   http://www.chem.ed.ac.uk/sites/default /files/outreach/experiments/indigoteach.pdf.

[2A] T. S. Briggs and W. C. Rauscher. *J. Chem. Educ.*, **1973**, *50*, 496.

[3A] http://www.rsc.org/images/mirror_tcm18–188809.pdf

[4A] R. G. Silberman, *J. Chem. Educ.*, **1983**, *60*, 996.

[5A] https://www.flinnsci.com/media/621822/91737.pdf

[6A] J. P. Birk and A. E. Lawson, *J. Chem. Educ.*, **1999**, *76*, 914.

EPILOGUE

I tried to present in this textbook a new system of teaching chemistry that may appeal to chemists and nonchemists alike, as it projects the beauty and aesthetics of the molecular universe, and it teaches quite a few complex topics in a fun and chemistry-loving manner.

Yet, I am fully aware that this is a somewhat unusual textbook of chemistry. Firstly, not every teacher may feel at ease to use the "click" or Nirvana" words standing in front of the students. But to share my experience with teaching this course for many years, I can tell the prospective teacher that after a single lecture, saying "click" and/or "Nirvana" becomes a second nature. And more importantly, the students enjoy this and find it playful and a good device for learning chemistry.

Secondly, many/some prospective teachers may feel that it is impossible to teach chemistry without "this or that topic," and certainly not without the quantitative, more rigorous part of chemistry. These omissions were purposeful. I can only share again my experience with you: *If you first teach your students the beauty of molecules, their structures, and their architectural-shaping interactions, the students will be more willing and attentive to learn the quantitative aspects of chemistry in advanced courses.* They will then better understand the motivation for doing all the calculations of moles and pKa, pH, equilibrium constants, usage of energy and rate theory, or quantum theory. If you are still not convinced and not even willing to try, then there may be no way I can convince you.

Chemistry as a Game of Molecular Construction: The Bond-Click Way, First Edition. Sason Shaik.
© 2016 John Wiley & Sons, Inc. Published 2016 by John Wiley & Sons, Inc.

In any event, I hope that the merits of this textbook will outweigh its deficiencies, for which none of my contributors and those whom I acknowledged should be held responsible.

SASON SHAIK

Jerusalem, June 2, 2014

ANSWERS TO PROBLEM SETS

2.1 The periodic table arranges the elements according to their atomic number and groups them into families of related atoms. The rows represent the periods within which the elements are arranged according to increasing atomic number and the columns gather elements into families.

2.2 (a) 15. (b) 15. (c) Phosphorus is located in the fifth column (group 5A) and therefore it has five valence electrons. (d) 3. (e) 3.

2.3 (a) 5. (b) 5. (c) boron is located in third column (groups 3A) therefore it has three valence electrons (d) five electrons are needed to reach Nirvana (f) No. (g) It will generate an electron-deficient compound.

2.4 (a) O: Using the connectivity rule suggests that the simplest molecule will be O_2. Since oxygen is a double connector, and it contributes two electrons for bonding, it should form a double bond with another oxygen atom:

$$\ddot{\underset{\cdot\cdot}{O}}\colon \quad \colon\ddot{\underset{\cdot\cdot}{O}}\colon \quad \xrightarrow{\text{"click"}} \quad \colon\ddot{O}\!=\!\ddot{O}\colon$$

In this molecule, each oxygen atom has two bond pairs and two lone pairs, and hence eight electrons surround each O. As we shall learn in Retouches section 3.R.3 for Lecture 3, the molecule *is indeed* O_2, but the bonding of this molecule

Chemistry as a Game of Molecular Construction: The Bond-Click Way, First Edition. Sason Shaik.
© 2016 John Wiley & Sons, Inc. Published 2016 by John Wiley & Sons, Inc.

is exceptional. O_2 performs partial pairing and creates a magnetic molecule. Nevertheless, the present answer is OK for the time being ...

(b) He: Helium has two valence electrons, and it is located within the noble gas column in the periodic table. This means that it has already reached the state of Nirvana and as such will not tend to form molecules with covalent bonds.

(c) In the HCN molecule (which is also known as the poisonous hydrogen cyanide gas, which has a smell of almonds), we find the following connectivities, depicted in red, around the atoms:

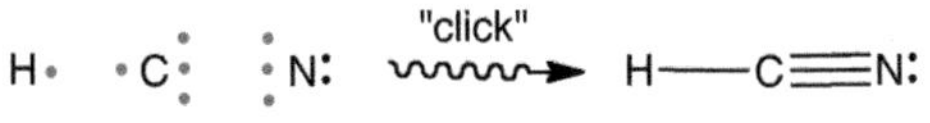

Remember, the term "connectivity" is the number of electrons the atom contributes to bonding.

- The hydrogen atom that has a single connectivity forms a single bond, and hence, it will be surrounded by two electrons.
- The carbon atom is a four connector. It forms four bonds, and hence, eight electrons will surround it in the molecule.
- The nitrogen atom is a triple connector. It forms three bonds and one lone pair, and hence, eight electrons surround it in the molecule.

(d) The beryllium atom is a double connector, and hence, it will require two Cl atoms, generating $BeCl_2$ as follows:

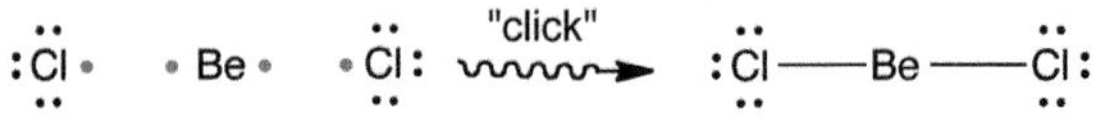

- In this molecule, Be is surrounded by only four electrons / two electron pairs. As we shall learn in section 3.2, chapter 3 that beryllium does not reach the state of Nirvana and forms an electron-deficient molecule.
- Each chlorine atom in the molecule has a single bond and is a single connector, donating one electron. As such, it reaches Nirvana with one bond pair and three lone pairs.

(e) Two molecules can be constructed by usage of two nitrogen atoms and hydrogen atoms:

Molecule 1: In the N_2H_2 molecule, each nitrogen atom is a triple connector. It will form three bond pairs and will possess in addition two nonbonding

electrons in a lone pair. As such, each N will be surrounded by eight electrons. Each hydrogen atom is a single connector, thus forming one bond, and will be surrounded by two electrons.

Molecule 2: In the N_2H_4 molecule (so-called hydrazine and serving among other things as rocket fuel), each nitrogen atom contributes three electrons for bonding. As such, each N has three bond pairs and one lone pair, and hence, eight electrons surround each N. Here, all the bonds are single bonds, in contrast to Molecule 1.

- Each hydrogen atom contributes one electron for bonding, and hence, it is surrounded by two electrons.

2.5 (a)

(b) The principle of conservation of the atom requires the following balanced equation:

$$CH_4 \; + \; 2O_2 \; \longrightarrow \; CO_2 \; + \; 2H_2O$$

(c) The combustion of the "natural gas" fuel releases energy in the form of heat (and fire). As explained in Lecture 2, making chemical bonds releases energy into the environment, whereas breaking chemical bonds consumes energy from the environment. Looking at the equation for the reaction, the bonds that are being broken are C—H bonds and an O—O bond. The bonds that are being made are O—H and C=O bonds. Given that the combustion reaction releases energy means that the bonds formed in the reaction are stronger than the bonds that are broken.

LECTURE 3: ANSWERS

3.1 BF_3 is electron deficient. It lacks a single electron pair. As such, the boron in BF_3 is a single-connector center. F^- has four lone pairs, so it can donate one of these to BF_3, and "click," we have a new bond:

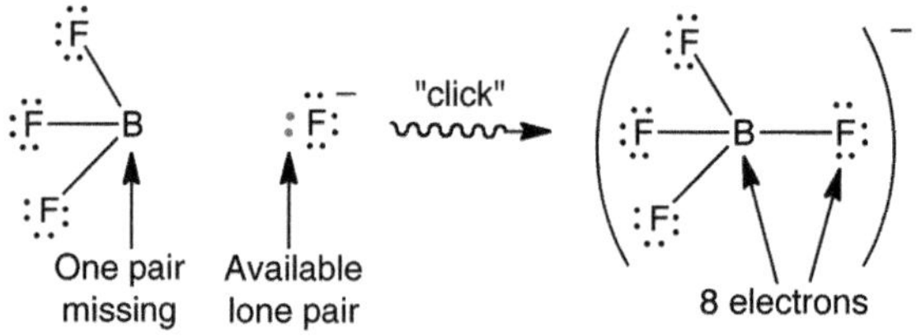

One pair missing   Available lone pair   8 electrons

In the BF_4^- ion, both B and F have attained Nirvana.

3.2 N is a triple connector, while O is a double connector, and hence, NO will not have an even number of electrons. It will be a free radical. This is apparent from the following drawing:

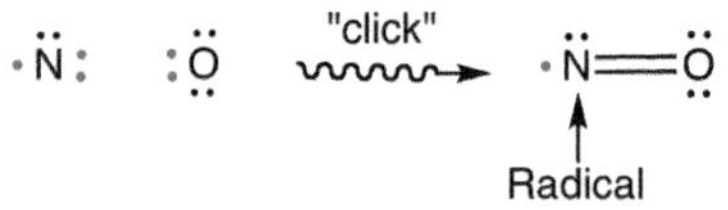

Radical

3.3 Since NH is a double connector, it requires two CH_3 fragments, and the molecule is constructed as follows:

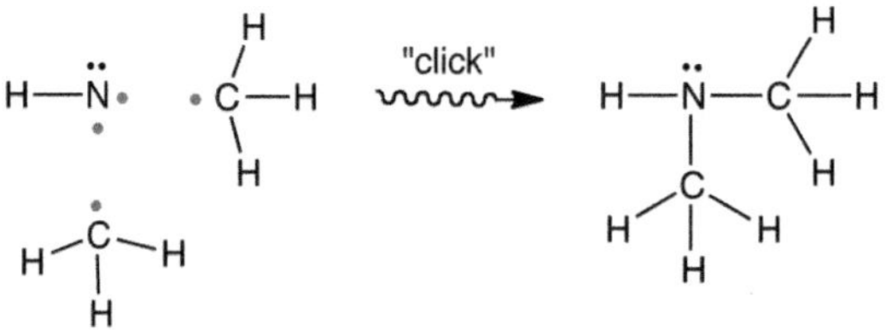

The resulting molecule is an amine.

3.4 Since H_2C and HN are both double connectors, the resulting molecules are the following:

(a)

(b)

3.5 Because both HC and N are triple connectors, the minimal molecule will have a triple bond as shown in the following drawing:

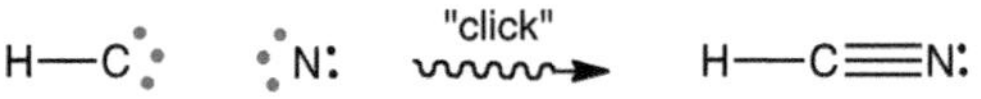

Note that this is the same deadly gas in Answer 2.4 (c). Only that now you are using the modular fragment HC, rather than starting from the atoms.

3.6 This is done by unclicking successive X–F bonds from these molecules (where X = C, N, and O). The so-generated fragments are depicted in the connectivity table below:

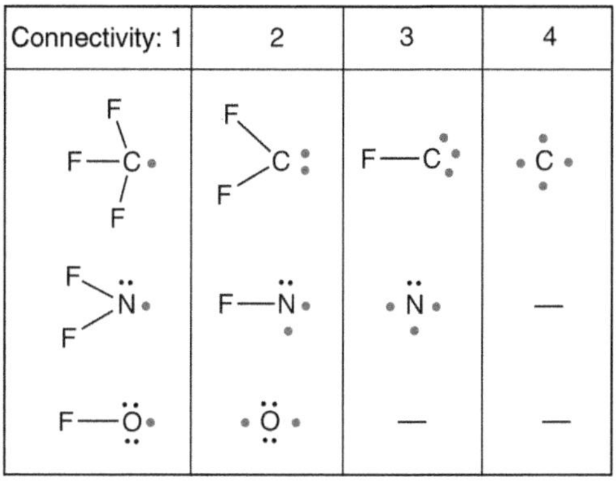

Thus, CF_4 is analogous to CH_4 and will give rise to four different fragments with connectivities of 1–4. NF_3 will generate three fragments with connectivities of 1–3, and OF_2 will give rise to two fragments with connectivities of 1–2.

From F_2C and O, which are double connectors, we generate the double-bonded $F_2C{=}O$ and $F_2C{=}CF_2$ molecules. From FC and N, which are triple connectors, we generate the triple-bonded $F{-}C{\equiv}N$ and $F{-}C{\equiv}C{-}F$ molecules.

3.7 By replacement of C, N, and O by their next family members, we get the following table of connectivity:

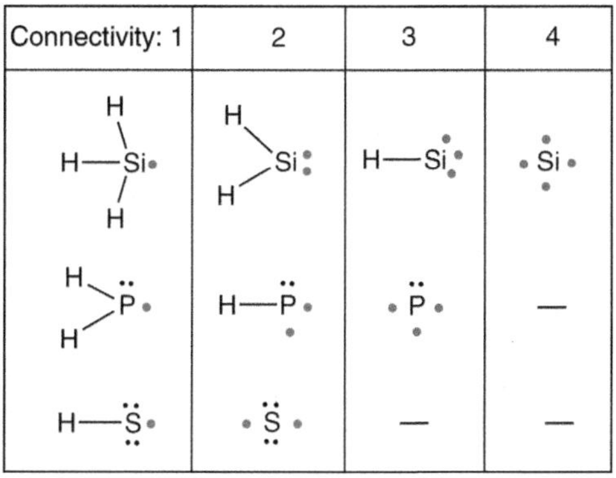

Try making molecules with these and other fragments. For example, S can combine with two $H_3C\bullet$ fragments to give $H_3C{-}S{-}CH_3$, and two $H_3CS\bullet$ fragments can combine into $H_3C{-}S{-}S{-}CH_3$ or with S into $H_3C{-}S{-}S{-}S{-}CH_3$. Similarly $HS\bullet$ and $H_3C\bullet$ can combined to $H_3C{-}SH$. Did you know that the latter two molecules help to give cheddar cheese its special odor?

3.8 A bonus question means that it is not a simple one. So do not feel alarmed if you could not solve it without looking at the answer. In BH_3, boron misses a pair of electrons compared with how many it needs to reach Nirvana. What happens is that two molecules cooperate and use one of their B—H bond pairs to be shared with the boron atom in the other molecule. These shared B—H bonds form bridged B—H—B moieties, which have electron pairs belonging to the three centers, as shown below:

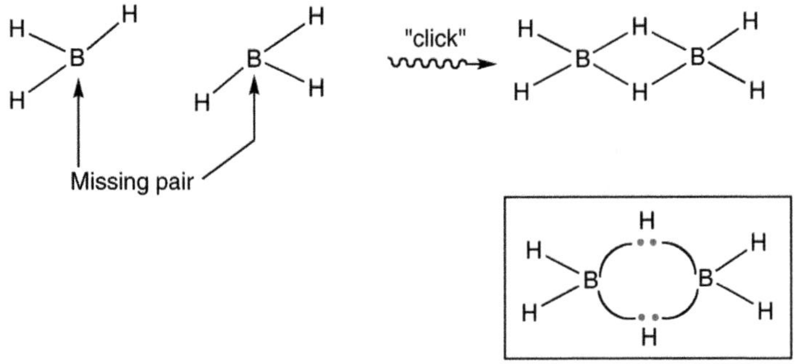

Here, each boron atom has two B—H bond pairs, and it shares two additional electron pairs with the bridging H's and the other boron atom (as shown by the curved line connecting the two boron atoms and the electron pair in the center). In such a way, boron sort of satisfies its octet. All the H's have electron pairs around them.

LECTURE 4: ANSWERS

4.1 The isomers are drawn below. In part (a), we show explicitly all the C—H bonds, while in (b), we simply write the groups, for example, CH_3.

To fully answer this question, we start from the straight chains, using a chain of CH_2 units, and two H_3C caps for the ends. Then, we replace one CH_2 by a CH fragment and add a H_3C cap. Then, we should continue to replace more CH_2 groups by CH fragments and add a H_3C cap for each such replacement. Finally, we replace a CH_2 by a C fragment, and then, we need three H_3C caps, and so on. This is done in the drawings below:

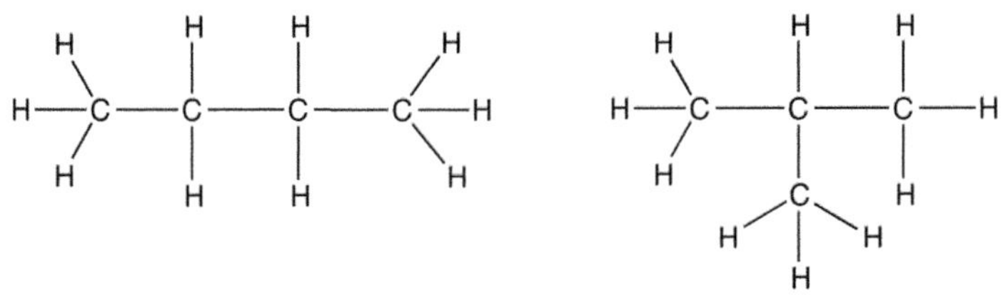

(b)

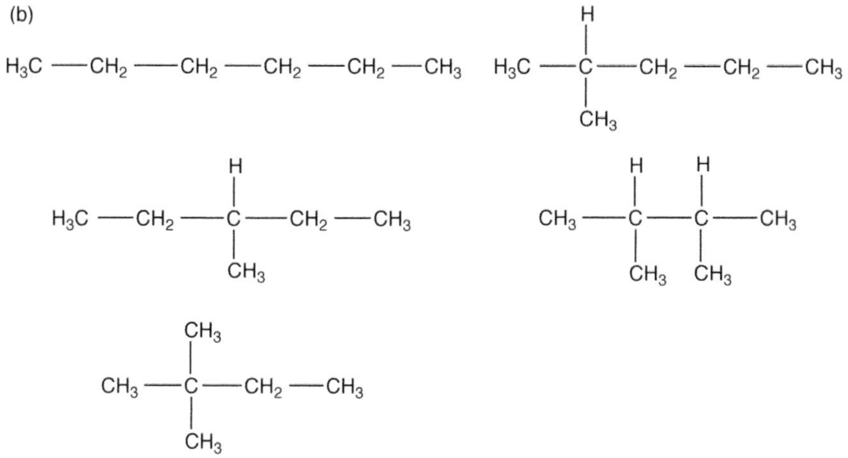

For (a), there are only two isomers, one with a straight chain, the other branched and arising from the replacement of $H_2C \rightarrow HC$. For part (b), there are three molecules corresponding to such replacements (including one molecule arising from the replacement of two CH_2 groups with CH groups) and one $H_2C \rightarrow C$ replacement.

4.2 Clicking CH fragments leads to two major isomers, one with alternating C=C and C–C bonds, and one a cage type [$(CH)_8$ has more isomers. Try …]. These are shown below:

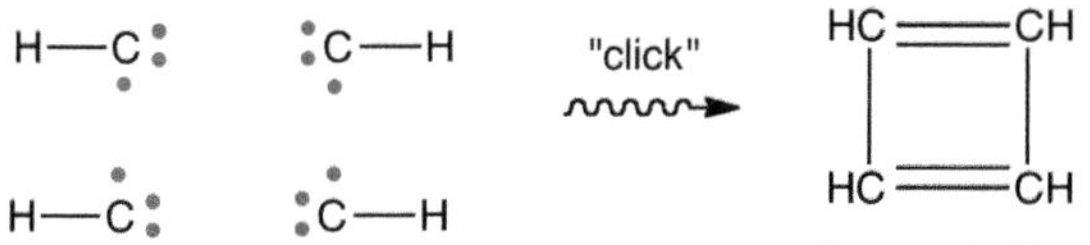

Cyclobutadiene
(C_4H_4)

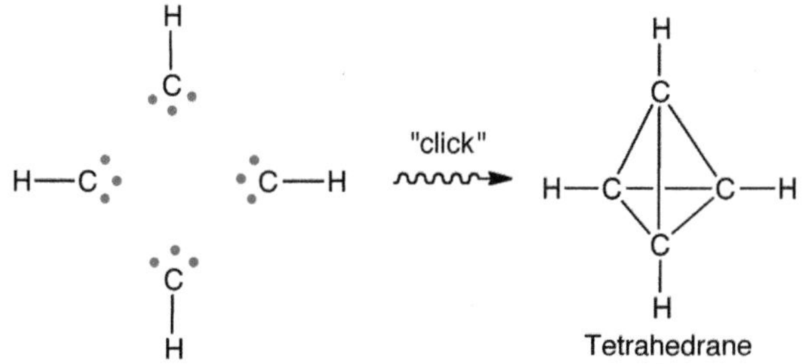

Tetrahedrane
cage
$(CH)_4$

Similarly

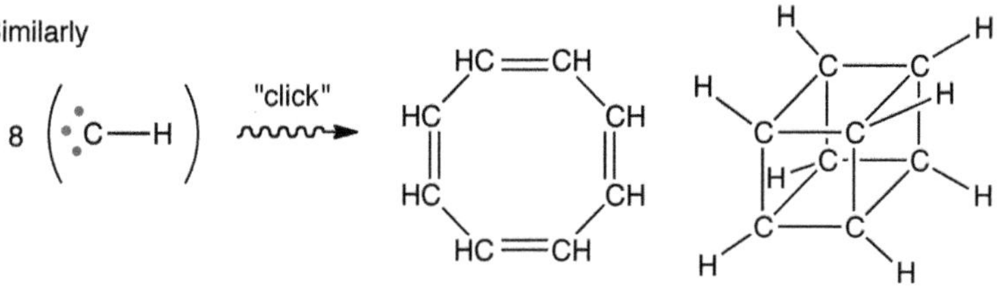

4.3 Clicking CH and C fragments into fused hexagons requires an even number of C fragments, which occupy the ring fusion positions. The HC fragments will occupy the rest of the positions. The so-constructed molecules are drawn below:

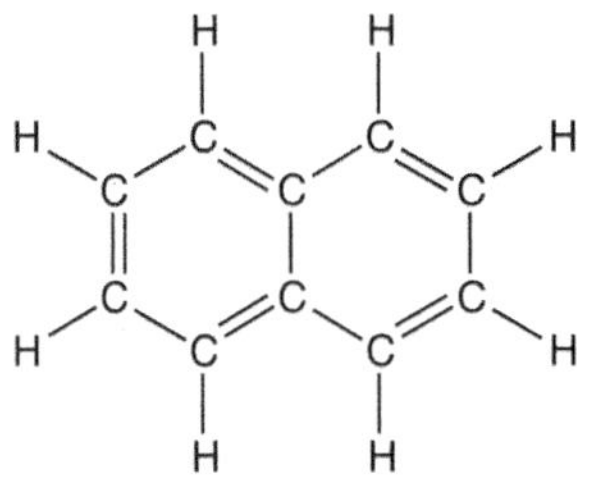

Naphthalene

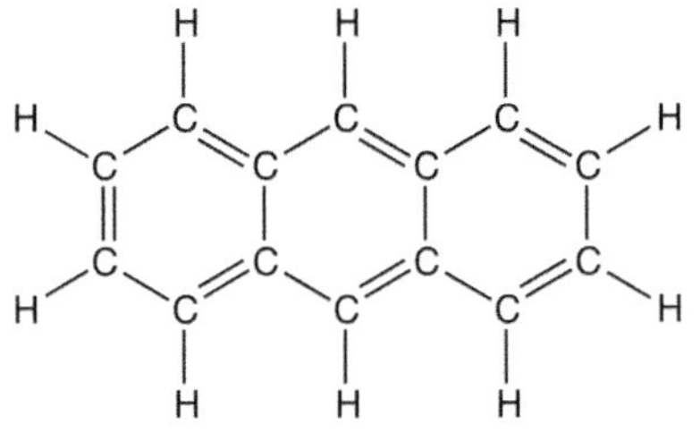

Anthracene

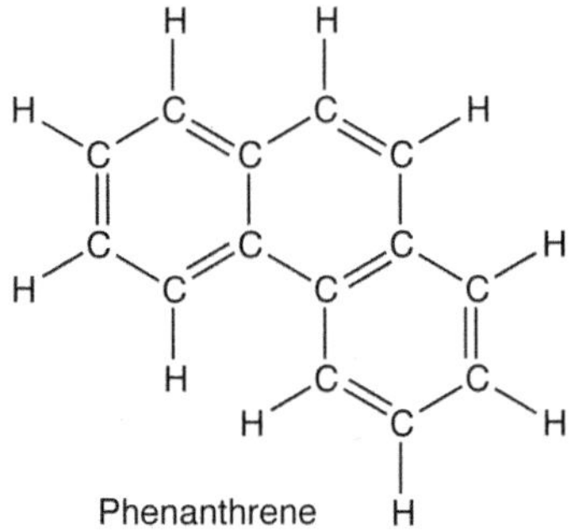

Phenanthrene

The first molecule, so-called naphthalene, is $C_{10}H_8$, which possesses a single C—C fusion unit connecting four HC fragments on each side of the fusion unit. For $C_{14}H_{10}$, we have two C—C fusion units. If we align them, we create three hexagons in a line, in the molecule called anthracene. If we line the fusion units skewed, the resulting hexagons will form a curved molecule, so-called phenanthrene. You can think of the entire family of these molecules made of polyhexagonal molecules and construct straight and curved ones by abstracting, successively, two C—H bonds and adding C_4H_4 units. Many of these molecules, which are known by the name "polynuclear aromatics," exhibit like benzene a carcinogenic (cancer-causing) activity.

4.4 Drawing these molecules is definitely not easy. Do not be discouraged. Instead simply try to follow the answer.

Since these structures were constructed in the text, we use them here as starting points. Keeping the directionality of the connectivity as in Scheme 4.10b and 4.10c, we click 12 and 16 H• fragments, respectively, and make the beautiful molecules shown below:

The first is the chair structure, C_6H_{12}, and the second is the cage structure $C_{10}H_{16}$. The latter is a small diamondoid fragment. Note that we could have constructed C_6H_{12} from six double-connector CH_2 fragments, and $C_{10}H_{16}$ from six CH_2 fragments and four triple-connector CH fragments.

4.5 The first fragment in Scheme 4.13 is a single connector, while the other two are double connectors. Another fragment of interest, not explicitly shown in Scheme 4.13, is a triple connector. Clicking the connectivities, we can generate the following four molecules:

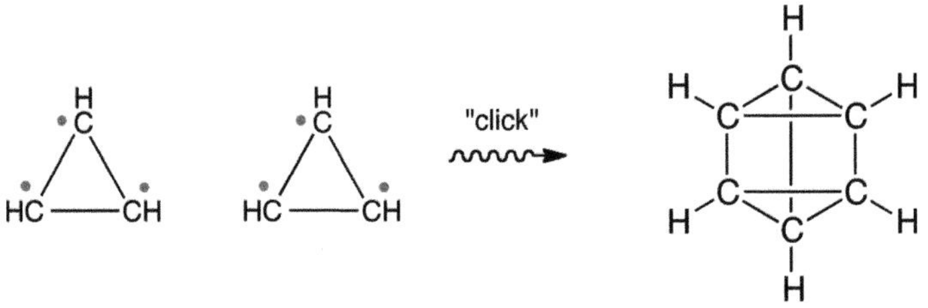

4.6 This is similar to the previous answer. By clicking the connectivities of the three fragments, each with an identical fragment, we get the following molecules:

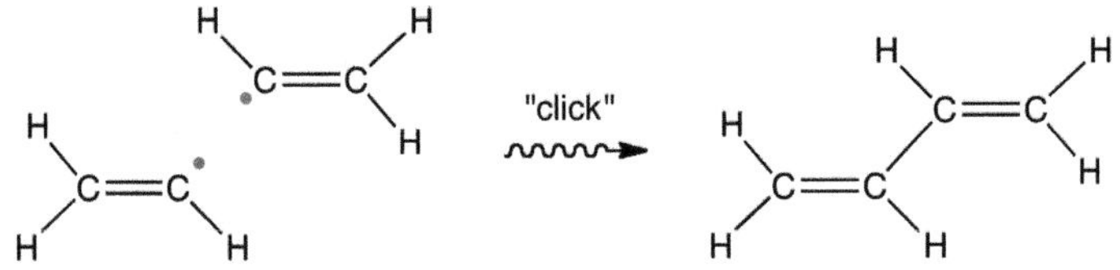

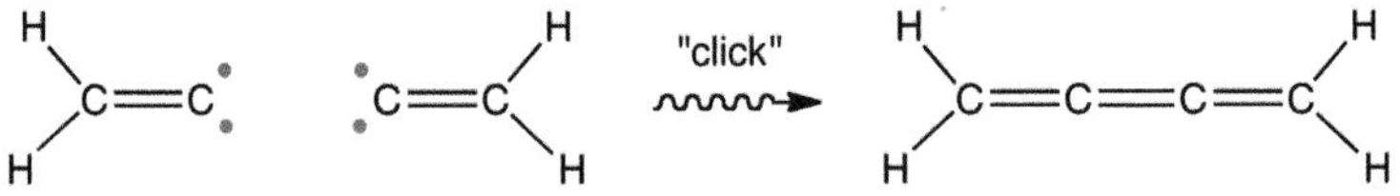

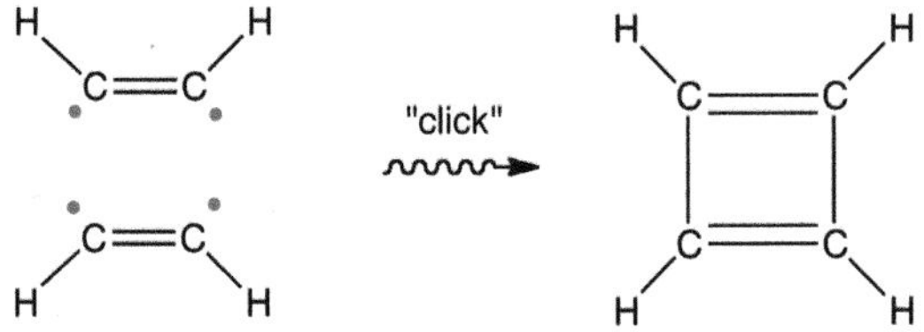

4.7 $2C_nH_{2n} + 3nO_2 \quad 2nCO_2 + 2nH_2O$

4.8 If you tried, you probably understand now C_{60}'s structure quite well!

LECTURE 5: ANSWERS

5.1 We use the fragments in the two schemes and adopt the implicit representation without C and H atoms in the skeleton. Thus, we can generate one monophenol having a single OH group on the benzene ring, and three different isomers of diphenols, as shown below:

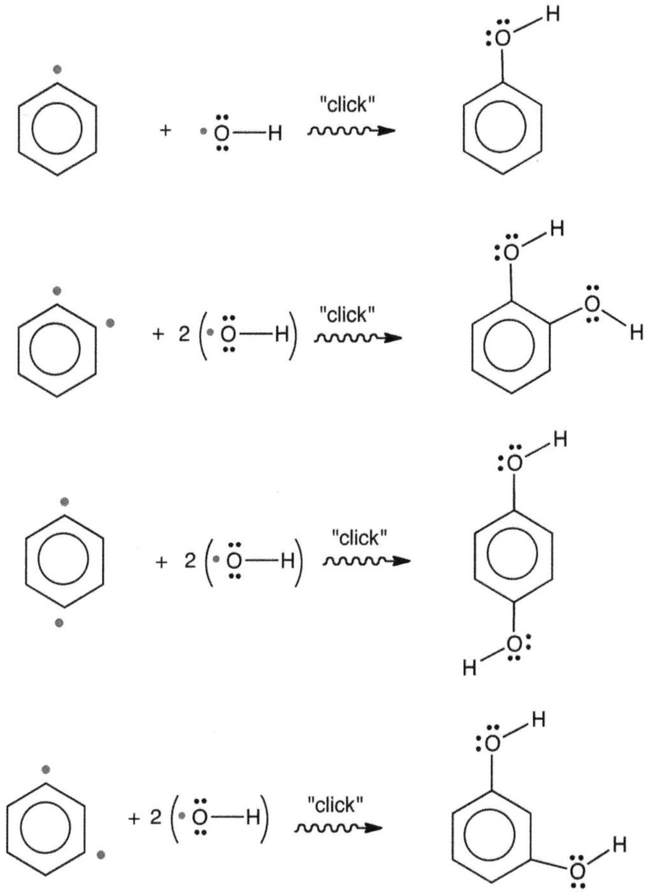

Searching on Google, one can find many uses of phenols. Some are used as disinfectants, others for making polymers. They make up the polyphenols in red wines. The Chinese lacquer coating is made from a polyphenol …

5.2 The key step is to identify the constituent fragments. Then we assemble them and click their connectivities. This is shown below, where we omit C and H atoms on the benzene ring:

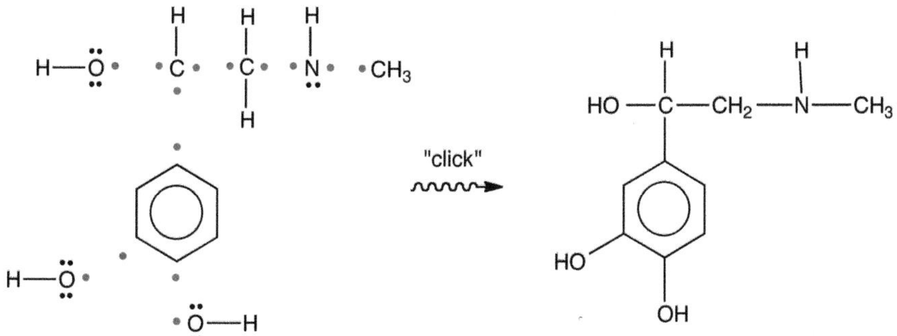

Adrenaline

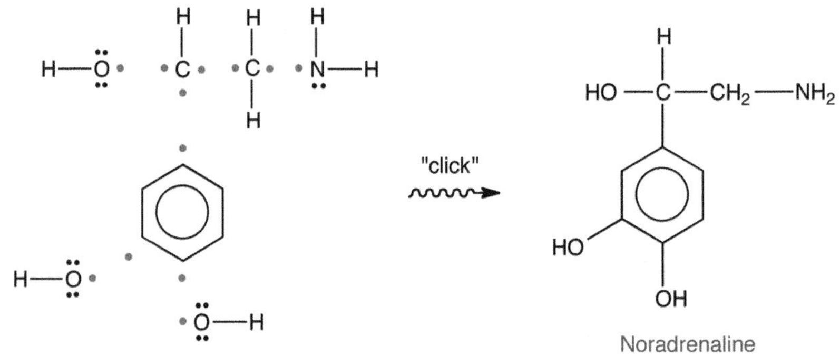

Noradrenaline

Note also that in the final drawings, we did not add the lone pairs. This is what chemists usually do when they consider the details to be too obvious to be included in the drawing.

Adrenaline and noradrenaline are part of the body's reaction to stress.

5.3 To exemplify, we use melatonin in the drawing below:

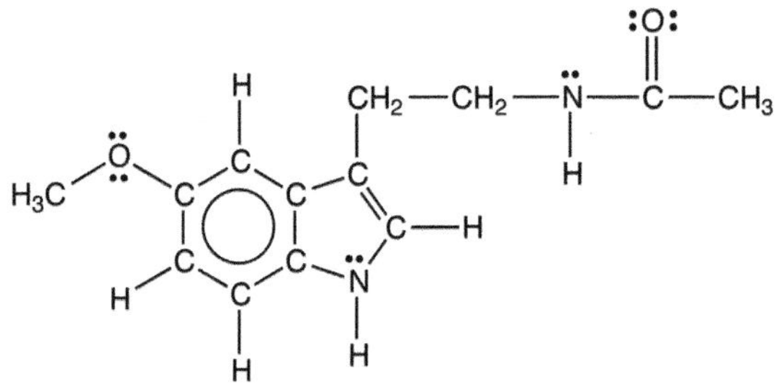

Melatonin

Please complete the rest on your own.

5.4 To exemplify, we use vanillin in the drawing below:

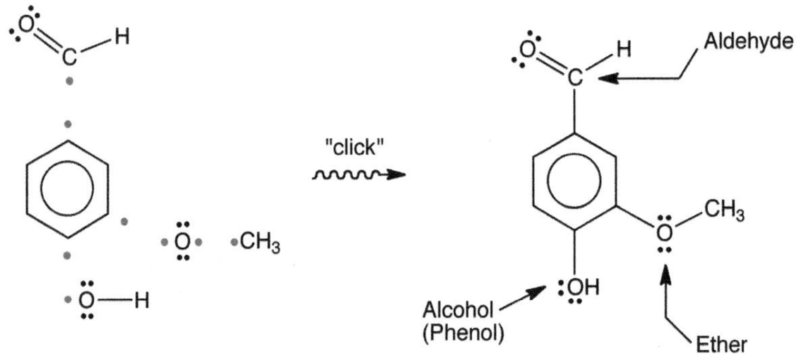

Vanillin

Please complete the rest on your own.

5.5 To exemplify, we use palmitic acid in the drawing below:

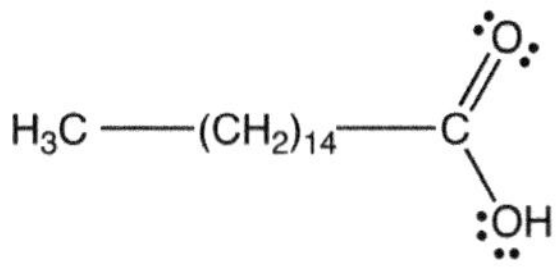

Palmitic acid

Please complete the rest on your own.

5.6 To exemplify, we use alpha-linolenic acid in the drawing below:

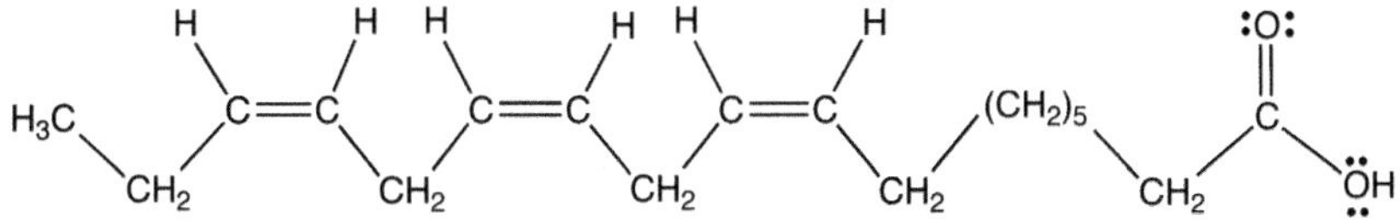

Alpha-Linolenic acid

Please complete the rest on your own.

5.7 The ester formation process involves excision of a water molecule from acetic acid and ethyl alcohol, as in the following drawing:

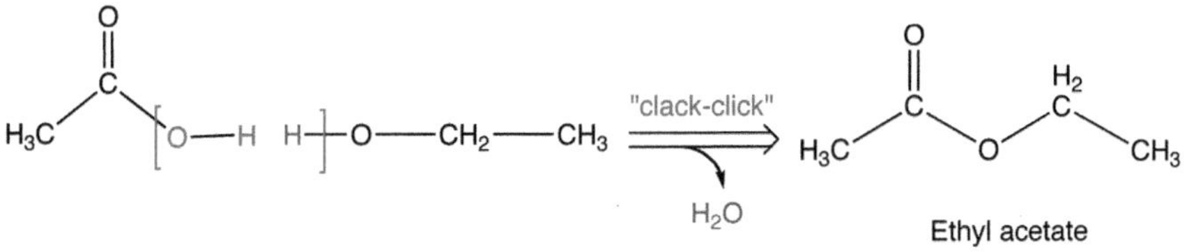

Ethyl acetate

5.8 Getting the components of a triglyceride is the reverse process of its formation. As such, we add a water molecule to each of the C–O bonds, and we obtain the glycerin molecule and the three fatty acids, as shown below:

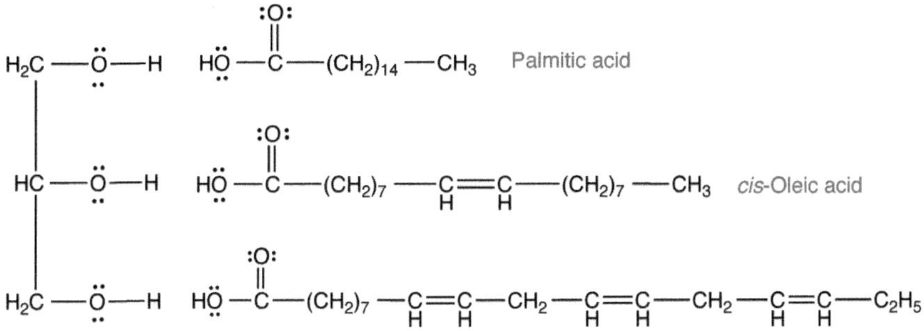

Glycerin   Alpha-Linolenic acid

5.9 The triglyceride is the ester shown below:

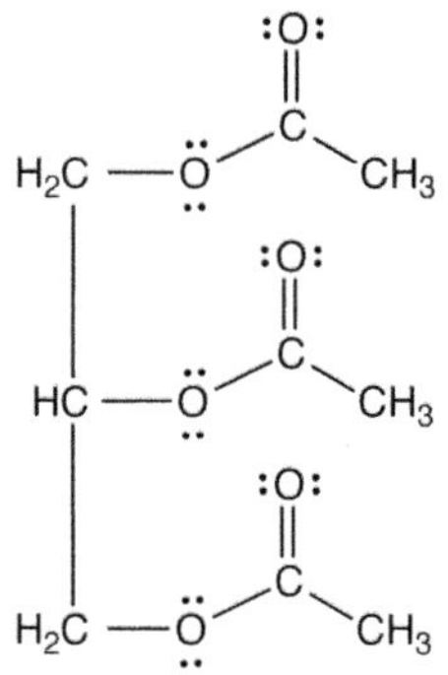

5.10 The proline ethyl ester is generated from the amino acid proline and the alcohol ethanol. The ester is shown below:

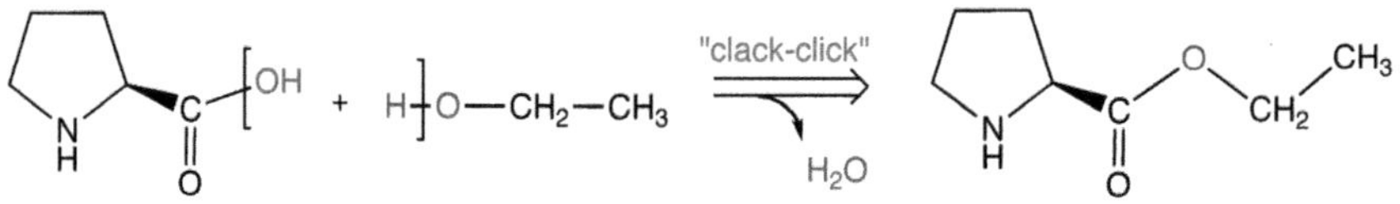

5.11 There are four different dipeptides. Two are made from identical amino acids, and the other two from one proline and one alanine. The four molecules are drawn below along with the names taken from Appendix 5.A.1:

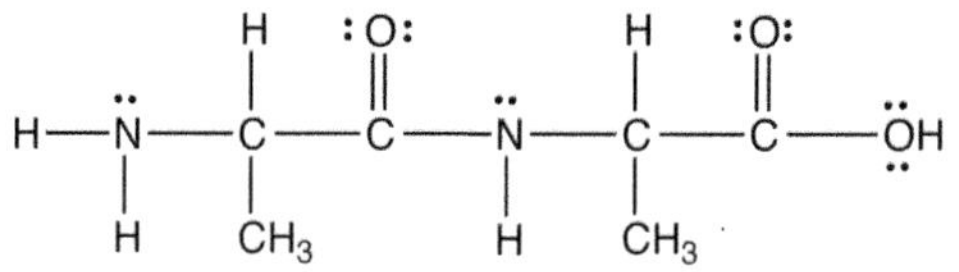

Ala-Ala (AA)

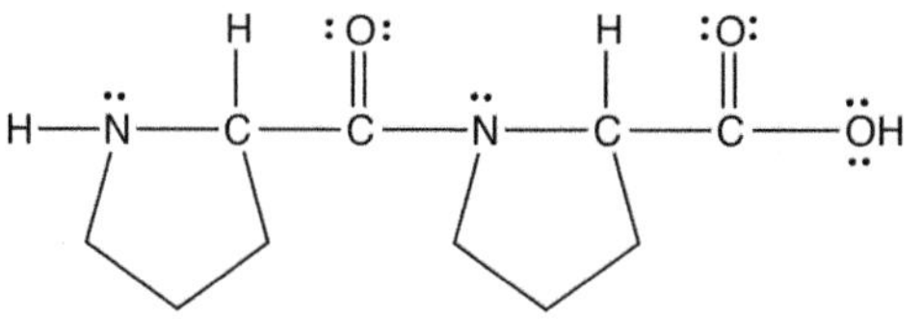

Pro-Pro (PP)

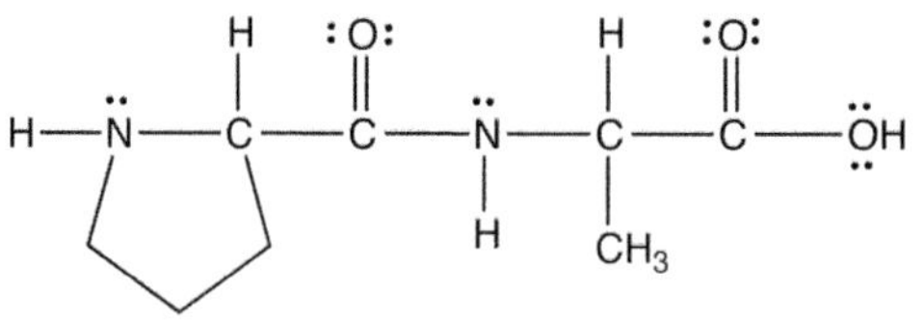

Pro-Ala (PA)

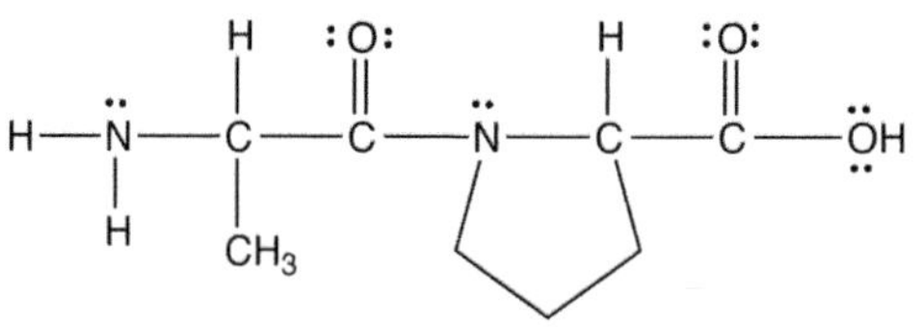

Ala-Pro (AP)

Note that the latter two dipeptides differ in the identities of the two ends. If you construct these molecules with the "clack-click" method, this difference will be apparent.

5.12 The easiest way is to draw the molecule in three steps as shown in (i)–(iii) below:

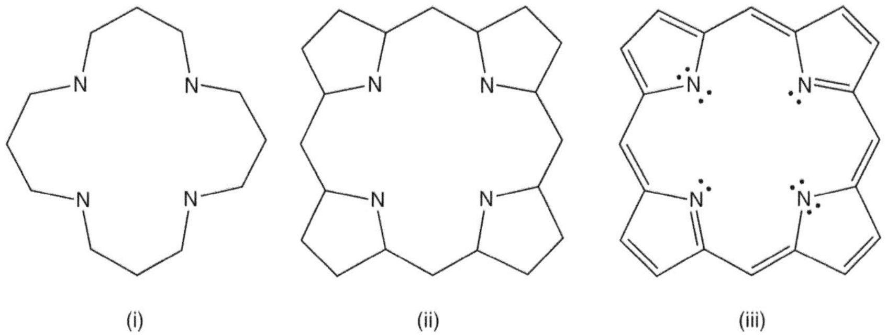

(i)   (ii)   (iii)

In the last step, we add the double bonds and the lone pairs on the nitrogen atoms.

Note that two of the nitrogen atoms must carry a negative charge (count electrons to prove to yourself). The neutral porphyrin has two N–H bonds. The H's are removed as protons when porphyrin makes the heme molecule.

LECTURE 6: ANSWERS

6.1 According to the qualifications, all the oxygen atoms must be linked to the sulfur atom. Accordingly, as we can see below, in all the molecules, S attains an electron-rich environment:

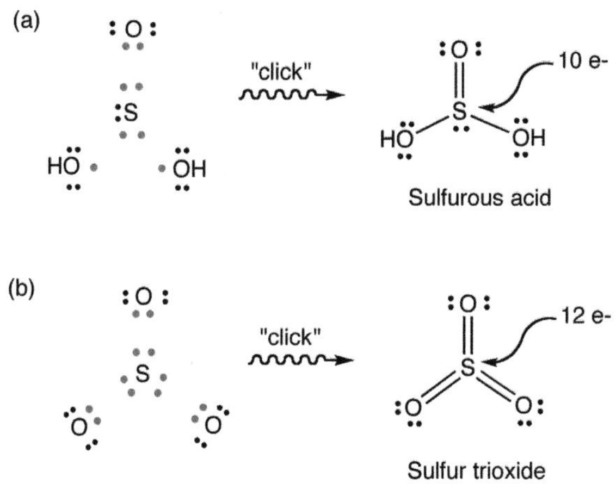

(a)

"click"

Sulfurous acid — 10 e-

(b)

"click"

Sulfur trioxide — 12 e-

(c)

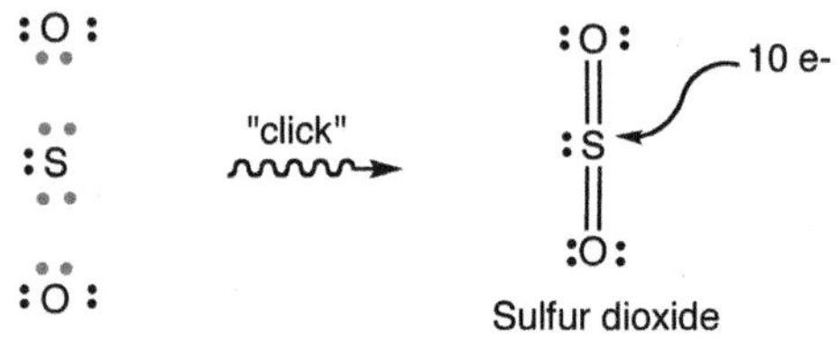

Sulfur dioxide

6.2 Based on the qualifications, all oxygen atoms must be bonded to Cl. Therefore, Cl assumes the electron-rich environment (with 14 electrons around it) shown below:

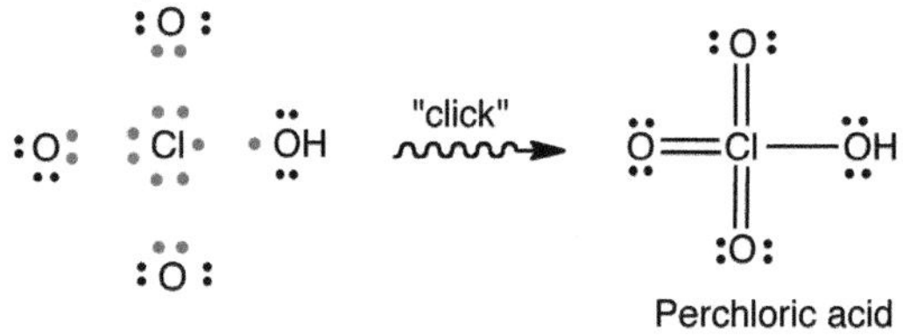

Perchloric acid

F strictly obeys the octet rule. The analogous HFO_4 molecule does not exist.

6.3 The best way to enumerate all the combinations is by using a table where the columns and rows are marked by the four nucleotides, A, C, G, and T, while all the squares inside the table have combinations of the two nucleotides. This table is shown below:

	A	C	G	T
A	AA	AC	AG	AT
C	CA	CC	CG	CT
G	GA	GC	GG	GT
T	TA	TC	TG	TT

It is seen that on the diagonal of the table, we have the dinucleotides made from identical units, for example, AA. In all other squares, we have nucleotides made from two different units, for example, AC. Note that, in the latter case, nucleotides appear in two orderings, for example, AC and CA. These are different molecules as may be seen from the example of AC and CA below:

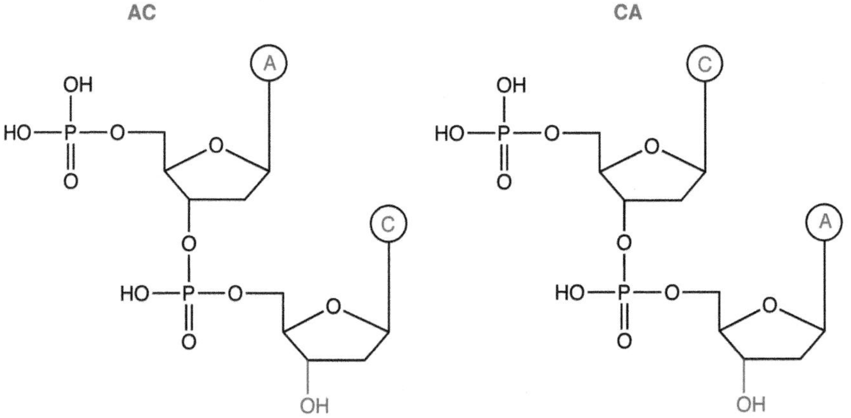

AC

CA

6.4 The fragments needed to make an artificial nucleotide are shown below:

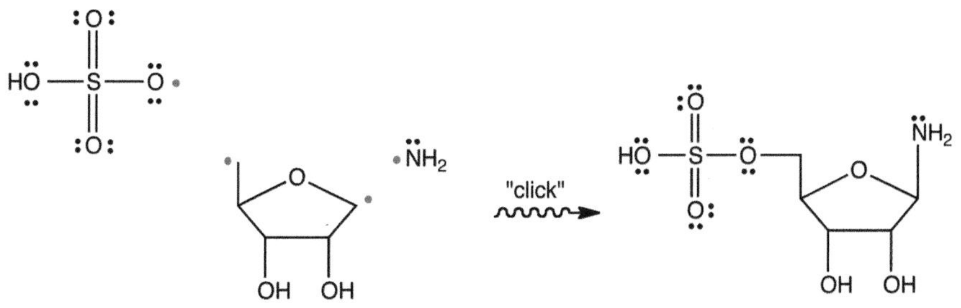

Note that as a base, we have chosen the NH_2 fragment derived from ammonia. There is an infinite number of bases that could have been used.

6.5 Using phosphoric acid and ethanol, we can make the following ester:

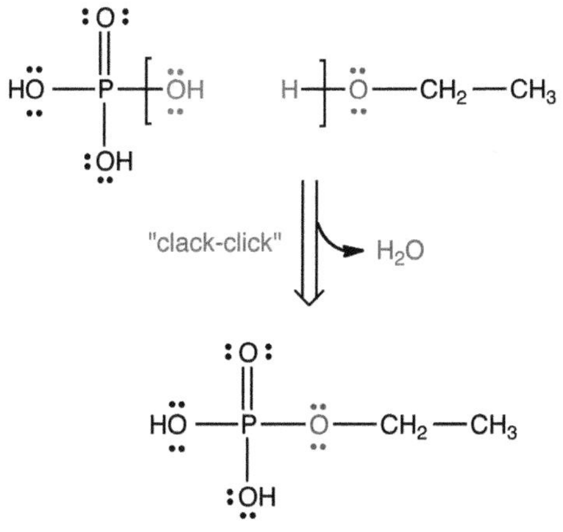

6.6 The construction of acrylonitrile and its radical polymerization are shown below:

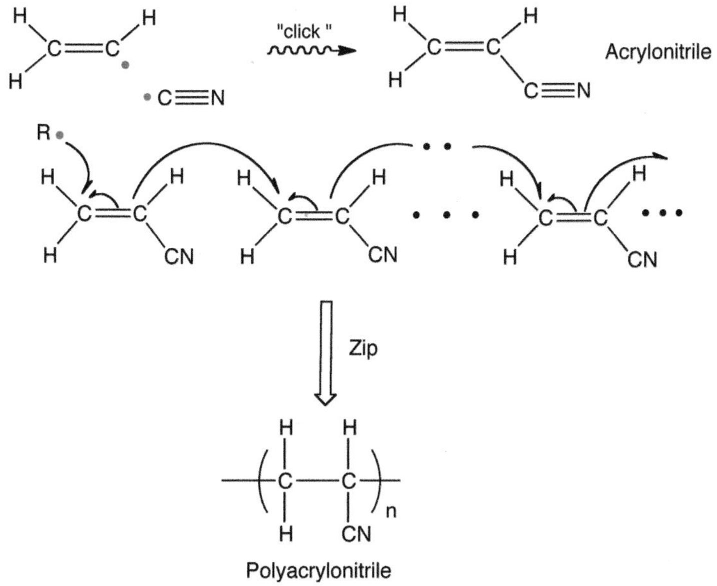

Polyacrylonitrile

Note that here and thereafter, we do not write the end groups of the polymer (e.g., the radical initiator and the terminal group that stops the polymerization, e.g., an abstracted H atom).

6.7 The initiator loses an N_2 molecule and thereby creates a free radical, as shown in the first chemical equation. Subsequently, the radical initiates the polymerization of styrene, as shown in the second equation.

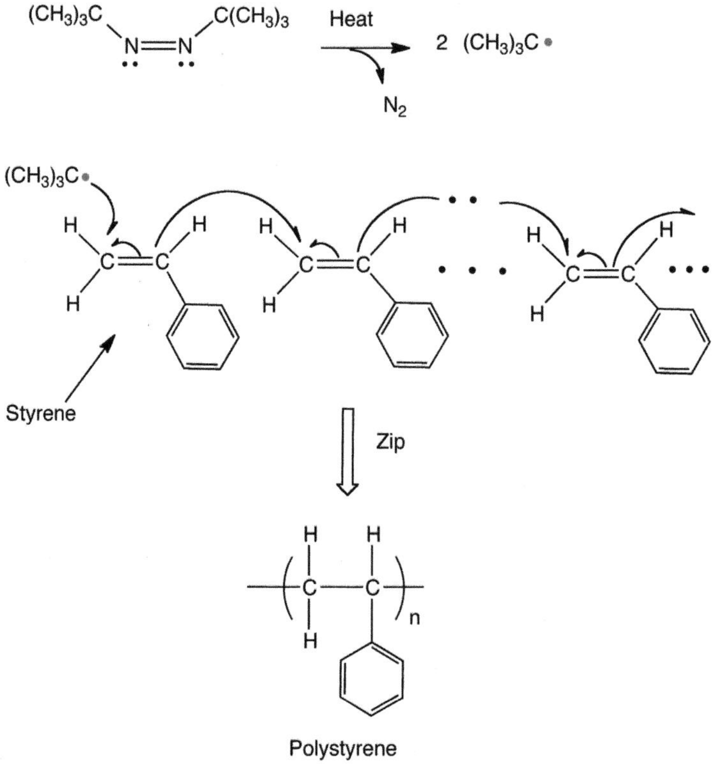

Styrene

Zip

Polystyrene

6.8 The copolymer is formed by extrusion of water molecules, forming ester bonds of the acid with glycol. This is exemplified for one acid, H_2CO_3, below:

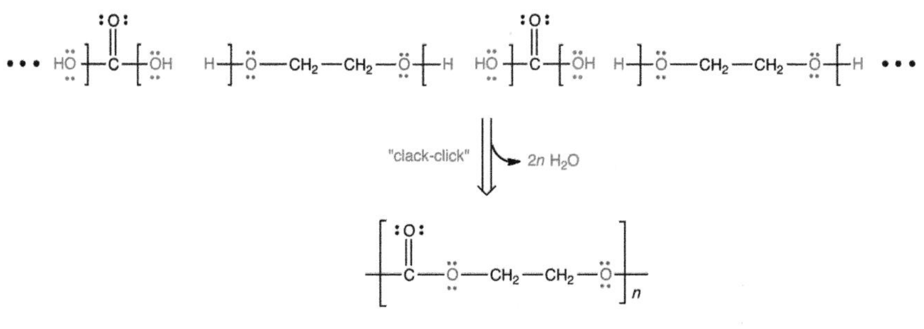

"clack-click" $2n\ H_2O$

Please complete the H_2SO_4 component of this problem on your own.

LECTURE 7: ANSWERS

7.1 Be is a double connector and hence it forms $BeCl_2$. But since the isolated $BeCl_2$ molecule is electron deficient, it will aggregate by accepting lone pairs form the Cl atoms of another molecule. The two structures are illustrated below:

(a)

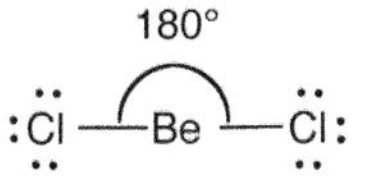

$$180°$$

:Cl —Be —Cl: Linear

(b)

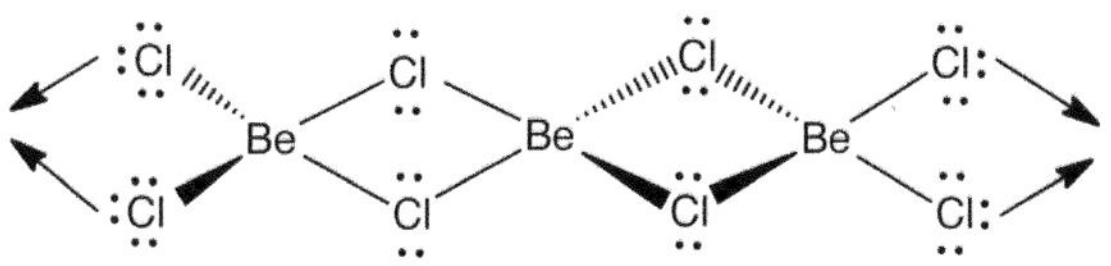

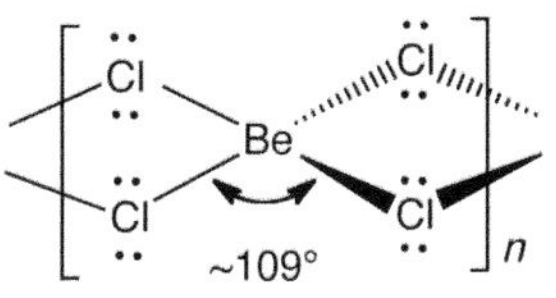

Tetrahedral

~109°

7.2 B is a triple connector and hence it forms BH_3, which is electron deficient. By accepting H^-, it attains Nirvana. The two structures are shown below:

(a)

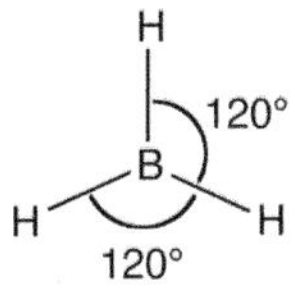

H

120°   Planar

B

H   H

120°

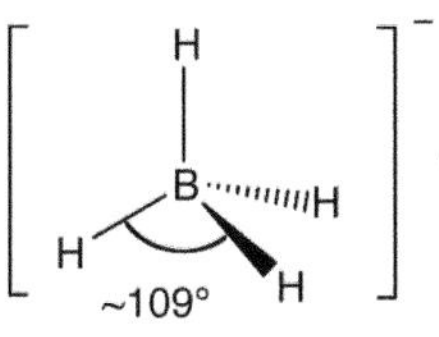

(b)

H

B······H   Tetrahedral

H

~109°   H

7.3 The two molecules are SF_2 and PBr_3. Since the central atoms in both molecules have four electron pairs, the structures will derive from a tetrahedron, as drawn below:

(a)

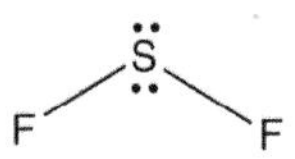

S

F   F

A planar molecule with a structure that derives from a tetrahedron

(b)

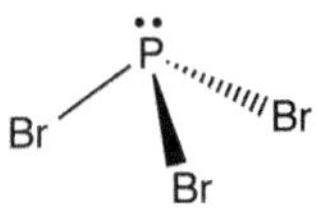

A pyramidal molecule with a structure
that derives from a tetrahedron

Note that the bond angles (although not indicated) in both molecules are smaller than the ideal angle of 109° (although we did not elaborate on this effect, it is due to the strongly repulsive nature of the lone electron pair on the central atom), while if we were to neglect the lone pairs in our considerations, the angles would have been 180° for SF_2 and 120° for PBr_3.

7.4 The two molecules are SF_4 and SF_6 for (a), and the one molecule is PBr_5 for (b), and their structures are shown below:

(a)

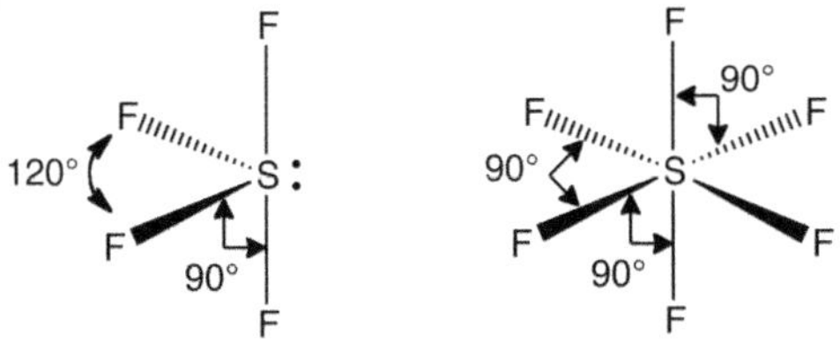

Derived from a
trigonal bipyramid

Octahedral

(b)

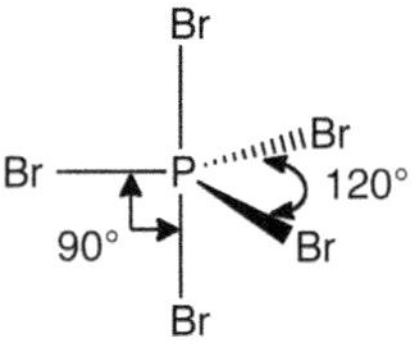

Trigonal bipyramidal

7.5 By replacements of two H's by two CH_3 groups, we generate *syn*, *cis*, and *trans* isomers. The *cis* and *trans* isomers are shown below. In both molecules, the C(H)C=C(H)C part forms a plane and the CH_3 groups are tetrahedral centers. All specified angles are ideal rather than precise angles.

cis

trans

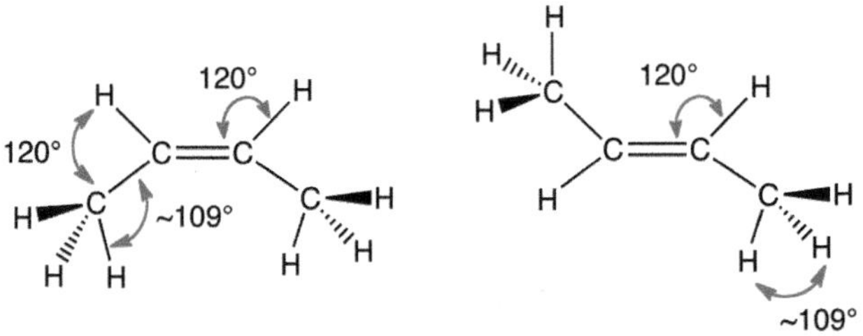

7.6 These molecules are all electron rich. Their structures are depicted below with idealized angles:

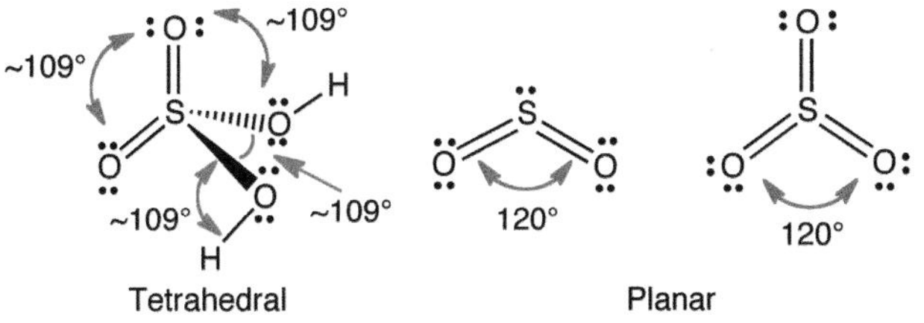

Tetrahedral   Planar

7.7 By adding the missing hydrogen atoms, one can identify three chiral centers, as shown below by the asterisks:

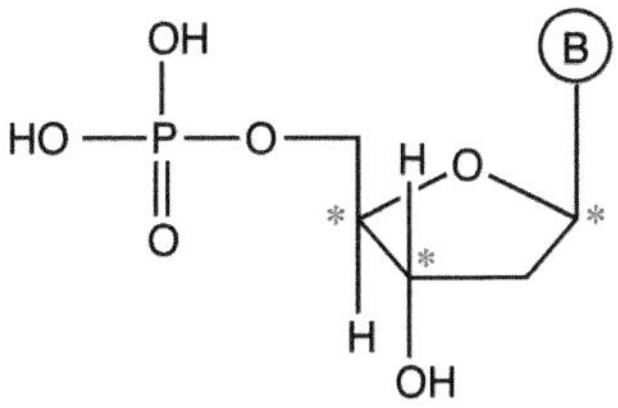

7.8 Using alanine (R=CH$_3$) as an example, we mark below the chiral center with an asterisk and draw the enantiomers as an object and its mirror image, and the senses of rotations within the enantiomers underneath the corresponding structures:

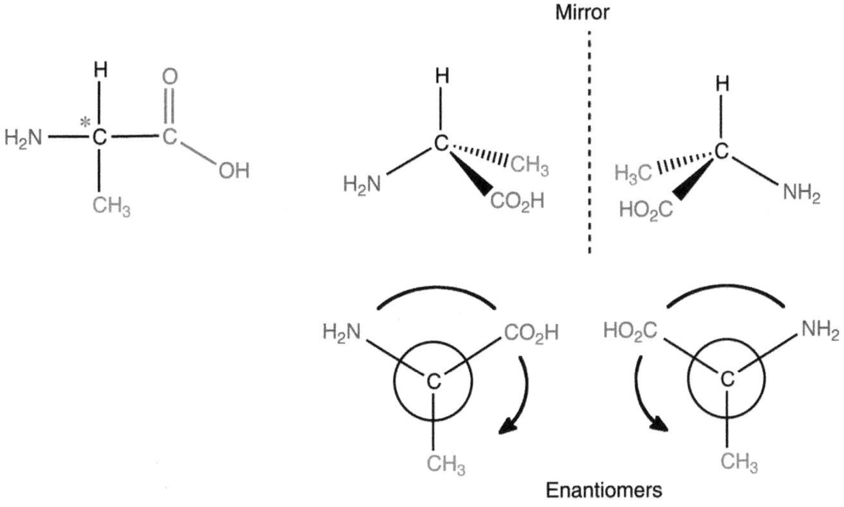

Enantiomers

7.9 By replacing three of the C atoms with three different group 4A atoms, we create a tetrahedral structure that cannot be superimposed on its mirror image, and is therefore chiral. The drawing below shows a replacement of three C atoms to Si, Ge, and Sn:

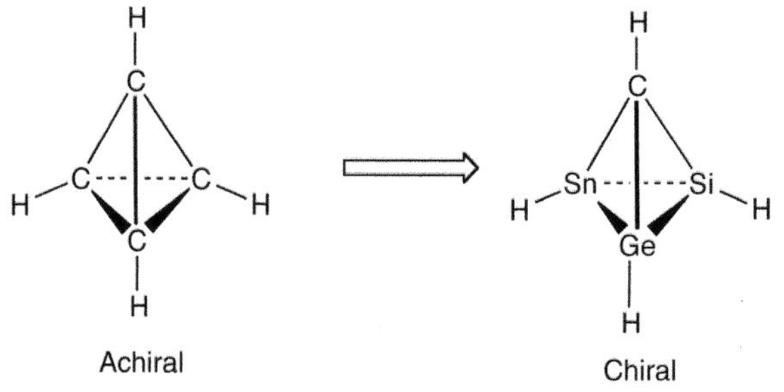

Achiral   Chiral

7.10 The two structures are shown below. Note that even though the structure $C_4FClBr(CH_3)$ has four different groups, it has a mirror of symmetry in the plane of the page and thus is not handed.

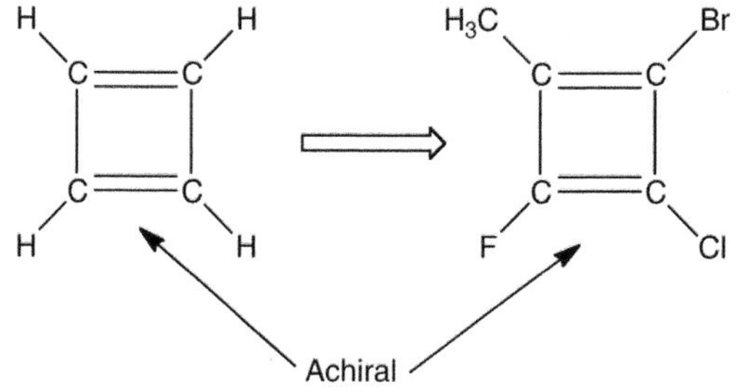

Achiral

7.11 The dimer drawn below has two hydrogen bonds (H-bonds). The O•••H—O angle of the H-bond is linear.

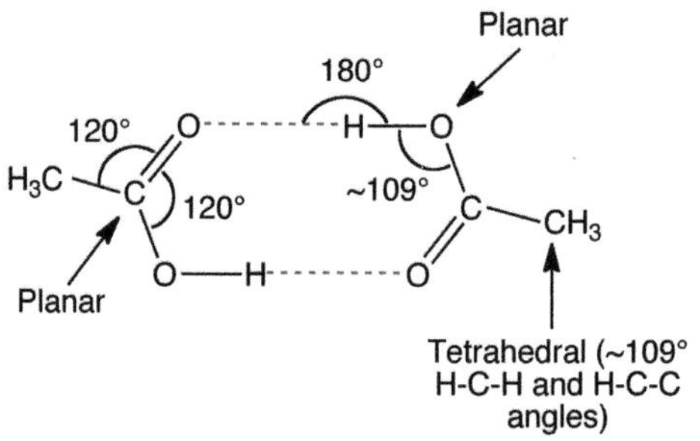

Planar

180°

120°

120°

~109°

Planar

Tetrahedral (~109° H-C-H and H-C-C angles)

7.12 The bond dipole can be assigned by looking at the electronegativity (EN) differences in Figure 7.A.1. As seen below, the differences are 0.4 for C—H versus 0.3 for all the others, so the bond polarity is slightly larger for the C—H bond. Note, however, that since C has a larger EN than H, the sense of polarity for the C–H bond (from δ^+ to δ^-) is opposite compared with all other E–H bonds for which the EN of H is higher. Since all the EH_4 molecules are symmetrical, the molecular dipole is zero for all of them.

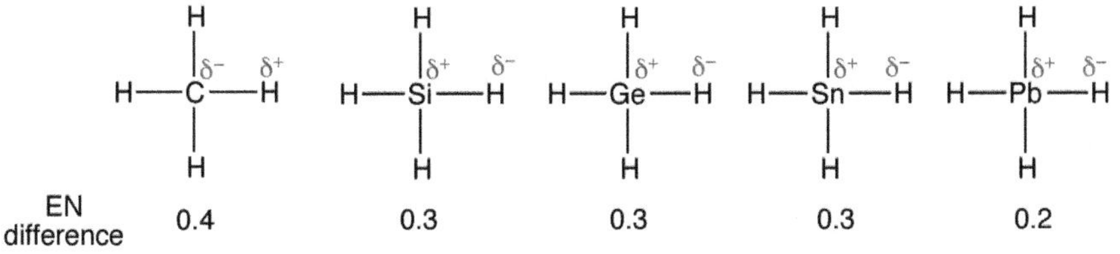

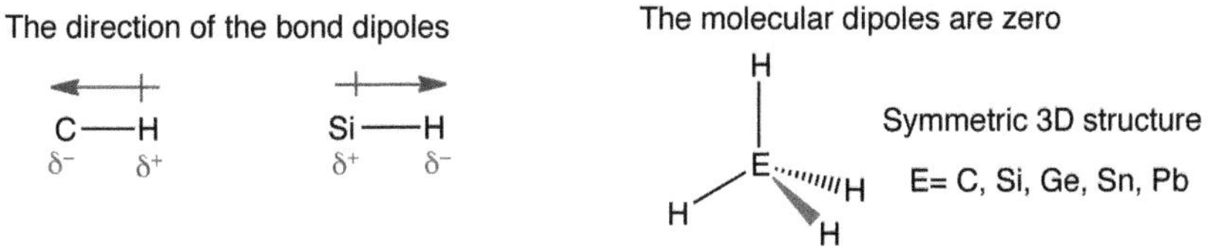

The direction of the bond dipoles

The molecular dipoles are zero

Symmetric 3D structure

E= C, Si, Ge, Sn, Pb

7.13 As shown below, benzene is planar (the double bond counts as a single spatial unit). Replacing the two H's in positions 1 and 4 creates a dichlorobenzene, which does not have an overall molecular dipole, since the individual bond dipoles cancel one another:

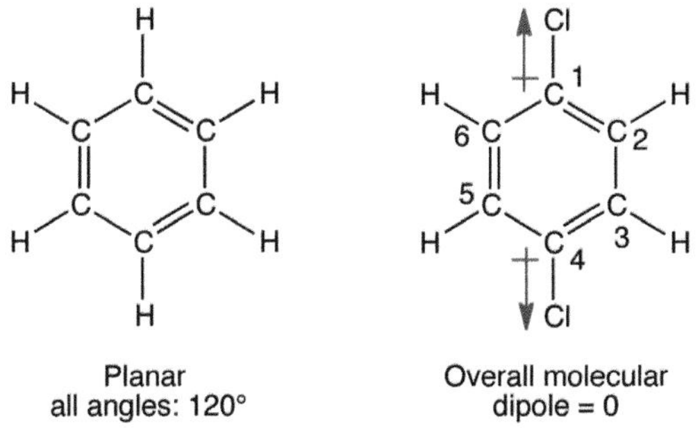

Planar
all angles: 120°

Overall molecular
dipole = 0

7.14 The HF molecules are glued together by H-bonds. The F•••H–F angle is 180° while the H–F•••H angle is roughly tetrahedral, which is how the H is oriented toward the lone pair on the F to which it is covalently bound. See below:

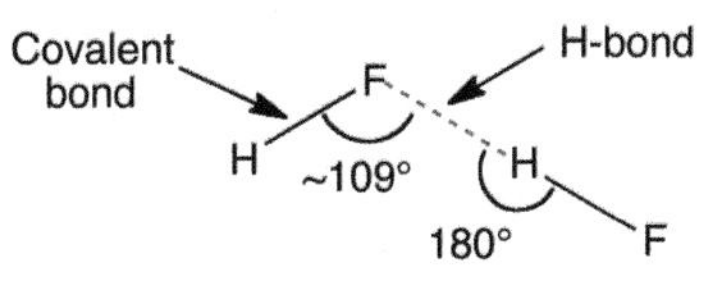

7.15 The most stable conformer is the one with the extended carbon chain. It is drawn below as instructed in the text (Scheme 7.9).

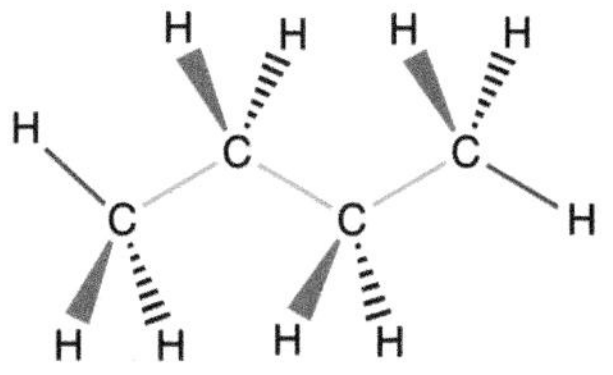

7.16 It is easier to start from the achiral isomer. If we cut the C—C bond of this isomer at the midpoint, the resulting two halves will be identical (see the left-hand side isomer). One can further ascertain that the sense of rotation between the groups bonded to one carbon is exactly opposite to that for the other carbon (try moving from OH to H via COOH or use plastic models to ascertain the sense of rotation) such that the chiralities of the two centers cancel one another. The other two isomers shown on the right-hand side are chiral; they are an object and its mirror image that are not superimposable.

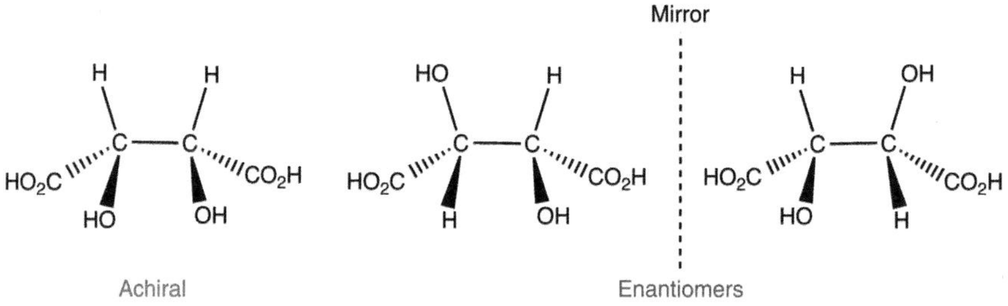

LECTURE 8: ANSWERS

8.1 The equations are written below:

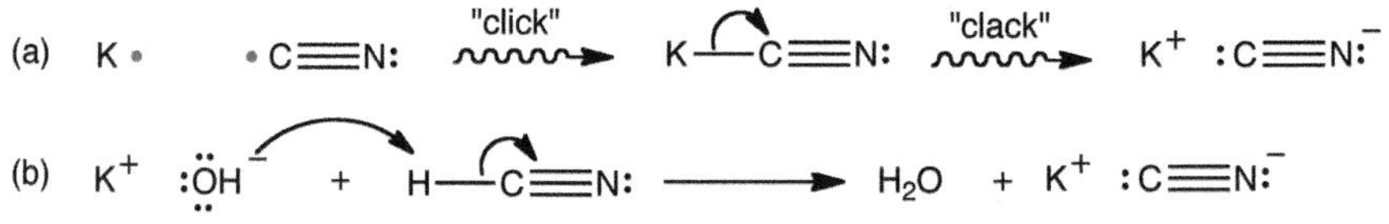

8.2 Using the principle that the ions should have octet in their valence shells, you can assign their charges. Then construct the following table that allows you to write all the possible ionic combinations. The idea is that the total charge must be zero, for example, using a 3+ cation with a 2− anion requires two cations and three anions, etc.

	Na$^+$	Mg^{2+}	Al^{3+}
O^{2-}	(Na$^+$)$_2$ O^{2-}	Mg^{2+} O^{2-}	(Al^{3+})$_2$ (O^{2-})$_3$
Cl$^-$	Na$^+$ Cl$^-$	Mg^{2+} (Cl$^-$)$_2$	Al^{3+} (Cl$^-$)$_3$
SO$_4$$^{2-}$	(Na$^+$)$_2$ SO$_4$$^{2-}$	Mg^{2+} SO$_4$$^{2-}$	(Al^{3+})$_2$ (SO$_4$$^{2-}$)$_3$
PO$_4$$^{3-}$	(Na$^+$)$_3$ PO$_4$$^{3-}$	(Mg^{2+})$_3$ (PO$_4$$^{3-}$)$_2$	Al^{3+} PO$_4$$^{3-}$

8.3 The NH_4 and $N(CH_3)_4$ species are cations with 1+ charge, while all the others are anions with 1− charge. The following table enumerates all the possible ionic compounds that can be made from these cations and anions:

	NH$_4$$^+$	(CH$_3$)$_4$N$^+$
BF$_4$$^-$	NH$_4$$^+$ BF$_4$$^-$	(CH$_3$)$_4$N$^+$ BF$_4$$^-$
Br$^-$	NH$_4$$^+$ Br$^-$	(CH$_3$)$_4$N$^+$ Br$^-$
CH$_3$O$^-$	NH$_4$$^+$ $^-$OCH$_3$	(CH$_3$)$_4$N$^+$ $^-$OCH$_3$

8.4 Based on the identity of the constituent atoms, the molecule is ionic. N is a triple connector and Li is a single connector. So we click N with 3Li and then we clack to form $(Li^+)_3N^{3-}$ as shown below:

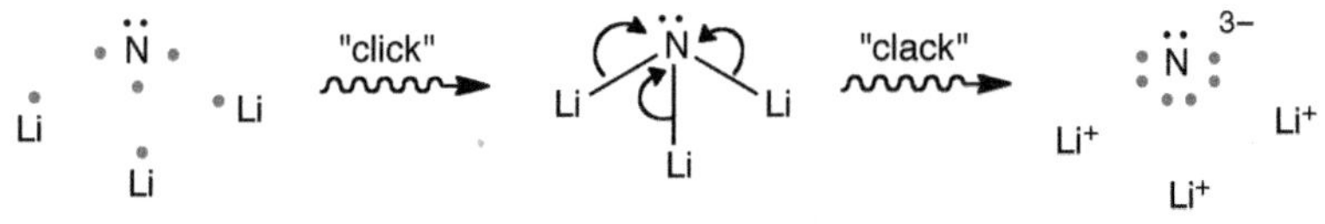

8.5 In the acid $HClO_4$, Cl adopts an electron-rich valence shell with three Cl=O bonds and one Cl–OH bond. Removing the H creates a single-connector fragment, $O_3ClO\bullet$, which is a very electronegative fragment, and hence, will form an ionic bond with K. Thus, we use is with K• and O_3ClO• to click the bond and then we clack to create the ionic bond:

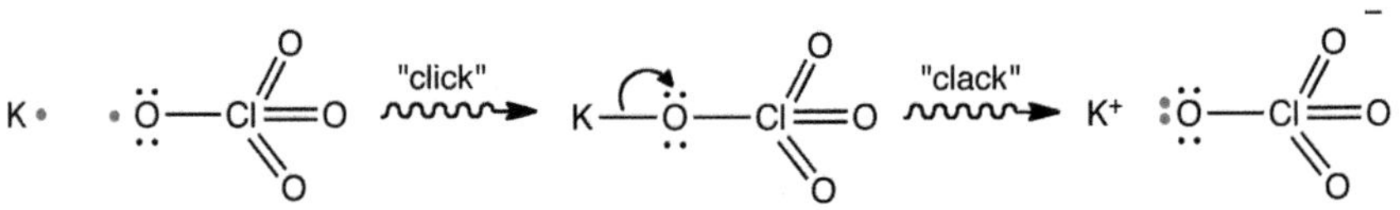

8.6 A possible ionic liquid must protect the ions from approaching too closely. One possible candidate is the one shown below, where the cation is shielded by the $C(CH_3)_3$ groups:

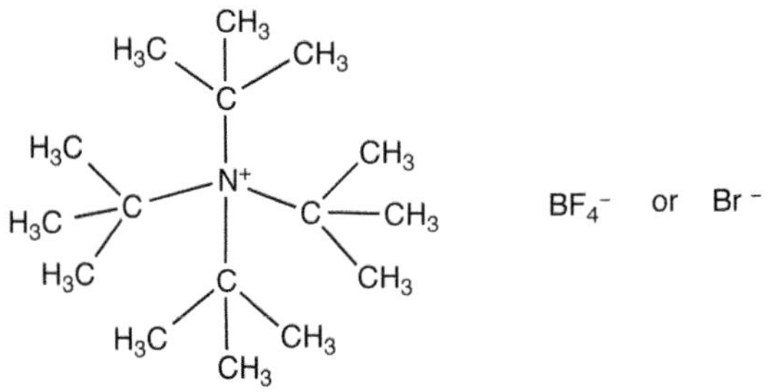

BF_4^- or   Br $^-$

8.7 Since in Na^+Cl^- the two ions are each coordinated by six counterions, we expect the $Ca^{2+}O^{2-}$ species to exhibit the same structural features. The ions and their immediate environments are accordingly depicted below:

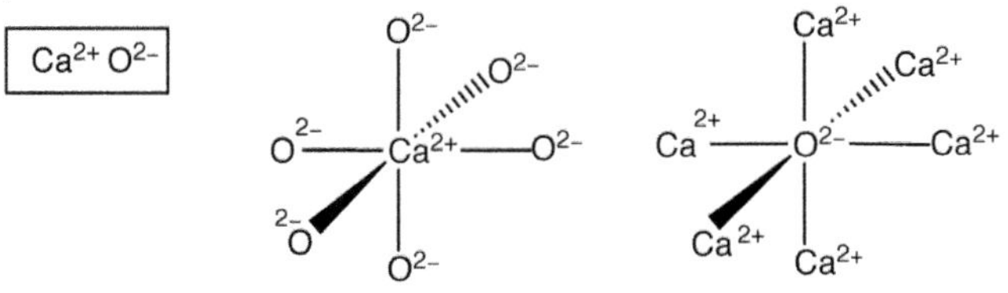

8.8 The reaction involves proton abstraction from HCl by the HO^- base, as shown below:

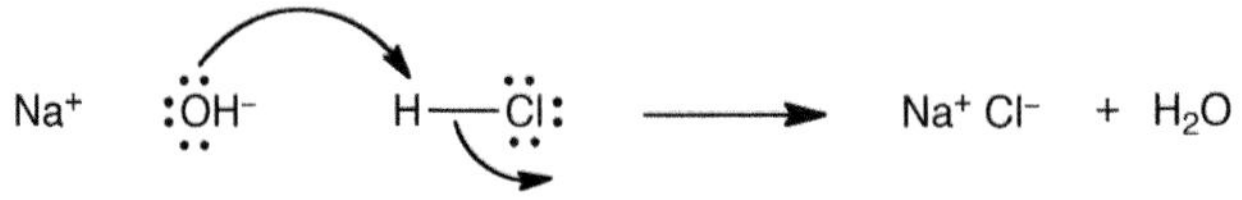

$$Na^+ \quad :\ddot{O}H^- \qquad H—\ddot{C}l: \qquad \longrightarrow \qquad Na^+ Cl^- \;+\; H_2O$$

8.9 Since HCl is a very strong acid, it easily donates a proton to the CO_3^{2-} anion. The anion requires two protons to form carbonic acid, H_2CO_3, which tends to split into water and CO_2, as shown below:

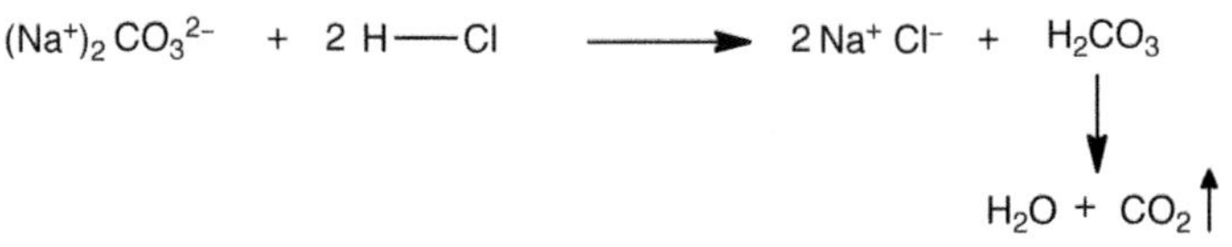

$$(Na^+)_2\, CO_3^{2-} \;+\; 2\,H—Cl \qquad \longrightarrow \qquad 2\,Na^+ Cl^- \;+\; H_2CO_3$$

$$\downarrow$$

$$H_2O + CO_2 \uparrow$$

The upward-pointing arrow near CO_2 indicates that the gas bubbles out of the solution. This is the origin of the bubbles.

8.10 When you mix the two ionic compounds, they exchange ions and generate $Ca^{2+}CO_3^{2-}$, which is not soluble in water, and hence, it precipitates as a solid. The downward-pointing arrow in the equation below indicates the precipitation:

$$(Na^+)_2 CO_3^{2-} \quad + \quad Ca^{2+}(Cl^-)_2 \quad \longrightarrow \quad Ca^{2+}CO_3^{2-}\downarrow \ + \ 2\,Na^+\,Cl^-$$

8.11 (a) pH > 7. (b) pH < 7. (c) pH > 7. (d) pH < 7.

LECTURE 9: ANSWERS

9.1 To determine the charge state of the transition mental (TM), we need to know the total charge of the complex, and the charge state of the ligands. For example, in $Fe(CN)_6^{4-}$ the total charge is 4− and the ligands are each 1−, so the total charge on the ligands is 6−. This means that we have Fe^{2+} and hence six electrons on the iron cation. Doing the same for all other complexes, we find the charge on the TM (TM charge = complex charge − ligand charge), as shown in the table below:

	(a)	(b)	(c)	(d)	(e)	(f)
ion	Fe^{2+}	Mn^+	Cr^0	Fe^{3+}	Zr^{4+}	Pd^{2+}
e-	6	6	6	5	0	8

9.2 Since the Pt^{2+} complex (with eight electrons on the TM center) is planar, there are *cis* and *trans* isomers, as shown below:

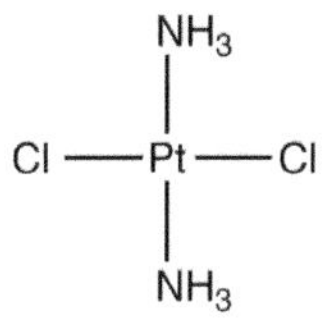

trans 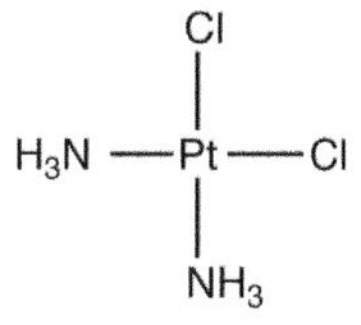   cis

9.3 The ligand $H_2N(CH_2)_2NH_2$ contributes four electrons to bonding with the iron. CH_3CN donates two electrons through the lone pair on N, and O^{2-} donates four electrons. A complex that obeys the 18e rule is shown below:

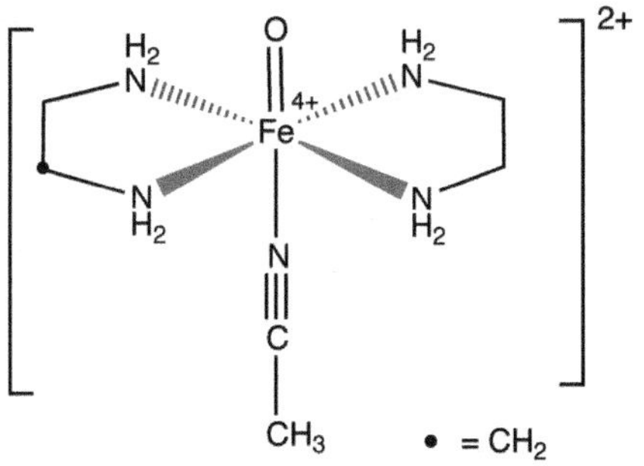

• = CH_2

Here, Fe^{4+} has four electrons, the two $H_2N(CH_2)_2NH_2$ ligands donate eight electrons, and the O^{2-} and CH_3CN donate together six electrons. The total electron count around Fe^{4+} is 18. Hence, the complex obeys the 18e rule. The total charge on the complex must be 2+ to keep the 18e count.

9.4 The trigonal bipyramidal $Fe(CO)_5$ complex has two unique positions, one on the OC–Fe–CO axis and one on the $Fe(CO)_3$ plane. Accordingly, the number of isomers expected from replacements of CO ligands by $P(CH_3)_3$ are the following:

(a)

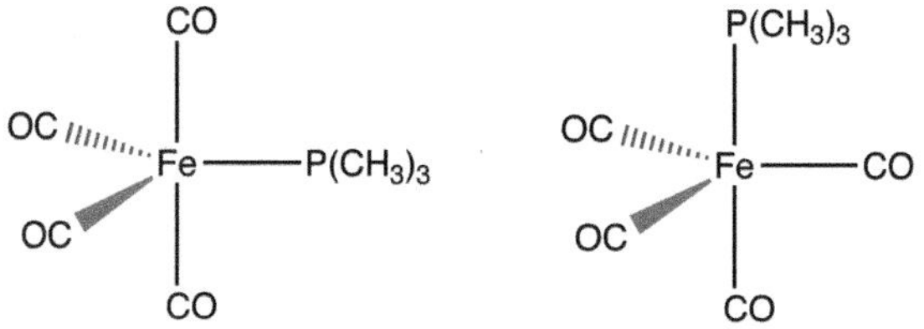

(b)

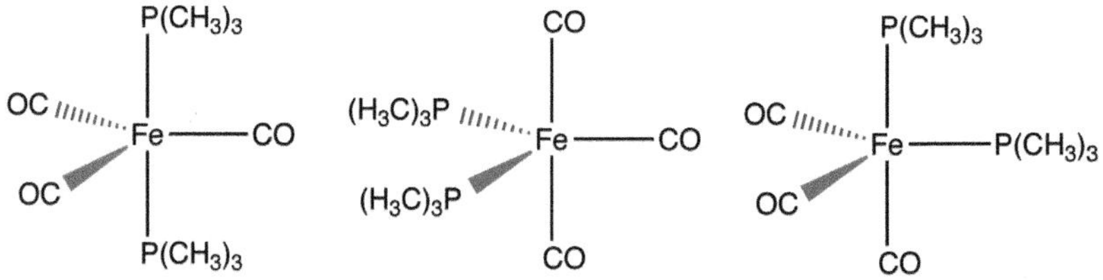

9.5 The complex has an octahedral ligand environment. The two $H_2N(CH_2)_2NH_2$ ligands can occupy four positions in one plane, or in two mutually perpendicular planes. The latter complex is chiral and has two enantiomers (try to superimpose of the object and its mirror image):

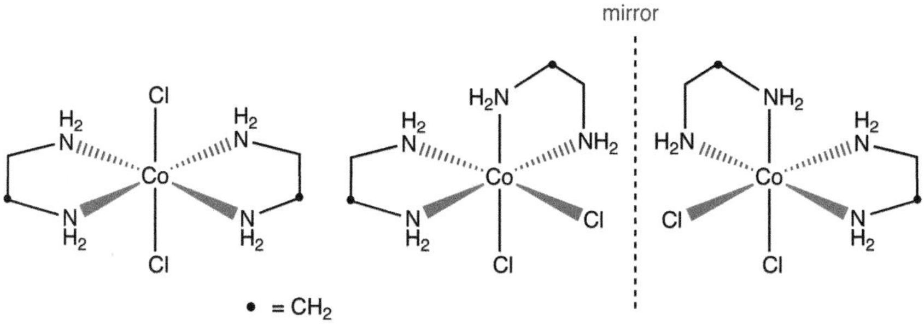

mirror

• = CH_2

9.6 $Ni(CO)_4$ is tetrahedral and all its positions are identical to one another. Hence, only one isomer is expected, and it is shown below:

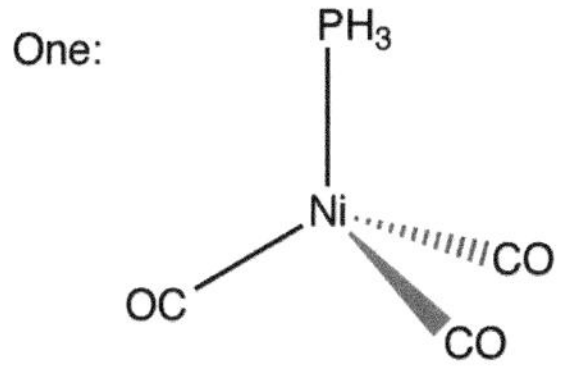

One:

9.7 To obey the 18e rule, the atom Fe (with eight electrons) will coordinate to one C_4H_4 ligand (donating four electrons) and three CO ligands (donating a total of six electrons). The complex is shown below:

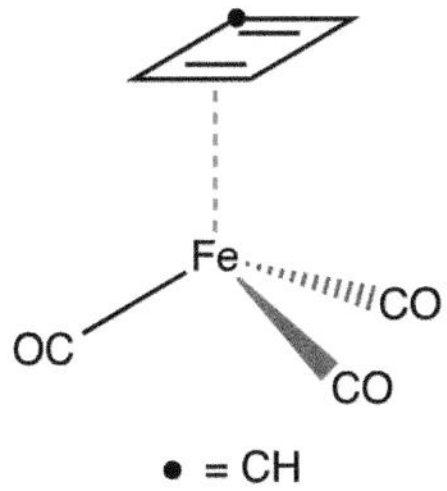

● = CH

9.8 $Fe(CO)_4$ is a double connector and is isolobal to CH_2. In principle, it can form a molecule with Fe=Fe double bond, which is an analogue of ethylene. It can form a three-membered ring $[Fe(CO)_4]_3$, which is an analogue of the $(CH_2)_3$ ring. It can form, in principle, many other rings.

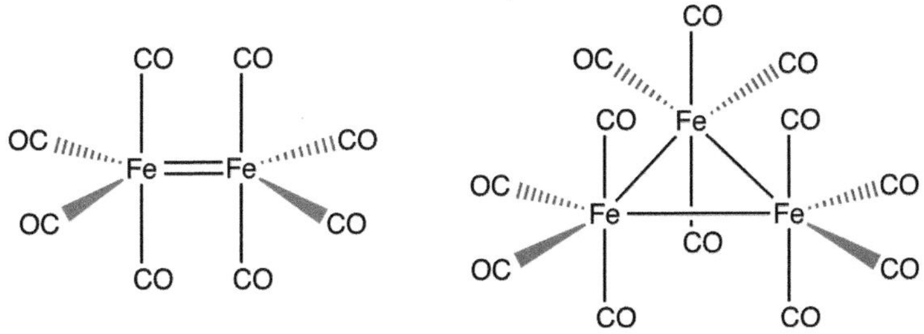

9.9 Using the covalent radii in Table 9.1, we can estimate the C–H and C–C bond lengths of the two molecules. These are shown below:

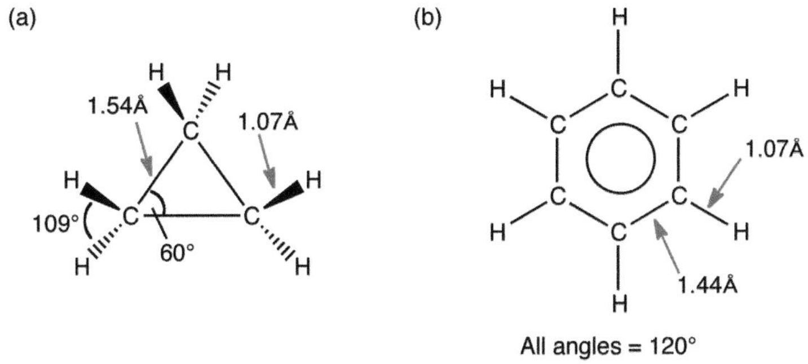

(a)

(b)

All angles = 120°

In C_3H_6 in (a), the carbons are tetrahedral centers, with C—C and C—H bond lengths of 1.54 Å and 1.07 Å, respectively. Note that the internal C—C—C angle is not a native tetrahedral angle, but is 60°, which is the constraint of the triangular structure. (This is why chemists say that cyclopropane exhibits "ring strain" and its bonds are highly bent.)

In C_6H_6, the structure is planar with angles of 120°. Since each CC bond is one time C=C and one time C—C (consult Retouches section 4.R.5), then the CC distance is an average of C—C and C=C bonds. This average distance ((1.54 + 1.34)/2) is 1.44 Å. The actual distance is 1.40 Å, and the discrepancy reflects the fact that the covalent radii in Table 9.1 represent average values, and hence sometimes, the predictions deviate from the experimentally determined values.

9.10 Using Table 9.1, the distances are determined as a sum of the covalent radii of C with H and with X. These distances and the three-dimensional (3D) information are shown below:

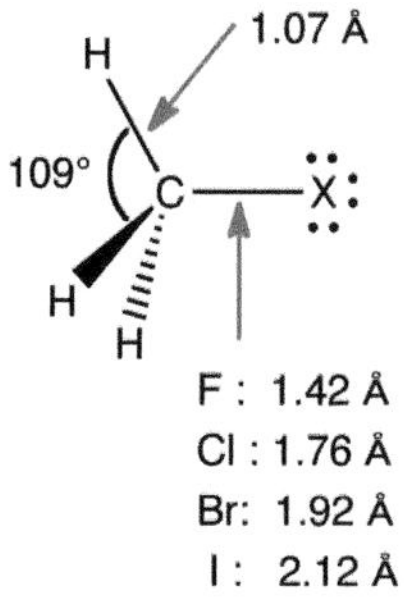

The bond length of the C—X bond increases with the size of the halogen.

9.11 Using Tables 9.1 and 9.A.1, the distances are determined as a sum of the covalent radii of Ni with C and of C with O. These distances and the three-dimensional (3D) information are shown below:

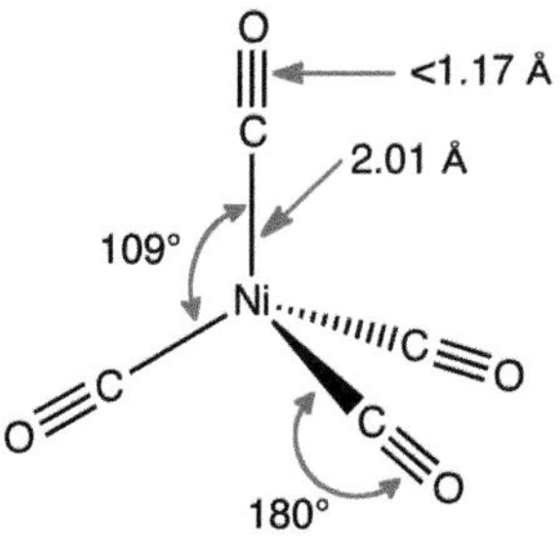

Note that since we do not have the covalent radius for a triple-bonded O, we simply write the CO distance as being smaller than 1.17 Å (i.e., <1.17 Å), which is the value determined by summing the covalent radius of triple-bonded C with the covalent radius of a double-bonded O.

9.12 The reaction energy is determined as the difference between the bond dissociation energy (BDE) sum of the reactants' bond energies minus the BDE sum of the products' bond energies. The rationale is the following: Since we are breaking the bonds of the reactants, we invest energy, and hence, we take these BDEs with a positive sign; and since we are forming new bonds of the product, we get back the corresponding bond energies, and hence, we must take these BDEs with a negative sign. The calculation is shown below:

$$CH_4 \ + \ 2O_2 \ \longrightarrow \ CO_2 \ + \ 2H_2O$$

$$\text{Reaction energy} = 4BDE_{CH} + 2BDE_{OO} - 2BDE_{C=O} - 4BDE_{OH}$$

Using BDEs from table 9.A.2, we calculate the following reaction energy:

$$\text{Reaction energy} = 4(98.7) + 2(116) - 2(191) - 4(111) = -199 \text{ kcal/mol}$$

It is seen that the reaction energy is negative, which means that the reaction releases energy. When we exercise, we say that we "burn" fat (e.g., fatty acids, etc.). This means that our fat is converted to H_2O and CO_2 by reacting with O_2, as shown in the above equation. Note that fat is not composed of CH_4, but it behaves similarly, in that it reacts with O_2 to yield carbon dioxide and water.

LECTURE 10: ANSWERS

10.1 Let us take the HO• radical. These radicals are formed in the cells, and their population may sometimes increase artificially. Taking a typical *trans* fatty acid, we notice that it has a CH_2 group in between two double bonds. The C—H bonds of this group have rather small bond dissociation energies (compared with other C—H bonds in the molecule. There is a good reason for this which you may learn about in advanced courses). Hence, the radical abstracts an H atom from these bonds, as shown in reaction (1) below. The new radical reacts with oxygen (reaction 2) to form a peroxy radical, C—OO•. The latter radical abstracts H from another molecule of the fatty acid (reaction 3) and forms a peroxide molecule and a new radical. This new radical goes back and repeats the H abstraction in reaction (1). Thus, the membrane molecules are oxidized to peroxides in a chain process that eventually causes great damage to the cell. The peroxide has adverse effects on the DNA (mutation of DNA bases and toxicity).

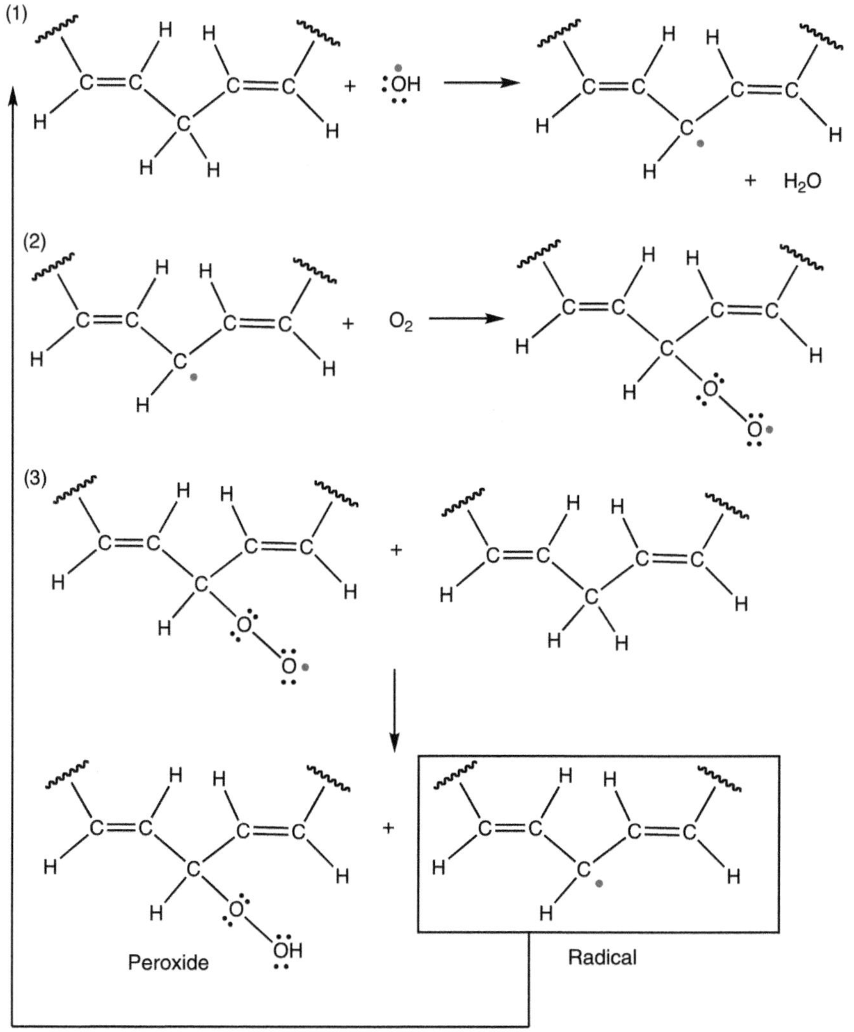

(1)

(2)

(3)

Peroxide

Radical

10.2 Limestone is made from $CaCO_3$. Acid rainfall (e.g., containing HCl) initiates the formation of a water-soluble $CaCl_2$ salt and carbonic acid, which falls apart giving H_2O and CO_2:

$$2HCl + CaCO_3 \rightarrow CaCl_2 + H_2CO_3 \rightarrow CaCl_2 + H_2O + CO_2$$

10.3 The reaction forms HCl as follows:

$$Cl\bullet + CH_4 \rightarrow HCl + CH_3\bullet$$

INDEX

Chemistry as a Game of Molecular Construction: The Bond-Click Way, First Edition. Sason Shaik.
© 2016 John Wiley & Sons, Inc. Published 2016 by John Wiley & Sons, Inc.